VOLUME SIX HUNDRED AND THIRTY EIGHT

METHODS IN ENZYMOLOGY

Chemical Tools for Imaging, Manipulating, and Tracking Biological Systems: Diverse Methods for Prokaryotic and Eukaryotic Systems

METHODS IN ENZYMOLOGY

VOLUME SIX HUNDRED AND THIRTY EIGHT

METHODS IN ENZYMOLOGY

Chemical Tools for Imaging, Manipulating, and Tracking Biological Systems: Diverse Methods for Prokaryotic and Eukaryotic Systems

Edited by

DAVID M. CHENOWETH

Roy and Diana Vagelos Laboratories,
Department of Chemistry,
University of Pennsylvania,
Philadelphia, PA, United States

ELSEVIER

ACADEMIC PRESS
An imprint of Elsevier

Academic Press is an imprint of Elsevier
50 Hampshire Street, 5th Floor, Cambridge, MA 02139, United States
525 B Street, Suite 1650, San Diego, CA 92101, United States
The Boulevard, Langford Lane, Kidlington, Oxford OX5 1GB, United Kingdom
125 London Wall, London, EC2Y 5AS, United Kingdom

First edition 2020

ISBN: 978-0-12-820145-9
ISSN: 0076-6879

For information on all Academic Press publications
visit our website at https://www.elsevier.com/books-and-journals

Publisher: Zoe Kruze
Editorial Project Manager: Shellie Bryant
Production Project Manager: James Selvam
Cover Designer: Greg Harris

Typeset by SPi Global, India

Contents

Contributors

Yasutomi Asano
Tri-Institutional Therapeutics Discovery Institute, New York, NY, United States

Ashley R. Brown
Department of Chemistry and Biochemistry, University of Delaware, Newark, DE, United States

Nathaniel W. Brown
Department of Chemistry, University of Minnesota, Minneapolis, MN, United States

Kevin E. Bruce
Department of Biology, Indiana University, Bloomington, IN, United States

Michael D. Burkart
Department of Chemistry and Biochemistry, University of California, San Diego, La Jolla, CA, United States

Zakey Yusuf Buuh
Department of Chemistry, Temple University, Philadelphia, PA, United States

Erin E. Carlson
Department of Chemistry, University of Minnesota; Department of Medicinal Chemistry, University of Minnesota; Department of Biochemistry, Molecular Biology and Biophysics, University of Minnesota, Minneapolis, MN, United States

Katherine Charov
Department of Chemistry and Biochemistry, University of California, San Diego, La Jolla, CA, United States

Hui Chen
Department of Cell and Developmental Biology, Perelman School of Medicine, University of Pennsylvania, Philadelphia, PA, United States

David M. Chenoweth
Roy and Diana Vagelos Laboratories, Department of Chemistry, University of Pennsylvania, Philadelphia, PA, United States

Clifford M. Csizmar
Department of Medicinal Chemistry, University of Minnesota, Minneapolis, MN, United States

Alexander Deiters
Department of Chemistry, University of Pittsburgh, Pittsburgh, PA, United States

Kristen E. DeMeester
Department of Chemistry and Biochemistry, University of Delaware, Newark, DE, United States

Juliel Espinosa
Laboratory of Chemical Biology and Microbial Pathogenesis, The Rockefeller University, New York, NY, United States

Sarah J. Fansler
Biological Sciences Division, Pacific Northwest National Laboratory, Richland, Washington, United States

Samir Gautam
Yale University School of Medicine, Department of Internal Medicine, Section of Pulmonary, Critical Care and Sleep Medicine; Yale University, Department of Chemistry, New Haven, CT, United States

Matthew C. Good
Department of Cell and Developmental Biology, Perelman School of Medicine; Department of Bioengineering, School of Engineering and Applied Science, University of Pennsylvania, Philadelphia, PA, United States

Catherine Leimkuhler Grimes
Department of Chemistry and Biochemistry; Department of Biological Sciences, University of Delaware, Newark, DE, United States

Tezcan Guney
Chemical Biology Program, Sloan Kettering Institute, Memorial Sloan Kettering Cancer Center, New York, NY, United States

Howard C. Hang
Laboratory of Chemical Biology and Microbial Pathogenesis, The Rockefeller University, New York, NY, United States

Rebecca Howell
Yale University, Department of Chemistry, New Haven, CT, United States

Maulik S. Jani
Department of Chemistry, University of Chicago; Grossman Institute of Neuroscience, Quantitative Biology and Human Behavior, University of Chicago, Chicago, IL, United States

Cheng Ji
Chemical Biology Program, Sloan Kettering Institute, Memorial Sloan Kettering Cancer Center, New York, NY, United States

Zachary S. Jones
Department of Chemistry and Biochemistry, University of Delaware, Newark, DE, United States

Byungchul Kim
Laboratory of Chemical Biology and Microbial Pathogenesis, The Rockefeller University, New York, NY, United States

Taehan Kim
Yale University, Department of Chemistry, New Haven, CT, United States

Nihar Kinarivala
Chemical Biology Program, Sloan Kettering Institute, Memorial Sloan Kettering Cancer Center, New York, NY, United States

Yamuna Krishnan
Department of Chemistry, University of Chicago; Grossman Institute of Neuroscience, Quantitative Biology and Human Behavior, University of Chicago, Chicago, IL, United States

Michael A. Lampson
Lynch Laboratories, Department of Biology, University of Pennsylvania, Philadelphia, PA, United States

Hai Liang
Department of Chemistry and Biochemistry, University of Delaware, Newark, DE, United States

Kevin N. Little
Department of Chemistry and Biochemistry, Florida State University, Tallahassee, FL, United States

Miguel Macias-Contreras
Department of Chemistry and Biochemistry, Florida State University, Tallahassee, FL, United States

Robert Maloney
Department of Chemistry, Temple University, Philadelphia, PA, United States

Carson J. Miller
Biological Sciences Division, Pacific Northwest National Laboratory, Richland, Washington, United States

Johannes Morstein
Department of Chemistry, New York University, New York, NY, United States

Naoyoshi Noguchi
Tri-Institutional Therapeutics Discovery Institute, New York, NY, United States

Shabnam Sharifzadeh
Department of Chemistry, University of Minnesota, Minneapolis, MN, United States

Joshua D. Shirley
Department of Medicinal Chemistry, University of Minnesota, Minneapolis, MN, United States

David A. Spiegel
Yale University, Department of Chemistry, New Haven, CT, United States

Lisa C. Standke
Pharmacology Program, Weill Cornell Graduate School of Medical Sciences, Memorial Sloan Kettering Cancer Center, New York, NY, United States

Andrea K. Steiger
Biological Sciences Division, Pacific Northwest National Laboratory, Richland, Washington, United States

Derek S. Tan
Chemical Biology Program, Sloan Kettering Institute; Pharmacology Program, Weill Cornell Graduate School of Medical Sciences; Tri-Institutional Research Program, Memorial Sloan Kettering Cancer Center, New York, NY, United States

Dirk Trauner
Department of Chemistry, New York University, New York, NY, United States

Ophelia I. Ukaegbu
Department of Chemistry and Biochemistry, University of Delaware, Newark, DE, United States

Aneesh T. Veetil
Department of Chemistry, University of Chicago; Grossman Institute of Neuroscience, Quantitative Biology and Human Behavior, University of Chicago, Chicago, IL, United States

Carston R. Wagner
Department of Medicinal Chemistry, University of Minnesota, Minneapolis, MN, United States

Rongsheng E. Wang
Department of Chemistry, Temple University, Philadelphia, PA, United States

Joshua S. Wesalo
Department of Chemistry, University of Pittsburgh, Pittsburgh, PA, United States

Christopher Whidbey
Department of Chemistry, Seattle University, Seattle, Washington, United States

Malcolm E. Winkler
Department of Biology, Indiana University, Bloomington, IN, United States

Richard Wombacher
Institute of Pharmacy and Molecular Biology, Ruprecht-Karls-Universität Heidelberg, Heidelberg, Germany

Aaron T. Wright
Biological Sciences Division, Pacific Northwest National Laboratory, Richland; The Gene and Linda Voiland School of Chemical Engineering and Bioengineering, Washington State University, Pullman, Washington, United States

Daniel Z. Wu
Roy and Diana Vagelos Laboratories, Department of Chemistry, University of Pennsylvania, Philadelphia, PA, United States

Yue Zhao
Department of Chemistry, Temple University, Philadelphia, PA, United States

Lei Zhu
Department of Chemistry and Biochemistry, Florida State University, Tallahassee, FL, United States

Michael J. Ziegler
Institute of Pharmacy and Molecular Biology, Ruprecht-Karls-Universität Heidelberg, Heidelberg, Germany

Preface

The development of new chemical tools for imaging, manipulating, and tracking biological systems is critical to our future understanding of the living world. The ability to visualize complex processes over a multitude of time and space scales as well as perturb these processes with varying levels of control can provide valuable insight into the complexities and nuances of detailed biological processes. The fundamental insights exposed by chemical tools provide new knowledge with often far reaching impacts on human health and medicine. New chemical tools catalyze new ideas, inspire new experiments, usher in new methods, and ultimately lead to new scientific discoveries. The development of a new chemical tool is not the endpoint, but rather the starting point in a cycle of inspiration leading to impactful scientific methods for gleaning insight into natural systems. These new tools are often at the forefront of scientific discovery, and it is the goal of this *Methods in Enzymology* series to broadly highlight new advances in chemical tools in an effort to catalyze new ideas and inspire future advances.

In this first volume, Grimes demonstrates the power of a chemoenzymatic approach for the rapid synthesis of uridine diphosphate-*N*-acetyl muramic acid (UDP-NAM) chemical tools (Chapter 1). Fluorescent β-lactone chemical tools for selective profiling of penicillin-binding proteins (PBP) are detailed by Carlson, enabling direct visualization of PBP activity (Chapter 2). In situ fluorescent chemical probes for quantification and localization of peptidoglycan crosslinking developed by Speigel and Gautam provide a versatile new tool for analysis of live bacteria by flow cytometry and super-resolution microscopy (Chapter 3). An efficient multigram-scale preparation of the antituberculosis lead compound salicyl-AMS by Tan demonstrates the critical importance of synthetic chemistry in the chemical tool development process (Chapter 4). Wright demonstrates the ability of new chemical tools to enable novel approaches for function-dependent cell sorting of the microbiome (Chapter 5). Methods for the preparation of large quantities of purified fully functional NlpC/p60 peptidoglycan hydrolases developed by Hang provide a much-needed starting point for the development of tools to study hydrolase function and regulation (Chapter 6). Krishnan has developed DNA nanocapsule-based tools for the chemically triggered and spatiotemporally controlled release of steroids, facilitating studies of membrane-initiated signaling in endothelial cells (Chapter 7). Fluorophore labeling of 5-ethynyl uridine-based chemical probes post incorporation into newly transcribed RNAs allowed Good to develop a protocol for direct imaging of nascent transcription and zygotic genome activation in vertebrate embryos (Chapter 8). Chemical tools and techniques for engineering reversible cell–cell interactions developed by Wagner provide a new starting point for a host of applications ranging from tissue engineering to immunotherapy (Chapter 9). Deiters' development of phosphine-based methods for control of protein function in live cell environments provides an orthogonal alternative to light-based systems (Chapter 10). Azobenzene photoswitches continue to provide a valuable motif for the development of chemical tools as demonstrated by the photoswitchable sphingosine-1-phosphate analog, PhotoS1P, developed by Trauner (Chapter 11). Zhu has expanded the substrate selectivity of the SNAP/CLIP chemical protein tagging system (Chapter 12). Orthogonal light-induced protein proximity systems based on gibberellic acid derivatives have been developed by Wombacher and utilized in living cells (Chapter 13). Our laboratory has developed photoactivatable trimethoprim-based chemical tools for controlling biological processes in a spatially defined and temporally controlled manner (Chapter 14). Targeted imaging tools based on site-specific antibody fragment conjugates have been developed by Wang (Chapter 15). Burkart has developed solvatochromic chemical probes for quantifying protein–protein interactions of the acyl carrier protein (Chapter 16).

Collectively, the chapters in this volume of *Methods in Enzymology* represent powerful examples of chemical tools for diverse applications in prokaryotic and eukaryotic systems. These methods highlight the diversity of chemical tools and demonstrate the breadth of applications and questions to be answered in a wide range of fields. The continued development of new chemical tools, such as those presented in this volume, will have far reaching impacts on a wide range of areas from fundamental biology to human medicine. I hope that this volume on chemical tools will appeal to the broad scientific community and foster the creation of new tools for future investigations at the frontier of science.

I thank all of the authors for their important contributions and sharing their detailed methods with the broad scientific community. I would like to thank the *Methods in Enzymology* editors, the Elsevier editorial team and the Elsevier production team for their guidance and expert assistance while assembling this volume. I would also like to thank Kimberly Chenoweth for her support and editing expertise during this project.

David M. Chenoweth
Department of Chemistry
University of Pennsylvania
Philadelphia, PA, United States

CHAPTER ONE

Utility of bacterial peptidoglycan recycling enzymes in the chemoenzymatic synthesis of valuable UDP sugar substrates

Ophelia I. Ukaegbu[a], Kristen E. DeMeester[a], Hai Liang[a], Ashley R. Brown[a], Zachary S. Jones[a], Catherine Leimkuhler Grimes[a,b,*]

[a]Department of Chemistry and Biochemistry, University of Delaware, Newark, DE, United States
[b]Department of Biological Sciences, University of Delaware, Newark, DE, United States
*Corresponding author: e-mail address: cgrimes@udel.edu

Contents

Abstract

Uridine diphosphate (UDP) sugars are essential precursors for glycosylation reactions in all forms of life. Reactions that transfer the carbohydrate from the UDP donor are catalyzed by glycosyltransferases (Gtfs). While the stereochemistry and negative physiological charge of UDP-sugars are essential for their biochemical function in the cell, these characteristics make them challenging molecules to synthesize and purify on scale in the laboratory. This chapter focuses on the utilization of a chemoenzymatic synthesis of muramyl UDP-sugars, key building blocks in the bacterial cell peptidoglycan. A scalable strategy to obtain UDP-*N*-acetyl muramic acid derivatives (UDP-NAM), the first committed intermediate used solely in peptidoglycan biosynthesis, is described herein. This methodology utilizes two enzymes involving the cell wall recycling enzymes MurNAc/GlcNAc anomeric kinase (AmgK) and NAM α-1-phosphate uridylyl transferase (MurU), respectively. The promiscuity of these enzymes allows for the unique chemical functionality to be embedded in bacterial peptidoglycan both in vitro and in whole bacterial cells for subsequent structural and functional studies of this important biopolymer.

Methods in Enzymology, Volume 638
ISSN 0076-6879
https://doi.org/10.1016/bs.mie.2020.02.014

1. Introduction

Uridine diphosphate (UDP)—sugars utilized in glycosylation reactions are critical molecules responsible for the regulation of biochemical transformations in many organisms (Decker & Kleczkowski, 2019; Goto et al., 2001). Glycosylation is a highly regulated posttranslational modification catalyzed by glycosyltransferases (Gtfs) that decorate substrate acceptor proteins, lipids or small molecules with a carbohydrate via a glycosidic bond forming a glycoconjugate (Lairson et al., 2008; Ohtsubo & Marth, 2006; Wagstaff et al., 2015) (Fig. 1). Early studies of cellular glycosylation processing focused on those that occur in the endoplasmic reticulum (ER) or Golgi of eukaryotes and rely on nucleotide sugar transporters (NSTs) to transport UDP-sugars from the cytosol and nucleus where they are synthesized (Goto et al., 2001; Hadley et al., 2014); these modifications ultimately get shuttled to the surface of the cell. More recently, it has been discovered that the addition of a *single* carbohydrate to substrates is important for the function of intracellular proteins (Hart, Housley, & Slawson, 2007). These glycoconjugates are involved in the regulation of numerous cellular events across multiple forms of life that include assisting in the stability and folding of proteins (Drake et al., 2018; Seyrek, Richter, & Lavrik, 2019; Solá & Griebenow, 2009) and modifying signaling proteins to restore their function (Seyrek et al., 2019). When these regulatory mechanisms fail, diseases can occur (Hart et al., 2007; Hwang et al., 2010). Understanding the small molecule details of these biochemical transformations is critical for the development of new therapeutics to treat disease (Compain & Martin, 2001).

Chemoenzymatic syntheses involving glycans are important in the production of biopharmaceuticals; they have recently shown utility in the production of glycoconjugate vaccines against clinically relevant bacterial infection such as *Neisseria meningitis* and *Streptococcus pneumonia* (Kay, Cuccui, & Wren, 2019). These glycoconjugate vaccines are composed of carrier proteins and polymers usually synthesized via chemoenzymatic glycosylation (McCarthy, Sharyan, & Sheikhi Moghaddam, 2018). To date, these glycoconjugates are one of the most effective vaccines against bacteria (Kay et al., 2019). Other glycosylated natural products used in the treatment of bacterial infections include antibiotics such as vancomycin (Fig. 1), eremomycin, and their lipophilic derivatives that target and inhibit the biosynthesis of most Gram-positive bacterial cell walls (Leimkuhler et al., 2005; Silver, 2003; Walsh, 2000; Watanakunakorn, 1984).

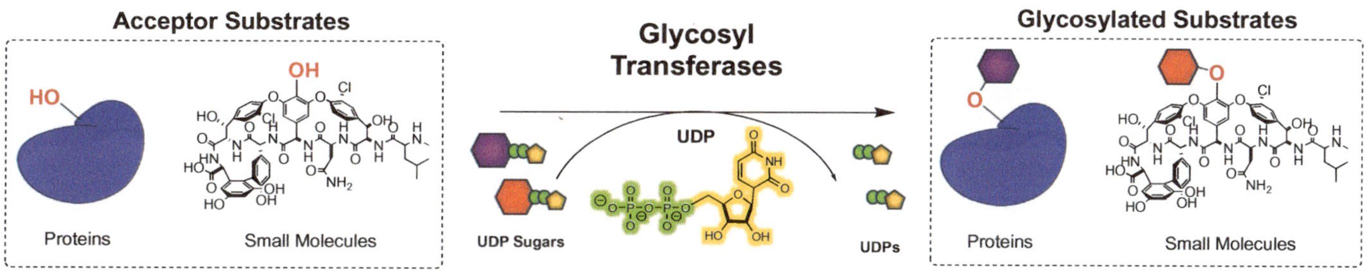

Fig. 1 Glycosyltransferase reaction mechanism using UDP sugar substrates. Glycosyltransferases (Gtfs) utilize activated UDP-sugar donor substrates to add carbohydrates onto acceptor substrates, proteins or small molecules scaffolds.

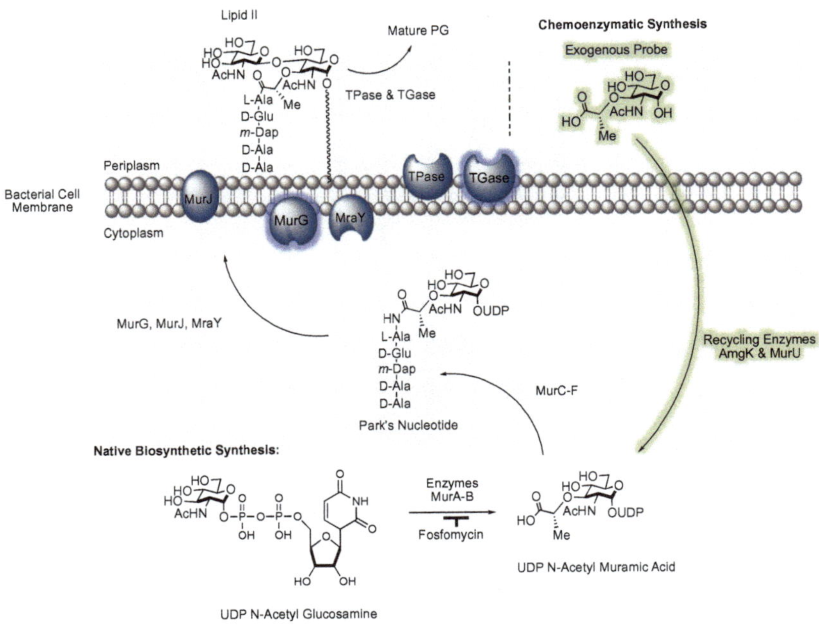

Fig. 2 Peptidoglycan biosynthetic and recycling pathway. The bacterial peptidoglycan biosynthetic pathway deploys over 13 unique enzymatic reactions to synthesize and recycle (green) the mature polymer. UDP sugars serve as key intermediates for the first two stages of this process.

The bacteria themselves also utilize critical glycosylation reactions and UDP donors in key steps during their cell wall biosynthesis including wall teichoic acid (WTA) (Brown et al., 2012; Swoboda et al., 2010), lipopolysaccharide (LPS) (Cote & Taylor, 2017), and peptidoglycan. The later relies on two types of Gtfs, MurG and other peptidoglycan transglycosylases (TGases) (Rod, FtsW, and penicillin binding proteins (PBPs) (Barrett et al., 2007; Scheffers & Pinho, 2005; Taguchi et al., 2019) (Fig. 2). Recently, TarM and TarS were identified as Gtfs involved in WTA biosynthesis, a major constituent of *Staphylococcus aureus* cell wall (Mistretta et al., 2019; Xia et al., 2010).

This methods paper focuses on the chemoenzymatic synthesis of bacterial peptidoglycan building blocks. The bacterial cell wall peptidoglycan is complex, consisting of repeating glycan units of β1–4 linked *N*-acetyl muramic acid (NAM) and *N*-acetyl glucosamine (NAG) that is further crosslinked through a pentapeptide chain bearing L and D amino acids on the NAM sugar (Alvarez et al., 2014; Schleifer & Kandler, 1972; Strominger, 2007) (Fig. 2). Nature adopted this conserved and complex biosynthetic mechanism involving multiple enzymatic steps in the construction of the bacterial cell wall peptidoglycan polymer (Johnson, Fisher, & Mobashery, 2013). The structure of peptidoglycan provides bacteria with a protective coat against environmental stresses. With peptidoglycan serving as a unique structure solely to bacteria, many traditional antibiotics are aimed at destroying or inhibiting the biosynthesis of this cell wall polymer (Johnson et al., 2013; Walsh, 2000). Understanding how peptidoglycan is made at the molecular level will provide insight into the development of new anti-infective therapeutics to combat antibiotic resistance.

While bacteria exert substantial chemical resources to construct its cell wall peptidoglycan, the polymer itself is a dynamic structure which constantly sheds components of its glycan and peptide units via amidases, muramidases, endopeptidases, and lytic transglycosylases (Humann & Lenz, 2009; Typas et al., 2010; Vermassen et al., 2019). Some bacteria have evolved the ability to reuse their NAM PG building blocks in a very sophisticated recycling program (Litzinger & Mayer, 2010; Mayer, 2012). This specific cell wall recycling machinery, which is absent in *Escherichia coli*, yet present in many bacterial species including *Pseudomonas putida*, includes the recycling enzyme AmgK that converts NAM into MurNAc 1-phosphate, which is then converted to UDP-MurNAc (UDP-NAM) by MurU (Scheme 1). These enzymes produce UDP-NAM, a critical precursor involved in the initial steps of the bacterial cell wall biosynthesis (Barreteau et al., 2008; Brown et al., 1995). Intrigued by the functionality of these two recycling enzymes, as well as the peptidoglycan biosynthetic enzymes, a modular synthetic strategy was implemented to screen AmgK and MurU substrate specificity and synthetic utility (Liang et al., 2017).

The promiscuity of these recycling and biosynthetic programs was utilized for the chemical labeling and fluorescent visualization of the bacterial peptidoglycan carbohydrate backbone (Liang et al., 2017). NAM derivatives bearing an alkyne or azide handle at the 2-*N*-acetyl position were fed to

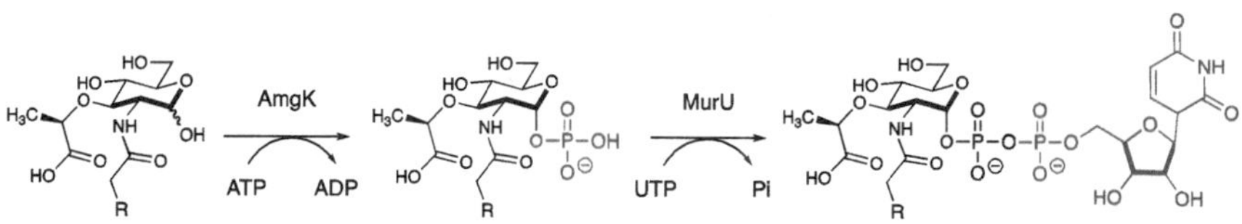

Scheme 1 Chemoenzymatic preparation of UDP-NAM. Enzymatic synthesis of muramic acid uridine diphosphate sugars with recycling enzymes AmgK and MurU.

genetically modified bacteria expressing the recycling genes *amgK* and *murU* (Gisin et al., 2013; Liang et al., 2017; Renner-Schneck et al., 2015). The modification permitted bio-orthogonal labeling and visualization of whole cells via click chemistry (Droumaguet et al., 2010; Kolb, Finn, & Sharpless, 2001; Liang et al., 2017; Rostovtsev et al., 2002). AmgK and MurU were used in situ to produce UDP-NAMs substrates which were subsequently used in in vivo labeling experiments (Liang et al., 2017). Upon completion of this work, we hypothesized that these enzymes could relieve the complex chemical synthetic efforts of UDP-NAM and its derivatives (Sadamoto et al., 2002), providing new small molecule probes to study bacterial PG structure and immune function at the carbohydrate level.

1.1 Rationale

Chemically synthesizing UDP-NAMs for use in a broad array of applications are challenging because these sugars contain two negative charges, making purification difficult. Additionally, one can expect a significant UDP-dimer side product in the activated coupling reaction that increasingly makes purification cumbersome resulting in low-yielding reactions (Hitchcock et al., 1998; Sadamoto et al., 2002). Here, we highlight that a shorter step, higher milligram quantity synthesis, and purification strategy of UDP-NAM derivatives is obtainable by chemoenzymatic synthesis utilizing the two major cell wall recycling enzymes AmgK and MurU, respectively (Gisin et al., 2013).

2. Materials and method

2.1 Materials, equipment and reagents

All chemicals were purchased from Sigma or ThermoFisher Scientific unless stated otherwise:

- DNA primers (Table 1) were synthesized by Eurofins Genomics LLC
- Protein purification column Bio-Rad
- Protease inhibitor tablets from Sigma-Aldrich

Table 1 List of primers.

Primer	Sequence
PpAmgK-For	5′GCTAGGATCCATGCCGGAACATGATG3′
PpAmgK-Rev	5′GATACTCGAGTTATGCACGTGCGCCTG3′
PpMurU-For	5′GCACTGGATCCATGAAAGCCATGATTC3′
PpMurU-Rev	5′GCATCCTCGAGTTATGCACGTTCACC3′

- GST Fast Flow resins were purchased from GE Healthcare Life Sciences
- Isopropyl β-D-1-thiogalactopyranoside (IPTG)
- Antibiotics were purchased from Gold Biotechnology or Sigma
- *E. coli* DH5α and BL21 (DE3) strains were from the Grimes laboratory stock
- *E. coli* strains were grown in liquid or solid Lysogeny broth (LB)
- Antibiotics: carbenicillin (100 mg/mL), kanamycin (5 mg/mL) or chloramphenicol (34 mg/mL)
- *P. putida* strain KT2440 was purchased from ATCC (ATCC number 47054)
- DNA ligase (NEB)
- Qiagen mini prep kit
- LB Agar culture plates
- PreScission Protease
- Bradford assay kit
- Glycerol
- Slick® Tubes (1.5 mL)
- Gel Imager
- Sonicator
- PCR machine
- Spectrophotometer
- Shimadzu Liquid Chromatography Mass Spectrometer
- Waters Autopurification System
- GE PD10 Desalting Columns
- Benchtop Centrifuge
- 5 kDa molecular weight spin filter (Corning)
- Bruker NMR

2.1.1 Reagents

- 1 M phosphate buffer pH, 7.9
- ATP 10 mM (Sigma-Aldrich)
- UTP 100 mM (Sigma-Aldrich)
- 100 mM Magnesium Chloride $MgCl_2$
- 10 mM NAM or NAM derivative
- Baker's yeast inorganic pyrophosphatase

2.2 Methods

2.2.1 Preparation of buffers (add DTT fresh into buffers)

2.2.1.1 GST lysis buffer

The buffer contains 150 mM NaCl, 50 mM Tris, 2 mM dithiothreitol (DTT) pH 7.0, containing one protease inhibitor cocktail tablet (add fresh before use; 1 tablet per 20 mL).

2.2.1.2 GST wash buffer

The wash buffer contains 500 mM NaCl, 50 mM Tris, 1 mM DTT, 1 mM EDTA pH 7.0.

2.2.1.3 GST elution buffer

The elution buffer contains 150 mM NaCl, 50 mM Tris, 1 mM DTT and 1 mM EDTA pH 7.0.

2.2.2 Cloning of AmgK and MurU constructs

Full-length *P. putida amgK* and *murU* were PCR amplified from *P. putida* genomic DNA (Gisin et al., 2013). The restriction sites used to generate the constructs are BamHI and *Xho*I for the forward and reverse primers, respectively (underlined in Table 1). The primers used to amplify AmgK and MurU are listed in Table 1 (PpAmgK-For/PpAmgK-Rev and PpMurU-For/PpMurU-Rev).

1. Perform PCR reaction and obtain PCR products of AmgK and MurU.
2. Clean up the PCR product by following the PCR purification protocol provided in the Qiagen mini prep kit.
3. In different tubes, digest the PCR product and pGEX-6P-1 vector with BamH1 and Xho1 restriction enzymes from NEB.
4. Run digestion products on a 1% DNA agarose gel, obtain DNA products by extracting the right gel band and purifying with Qiagen miniprep kit (follow the gel purification protocol provided in the kit).
5. Next, using NEB T4 ligase; ligate AmgK and MurU PCR product into pGEX-6P-1 vector to generate pGEX-PpAmgK and (in this step, ligate digested pGEX-6P-1 vector solely as a control) pGEX-PpMurU plasmids (allow ligation to proceed for approximately 2 h at room temperature).
6. Use the plasmid obtained after the ligation step to perform a transformation (by heat shock method) into BL21 (DE3) competent cells.
7. Spread the transformed cells on LB Agar plate (supplemented with 100 μg/mL total concentration of carbenicillin) and incubate at 37 °C overnight to produce colonies.
8. Pick colonies and incubate overnight in LB medium, harvest plasmids and confirm the gene inserts by PCR. Store cells with successfully cloned plasmid in 25% glycerol at −80 °C for long time storage.

2.2.3 Protein expression and purification

1. Make a 10 mL start-up cell culture by incubating cells containing pGEX-PpAmgK/pGEX-PpMurU plasmids into a 10 mL Fresh LB Medium that is supplemented with 100 μg/mL carbenicillin and allow to shake 200 rpm overnight at 37 °C.
2. Next day, use the start-up cell culture to inoculate 1 L fresh LB medium supplemented with carbenicillin antibiotics (add 100 μg/mL final concentration).
3. Shake the 1 L cell culture at 200 rpm at 37 °C until OD_{600nm} reaches 0.6.
4. Induce gene expression by adding 1 mM IPTG final concentration and continue incubation at 18 °C for 20 h or allow to shake in the incubator overnight.
5. Collect cells by centrifugation at 4000 *g* for 30 min at 4 °C (preset the centrifuge to 4 °C before spinning down cells).
6. Resuspend cells in 20 mL GST lysis buffer containing one protease inhibitor.
7. Next, vortex the cells until pellet is resuspended completely in solution with GST lysis buffer.
8. Sonicate the cells for 1 min, with a 1 min rest time (perform this step on ice) and repeat three times.
9. Centrifuge the lysed cells at 15,000 *g* for 40 min at 4 °C to remove the cell debris.
10. Pre-equilibrate the Glutathione Sepharose 4 Fastflow beads (GE Healthcare) with a column volume of lysis buffer.
11. Load the supernatant obtained from the previous step onto a protein purification column with pre-equilibrated Glutathione Sepharose 4 Fastflow beads.
12. Mix the lysed cell supernatant with the GST beads by shaking and incubate at 4 °C for 1 h while rotating in the cold.
13. After 1 h of rotation, collect a flow through and wash the column five times with 20 mL GST wash buffer (collect fractions in five different tubes).
14. After washing the beads, prepare to elute the protein of interest using a 10 mL GST elution buffer mixed with an appropriate amount of PreScission Protease and incubate at 4 °C overnight.
15. Next day, collect purified protein and analyze purity on SDS-Page gel.
16. Determine the concentration of the protein by Bradford assay. For long-term storage, glycerol can be added (20% final concentration)

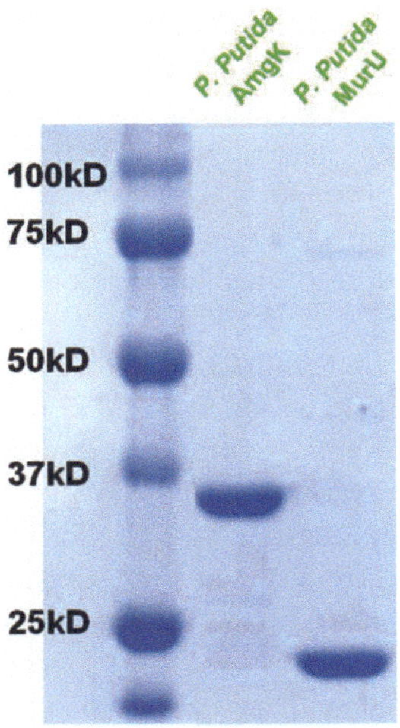

Fig. 3 Purified AmgK and MurU Enzymes. SDS page coomassie stained gel of purified AmgK (38 kDa) and MurU (24 kDa) enzymes.

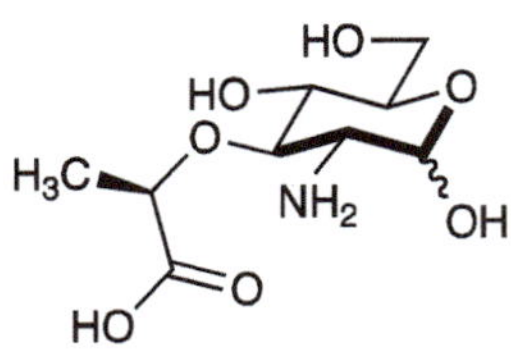

Fig. 4 Chemical structure of 2-amino muramic acid.

and the proteins stored at −80 °C. However, glycerol can interfere with purification of NAM products (see troubleshooting table). Avoid multiple freeze thaws of the purified protein (Fig. 3).

2.2.4 Synthesis of NAM derivatives

NAM derivatives with modifications at the 2-amino position were synthesized utilizing previously published protocol starting with commercially available glucosamine HCl as the common starting material (DeMeester et al., 2018, 2019; Liang et al., 2017). The modularity of this synthesis afforded large quantities (multi-grams) of the precursor, 2-amino muramic acid (DeMeester et al., 2019; Liang et al., 2017) (Fig. 4). This molecule was poised to functionalize using mild amide bond-coupling conditions with a variety of activated ester derivatives in order to generate a library of NAM compounds (DeMeester et al., 2018, 2019) (Scheme 2). This library includes multiple bio-orthogonal NAMs ranging from azides (compound **2**, Fig. 5A) and alkynes (compound **3**, Fig. 5A) for copper catalyzed azide-alkyne cycloaddition (CuAAC) or strain promoted azide-alkyne cycloaddition (SPAAC) "click" chemistry, ketone condensation intermediates (**5**) to inverse electron demand Diels–Alder (IEDDA) compounds (**4,9**) (Fig. 5A). Photoactivatable intermediates with diazirine and benzophenone functionality (**6,7,8**) were also synthesized in addition to difluoromethyl (**1**), fluorescein (**10**), and biotin tags **11,12**) (Fig. 5A) with the parent compound NAM shown as compound **13** (Fig. 5B).

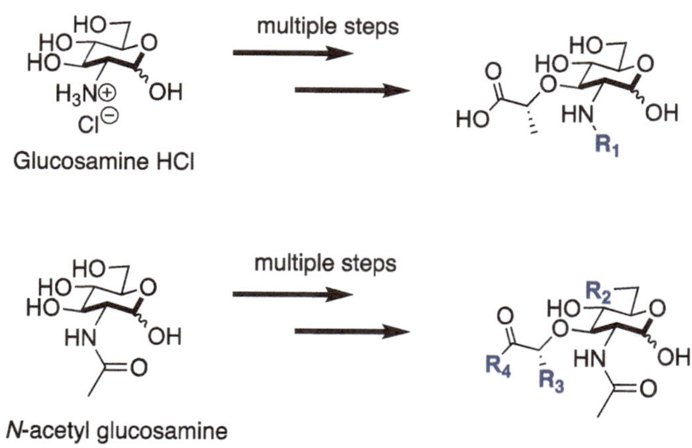

Scheme 2 Synthesis of 2-amino and *N* acetyl muramic acid derivatives. Beginning with commercially available glucosamine HCl, functionalization (R_1) at the 2-amino position was achieved. To prepare *N*-acetyl muramic acid derivatives with modifications at the 6-OH (R_2) muramic acid (R_3 and R_4), *N* acetyl glucosamine was used as starting material.

To assess if modifications around the 6-OH carbohydrate were tolerated a NAM derivative with an amine azide modification at the 6 position, 6-azido-NAM (**14, 15**), was synthesized (DeMeester et al., 2018; Schaefer et al., 2017). The final addition to this library of compounds included NAMs with modification at the 3-OH component of the sugar, extending off the muramic acid at both peptide (**16, 17, 18, 19**) and (*S*)-methyl (**20**) using previously reported syntheses (DeMeester et al., 2018; Melnyk et al., 2015; Schaefer et al., 2017). With these compounds in hand, we next screened the substrate specificity of the peptidoglycan recycling enzymes AmgK and MurU.

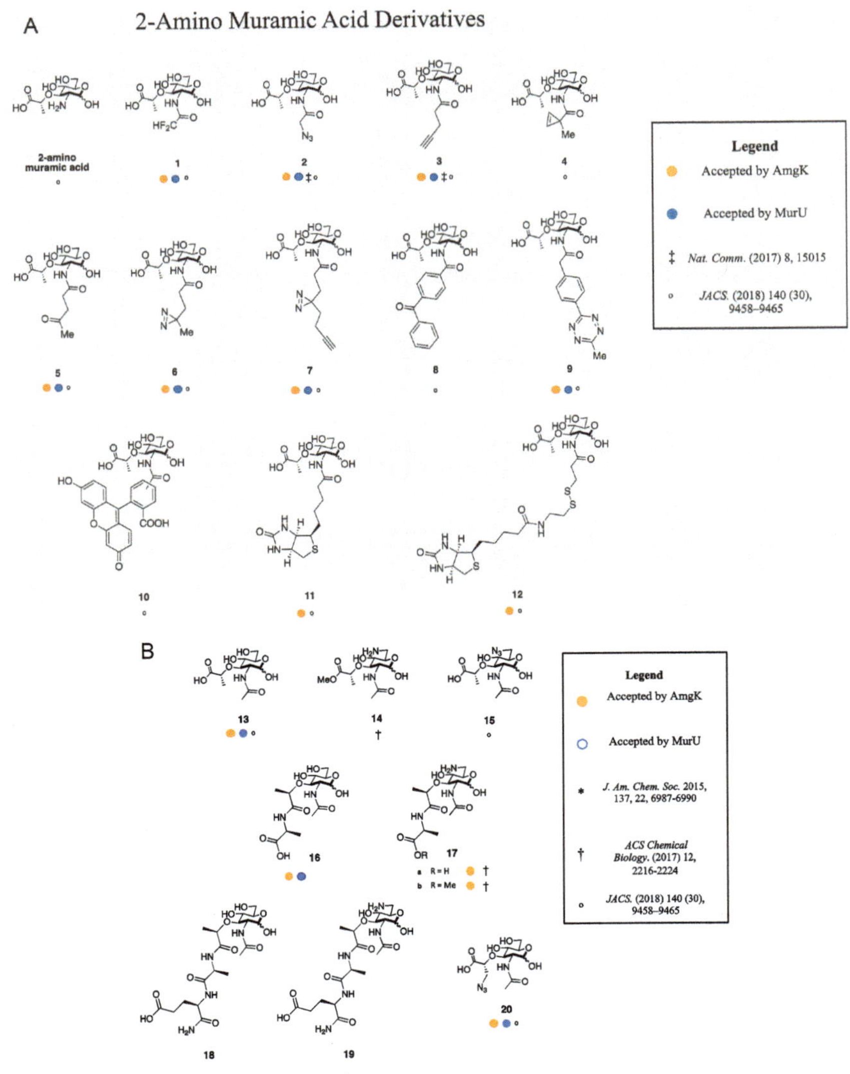

Fig. 5 Library of 2-amino (A) and *N*-acetyl (B) muramic acid derivatives.

2.2.5 Substrate specificity of AmgK and MurU

The substrate specificity of AmgK and MurU was probed initially with NAM derivatives (Fig. 5A and B) using liquid chromatography mass spectrometric detection system. The modifications were either on the

2-*N*-acetyl, 3, or 6-positions of the carbohydrate. Substrate specificity varied based on the modification. It was observed that AmgK and MurU accepted compounds **1–7, 9, 11, 12** that present small and moderately sized functionality at the 2-*N*-acetyl position as substrates (Fig. 5A). However, AmgK did not accept modifications where the alpha carbon did not contain any hydrogen atoms (Fig. 5A compounds **8, 10**). While the enzyme AmgK, accepted most modifications on the 2-*N*-acetyl position, it did not accept modifications on the 6-position (compounds **14, 15, 19**) unless the core sugar contained one L-alanine peptide unit on the 3-OH position (compound **17**). AmgK also tolerated modifications only present at the 3-OH position including both mono L alanine (**16**), and (*S*) methyl azide (**20**) but did not accept the dipeptide L-alanine D-isoglutamine (**18**). MurU appeared to be more promiscuous accepting all but three of the phosphate substrates (Fig. 5). All NAM derivatives accepted by both enzymes yielded the respective UDP carbohydrate products as confirmed by HRLC/MS. Overall, this preliminary substrate screen suggest that AmgK is "reading" the 6 position of the carbohydrate, as it will not accept any modification at this site, with the exception 6-amino (compounds **17a** and **b**, Fig. 5B); whereas it accepts any modification, even large biotin modifications, on the 2-position of the carbohydrate that contain the acetyl with a methylene unit (Fig. 5B). Once substrates have made it past the gatekeeper, AmgK, they are generally accepted by MurU, except for the biotin modification. This information provides insight into the nature of the AmgK-substrate binding and reactivity, as a crystal structure providing these details is only available to date for MurU (Renner-Schneck et al., 2015). We note that both enzymes could be engineered in future experiments to allow for larger modifications to be accepted.

To test substrate specificity of the recycling enzymes AmgK and MurU, please utilize the procedure below.

2.2.5.1 Protocol

1. Each NAM derivative was treated first with purified AmgK in a 100 μL reaction (Slick Tube).
2. Into a slick tube, add 66 μL milli-Q water, 5 μL of 1 M phosphate buffer, pH 7.9, 10 μL of 10 mM NAM derivative, 12 μL of 10 mM ATP, 3 μL of 10 mM $MgCl_2$ and 2 μL (20 μg/mL) of 1 mg/mL purified AmgK (Add enzyme last to start the reaction by carefully pipetting up and down).

3. Incubate the reaction at room temperature overnight.
4. Next day, prepare samples for mass spectrometry analysis; remove 10 µL into another tube and quench with 90 µL methanol).
5. Spin the quenched reaction at 8000 rpm on a benchtop centrifuge to precipitate the protein.
6. Then transfer supernatant to MS vial with insert for small volumes and proceed with MS analysis (the remaining volume will be carried forward into MurU reaction).
7. Into the remaining volume, add 12 µL of 10 mM UTP (Sigma-Aldrich), and 0.5 U of baker's yeast inorganic pyrophosphatase (Sigma-Aldrich).
8. Then 2.8 µL (20 µg/mL final concentration) of 0.71 mg/mL purified MurU enzyme to start the reaction.
9. Incubate the reaction at 37 °C overnight.
10. Remove 10 µL from the reaction and Quench with 90 µL methanol.
11. Spin the quenched reaction at 8000 rpm on a benchtop centrifuge to precipitate the protein.
12. Then transfer supernatant to MS vial with insert for small volumes and proceed with MS analysis.
13. Monitor product formation of each by HRLC/MS negative mode after quenching the reaction with methanol.

2.2.6 AmgK kinetic studies

The kinetic profile of the kinase, AmgK, in the presence of the unnatural NAM substrates were measured by the traditional NADH coupled assay. Briefly, ADP production is directly coupled to the oxidation of NADH to NAD^+ by pyruvate kinase and L-lactate dehydrogenase (LDH), respectively. It is well known that NADH strongly absorbs at 340 nm, while NAD^+ does not. Therefore, any reduction in concentration affects or decreases absorbance can be assessed by measuring absorbance at 340 nm. As is standard protocol, ATP was saturated at 5 mM, and the concentration of NAM substrates ranged from 0.05 to 5 mM. Utilizing the ATPase enzyme-coupled assay reported by Gisin et al. (2013), the catalytic efficiencies of **2** and **3** were 10-fold less than that of the natural substrate **13**, while compounds **5–7**, **9**, and **11** were 100-fold less efficient. However, the efficiency for **20** was similar to that of **13**, indicating that AmgK has some sensitivity at the 2-position (Table 2).

Table 2 Kinetic profile of AmgK with NAM derivatives.

Substrate	k_{cat} (s^{-1})	K_m (mM)	K_{cat}/K_m (M^{-1} S^{-1})
2[a]	14.07 ± 0.72	1.12 ± 0.16	$(1.26 \pm 0.19) \times 10^4$
3[a]	9.74 ± 0.45	0.55 ± 0.09	$(1.8 \pm 0.3) \times 10^4$
5[b]			3.46×10^3
6[b]			1.55×10^3
7[b]			1.02×10^3
9[b]			2.25×10^3
11[b]			2.95×10^3
13[a]	6.58 ± 0.21	$(3.04 \pm 0.58) \times 10^{-2}$	$(2.16 \pm 0.42) \times 10^5$
20[a]	4.67 ± 0.14	$(3.03 \pm 0.54) \times 10^{-2}$	$(1.54 \pm 0.28) \times 10^5$

[a]Kinetic parameters were calculated by fitting the experimental data into the Michaelis-Menten equation using program GraphicPad Prism-6.
[b]Apparent k_{cat}/K_m values were calculated by fitting the experimental data to the Lineweaver-Burk equation. Standard error (SE) was calculated based on three technical replicates of each sample concentration.

2.2.6.1 Protocol

1. In order to make a 200 μL reaction mixture, add 50 μL of 200 mM potassium phosphate buffer, pH 7.5, 40 μL of 10 mM $MgCl_2$, 2 μL of 0.1 M PEP, 4 μL of 10 mM NADH, 4 U Pyruvate kinase, 50 μL of 10 mM ATP, and 3 U (LDH) (all chemicals from Sigma-Aldrich).
2. Start the reaction by adding 1 μg/mL of AmgK enzyme at 25 °C. Measure the absorbance 340 nm every 5 s using a spectrophotometer.
3. Select the absorbance curve with a stable decrease and perform kinetic calculations using $6.2 \times 10^3\,M^{-1}\,cm^{-1}$ as the molar extinction coefficient of NADH at 340 nm.
4. Process the data according to the Michaelis-Menten Equation using GraphPad Prism 6 software or Lineweaver-Burk equation.

2.2.7 Synthesis and purification of clickable UDP-sugars

Compounds containing methyl, azide, alkyne, ketone and diazirine functionality (**13, 2, 3, 5**, and **6**) were subjected to a larger scale chemoenzymatic reaction with both AmgK and MurU to produce the desired UDP products (20–100 mg) (DeMeester et al., 2018). The aforementioned compounds were selected after confirming substrate specificity by mass spectrometry and obtaining kinetic evidence of AmgK reactivity. Representative reactions, purification strategies and NMR spectra are shown below.

2.2.7.1 Protocol

In order to make a 6 mL reaction sample, make the following dilutions from the stock solutions above and adjust volume as needed.

1. Into a tube, add about 1 mL milli-Q water, 0.4 mL of phosphate buffer (100 mM final concentration of phosphate buffer, pH 7.9), 0.8 mL of NAM or any NAM derivative, 1.6 mL of ATP, 40 μL of $MgCl_2$ and 270 μL of 1 mg/mL of purified AmgK for a 6 mL reaction sample (final concentration of enzyme is 45 μg/mL).
2. Incubate the reaction at room temperature overnight with gentle shaking.
3. Next day, remove 10 μL and quench with 90 μL methanol.
4. Spin the quenched reaction at 8000 rpm on a benchtop centrifuge to precipitate the protein.
5. Then transfer supernatant to MS vial with insert for small volumes and proceed with MS analysis (the remaining volume will be carried forward into MurU reaction).
6. Into the remaining volume, add 1.6 mL of UTP (Sigma-Aldrich), and 10 U of baker's yeast inorganic pyrophosphatase (Sigma-Aldrich).
7. Then add 378 μL (45 μg/mL final concentration) of purified MurU enzyme to start the reaction.
8. Incubate at 37 °C overnight.
9. Next day, centrifuged at 4 °C using a 5 kDa molecular weight spin filter (Corning).
10. Collect flow through and proceed to the purification of the UDP-NAM by HPLC.

2.2.8 HPLC purification of UDP-sugars

1. The supernatant from the enzymatic reaction after 5 kDa filtering was transferred directly to the purification system.
2. UDP-NAM purification is performed on the Waters 2767 Sample Manager with HPLC and SQD2 MS on a hydrophilic liquid chromatography (HILIC) preparatory column (Phenomenex) with the following method: flow rate 20 mL/min, injection 500 μL, 0.1% NH_4OH base in 92.5% HPLC grade acetonitrile and 7.5% Millipore H_2O as eluant A, and 0.1% NH_4OH base in 80% Millipore H_2O and 20% HPLC grade acetonitrile as eluant B. Inlet file (A/B): 0–1.0 min 95/5, 3.0–4.0 min 20/80, 4.3–5.0 min 95/5. Peaks were collected via mass and lyophilized (with water resuspension and lyophilization 2 ×) to yield purified products, UDP-2Az-NAM and UDP-2Alk-NAM on the 13.1 and 21.6 mg scale, respectively (Fig. 6).

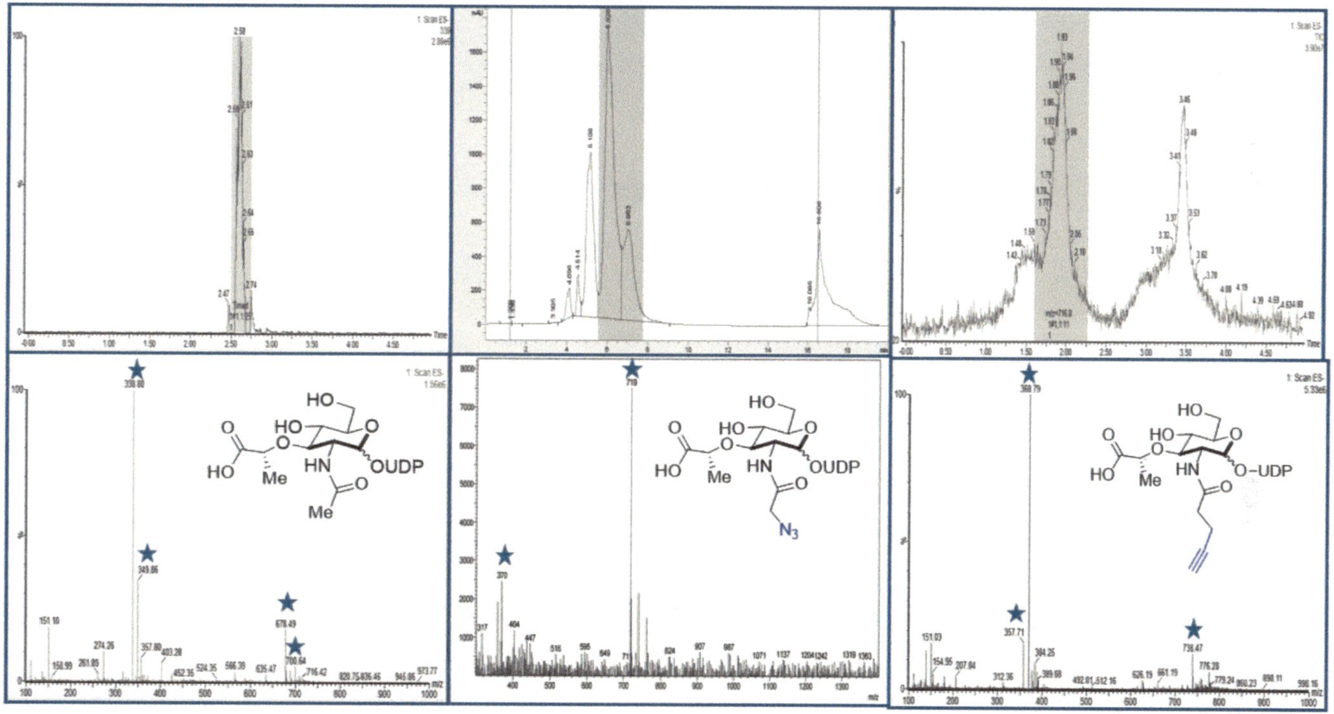

Fig. 6 HPLC purification of NAM UDP-sugars.

3. The products were passed through a PD10 (GE) desalting column as needed. Assigned NMR spectra for the aforementioned compounds are found below (Figs. 7–9). The yield obtained for purified UDP-2Az-NAM and 2Alk-NAM compounds are 19% and 89%, respectively (DeMeester et al., 2019).

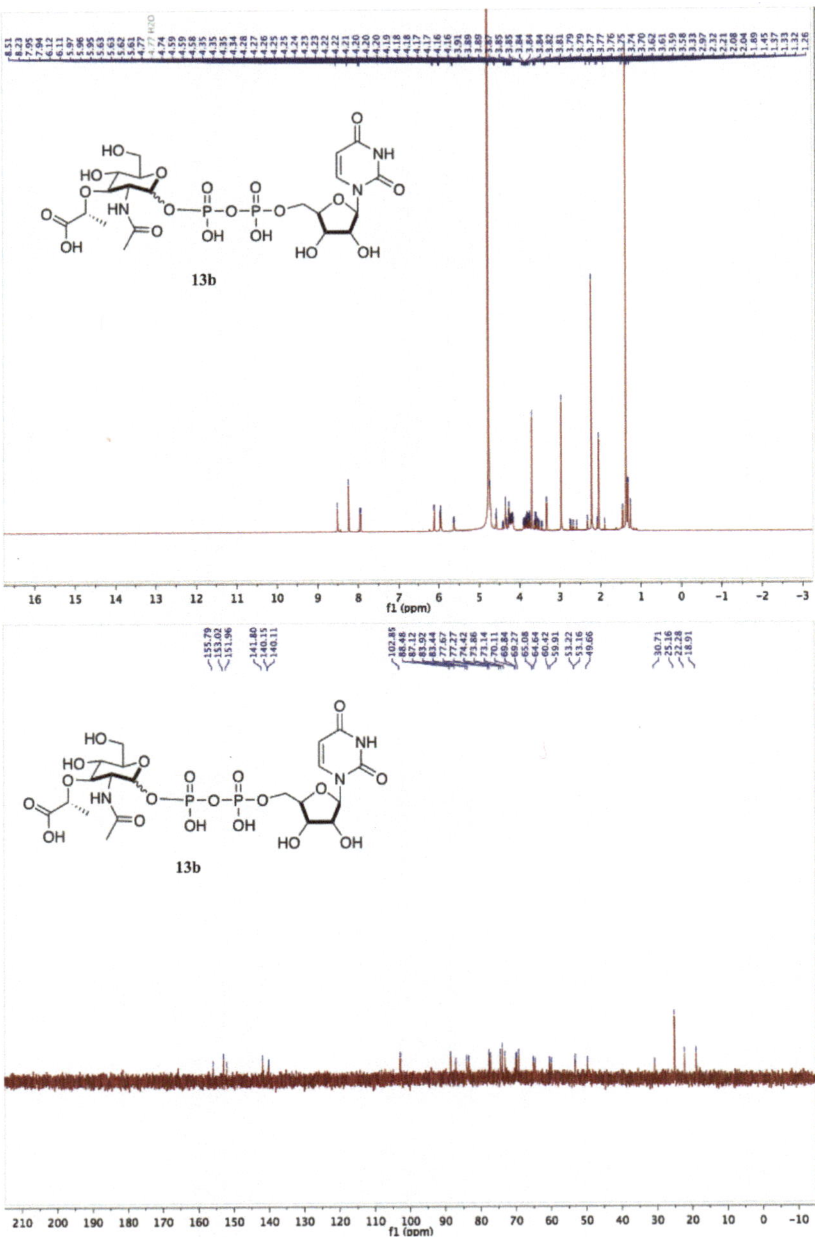

Fig. 7 ^{32}P NMR (top) and ^{1}H NMR (bottom) characterization of AmgK and MurU reaction with NAM to yield UDP NAM (compound **13b**).

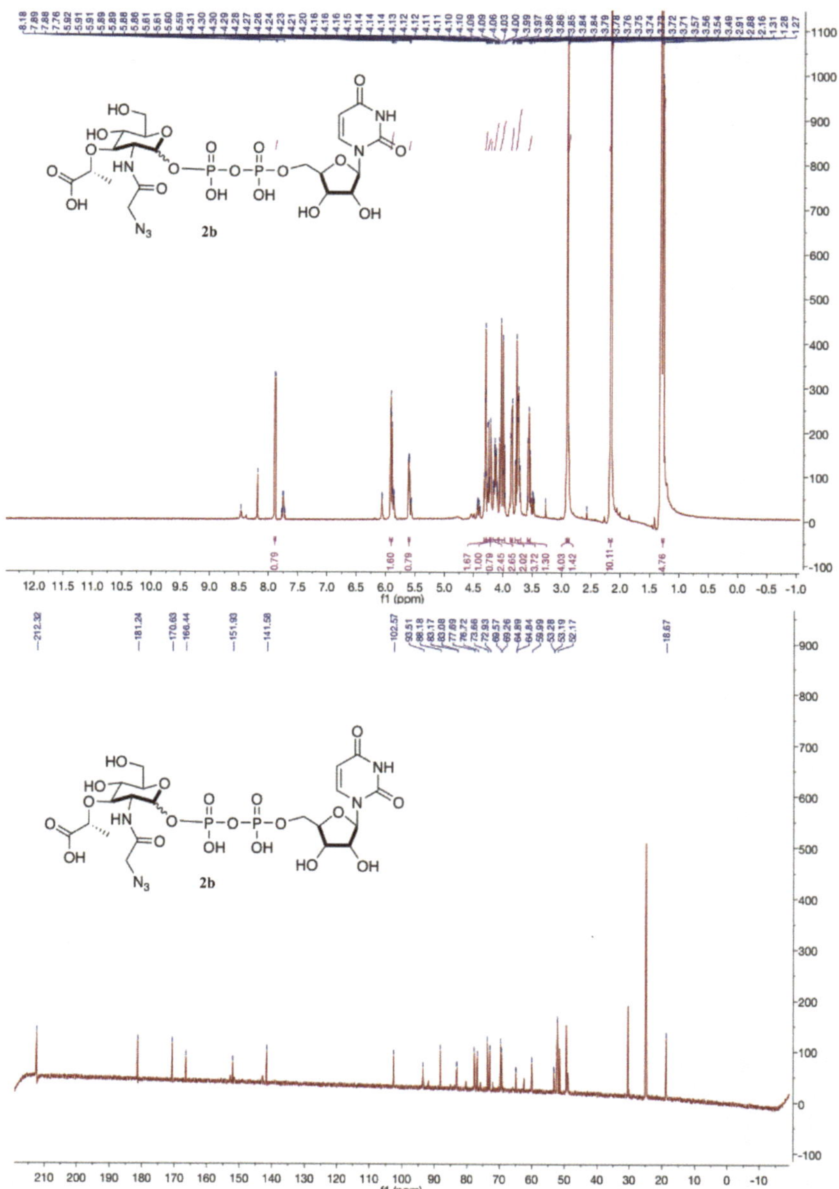

Fig. 8 ^{32}P NMR (top) and ^{1}H NMR (bottom) characterization of AmgK and MurU reaction with 2-azido NAM to yield UDP 2-azido NAM (compound **2b**).

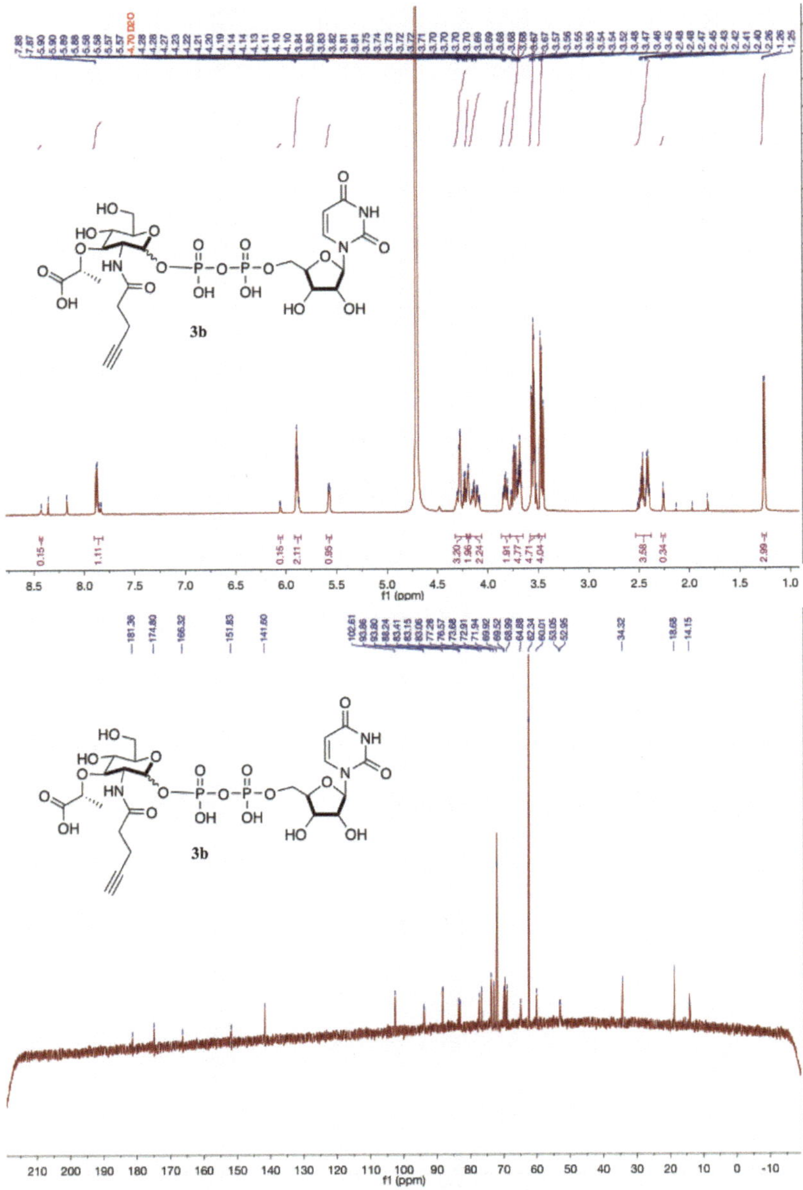

Fig. 9 ^{32}P NMR (top) and ^{1}H NMR (bottom) characterization of AmgK and MurU reaction with 2-alkyne NAM to yield UDP 2-alkyne NAM (compound **3b**).

Pros and cons

Pros	Cons
Chemoenzymatic synthesis of UDP-NAMs is less steps compared to organic synthesis and avoids complicated UDP dimer bi-products	Need molecular cloning and protein purification
Larger scale production of UDP-NAM	Enzyme activity may vary
Low cost, fast reaction with UDP products ready in 2 days	Main substrate (NAM or derivatives) needs to be synthesized chemically (2 weeks)
Most reagents are commercially available	NAM modifications on the 6-OH positions are not accepted by AmgK
Synthesis of different NAM derivatives with same enzyme reactions	Modifications with large functional groups on the 2-*N*-acetyl position are not accepted by the enzyme system

Troubleshooting and Optimization

Problem	Solution
Absence of product mass in the MS	• NAM or derivatives may be too dilute (slightly increase the concentration of NAM or derivative) • Make sure you are performing MS analysis in Negative mode • Check for the mass of the starting material (NAM or derivatives) • Check for the −2 charge state of the UDP products
Mass of starting material only	• For NAM derivative reactions, be sure to run a reaction with natural NAM in parallel to compare conversion • The enzymes may have lost function • Make sure the enzymes are still active and well folded (a circular dichroism can be used to characterize the enzyme folding state) • Avoid multiple freeze thaw cycles of the enzyme (make aliquots of the enzyme and store in the −80 °C)

Continued

Troubleshooting and Optimization—cont'd

Problem	Solution
	• Mix the reagents properly by pipetting up and down • Always use slick tubes to perform reaction
Low or no enzyme activity	• Check protein concentration and quality by SDS page gel • Test enzyme activity by running small reaction with NAM as substrate before each large-scale conversion
Slow reaction or low conversion	• Add more freshly prepared enzyme into reaction system. Also, check the hydrolysis of UTP or ATP substrates, and make sure the inorganic pyrophosphatase is added
Having trouble accessing clean NMR spectra, or your yield is greater than 100%	• Pass the purified product through PD10 desalting column and undergo multiple rounds of lyophilization

3. Summary

Glycosylation is a powerful enzymatic process that plays a pivotal role in all living organisms including prokaryotic cells. The modification controls numerous cellular processes including cell and protein stability and signaling (Ohtsubo & Marth, 2006). UDP NAM is an essential building block of bacterial peptidoglycan biosynthesis. While synthetic preparation of this molecule and derivatives thereof prove challenging, we outline here a robust and scalable step-by-step chemoenzymatic synthesis of UDP-NAM chemical probes. This process harnesses the promiscuity of bacterial cell wall recycling enzymes AmgK and MurU to generate a library of compounds that are of interest to the carbohydrate, microbiology and immunology communities studying nucleotide phosphates and their role in bacterial survival.

Acknowledgments

For financial support, this project was supported by a grant to C.L.G. from the Glycoscience Common fund (U01 CA221230); the Delaware COBRE program, with a grant from the National Institute of General Medical Sciences—NIGMS (P20GM104316). C.L.G. is a Pew Biomedical Scholar, Sloan Scholar, and Camille Dreyfus Scholar and thanks the Pew Foundation, the Sloan Foundation for Science Advancement and the Dreyfus foundation for support. O.I.U. and K.E.D. thanks the NIH Chemistry-Biology Interface Program (GM133395) and the University of Delaware Dissertation Fellowship for funding support. For instrumentation support, the Delaware COBRE and INBRE programs supported this project with a grant from the National Institute of General Medical Sciences–NIGMS (5 P30 GM110758, P20GM104316-01A1 and P20 GM103446) from the National Institutes of Health. We thank Papa Nii Asare-Okai, the director of the MS facility and Dr. Steve Bai the director of the NMR facility in the Chemistry and Biochemistry Department at the University of Delaware, with assistance in MS and NMR, respectively. We thank James Melnyk and Lauren Genova for the synthesis of muramyl monopeptide and muramyl dipeptide compound and Matthew Jensen for the 3-azido NAM compound. Also, we are grateful for the Fox lab at the University of Delaware for supplying the tetrazine compounds.

References

Alvarez, L., et al. (2014). Peptidoglycan remodeling by the coordinated action of multispecific enzymes. *Microbial Drug Resistance*, *20*(3), 190–198. https://doi.org/10.1089/mdr.2014.0047.

Barreteau, H., et al. (2008). Cytoplasmic steps of peptidoglycan biosynthesis. *FEMS Microbiology Reviews*, *32*(2), 168–207. https://doi.org/10.1111/j.1574-6976.2008.00104.x, John Wiley & Sons, Ltd (10.1111).

Barrett, D., et al. (2007). Analysis of glycan polymers produced by peptidoglycan glycosyltransferases. *Journal of Biological Chemistry*, *282*, 31964–31971. https://doi.org/10.1074/jbc.M705440200.

Brown, E. D., et al. (1995). MurA (MurZ), the enzyme that catalyzes the first committed step in peptidoglycan biosynthesis, is essential in *Escherichia coli*. *Journal of Bacteriology*, *177*(14), 4194–4197. https://doi.org/10.1128/jb.177.14.4194-4197.1995, American Society for Microbiology Journals.

Brown, S., et al. (2012). Methicillin resistance in Staphylococcus aureus requires glycosylated wall teichoic acids. *Proceedings of the National Academy of Sciences of the United States of America*, *109*, 18909–18914. https://doi.org/10.1073/pnas.1209126109.

Compain, P., & Martin, O. R. (2001). Carbohydrate mimetics-based glycosyltransferase inhibitors. *Bioorganic and Medicinal Chemistry*, *9*, 3077–3092. https://doi.org/10.1016/S0968-0896(01)00176-6.

Cote, J. M., & Taylor, E. A. (2017). The glycosyltransferases of LPS core: A review of four heptosyltransferase enzymes in context. *International Journal of Molecular Sciences*, *18*(11), 2256. https://doi.org/10.3390/ijms18112256, MDPI.

Decker, D., & Kleczkowski, L. A. (2019). UDP-sugar producing pyrophosphorylases: Distinct and essential enzymes with overlapping substrate specificities, providing de novo precursors for glycosylation reactions. *Frontiers in Plant Science*, *9*, 1822. https://doi.org/10.3389/fpls.2018.01822.

DeMeester, K. E., et al. (2018). Synthesis of functionalized *N*-acetyl muramic acids to probe bacterial cell wall recycling and biosynthesis. *Journal of the American Chemical Society*, *140*(30), 9458–9465. https://doi.org/10.1021/jacs.8b03304, American Chemical Society.

DeMeester, K. E., et al. (2019). Metabolic incorporation of *N*-acetyl muramic acid probes into bacterial peptidoglycan. *Current Protocols in Chemical Biology*, *11*(4), e74. https://doi.org/10.1002/cpch.74, John Wiley & Sons, Ltd.

Drake, W. R., et al. (2018). New use for CETSA: Monitoring innate immune receptor stability via post-translational modification by OGT. *Journal of Bioenergetics and Biomembranes*, *50*(3), 231–240. https://doi.org/10.1007/s10863-018-9754-z, Springer US.

Droumaguet, L., et al. (2010). (6) For selected references to CuAAC applications in 2010, see: (a) Sharma, P.; Moses. *Journal of the American Chemical Society*. https://doi.org/10.1021/jo101495k.

Gisin, J., et al. (2013). A cell wall recycling shortcut that bypasses peptidoglycan de novo biosynthesis. *Nature Chemical Biology*, *9*, 491–493. https://doi.org/10.1038/nchembio.1289.

Goto, S., et al. (2001). UDP–sugar transporter implicated in glycosylation and processing of Notch. *Nature Cell Biology*, *3*(9), 816–822. https://doi.org/10.1038/ncb0901-816.

Hadley, B., et al. (2014). Structure and function of nucleotide sugar transporters: Current progress. *Computational and Structural Biotechnology Journal*, *10*(16), 23–32. https://doi.org/10.1016/j.csbj.2014.05.003, Research Network of Computational and Structural Biotechnology.

Hart, G. W., Housley, M. P., & Slawson, C. (2007). Cycling of O-linked β-N-acetylglucosamine on nucleocytoplasmic proteins. *Nature*, *446*(7139), 1017–1022. https://doi.org/10.1038/nature05815.

Hitchcock, S. A., et al. (1998). The first total synthesis of bacterial cell wall precursor UDP−N-acetylmuramyl-pentapeptide (park nucleotide). *American Chemical Society*, *120*, 1916–1917. https://doi.org/10.1021/JA973172D.

Humann, J., & Lenz, L. L. (2009). Bacterial peptidoglycan-degrading enzymes and their impact on host muropeptide detection. *Journal of Innate Immunity*, *1*, 88–97. https://doi.org/10.1159/000181181.

Hwang, H., et al. (2010). Glycoproteomics in neurodegenerative diseases. *Mass Spectrometry Reviews*, *29*(1), 79–125. https://doi.org/10.1002/mas.20221.

Johnson, J. W., Fisher, J. F., & Mobashery, S. (2013). Bacterial cell-wall recycling. *Annals of the New York Academy of Sciences*, *1277*(1), 54–75. https://doi.org/10.1111/j.1749-6632.2012.06813.x, 2012/11/16.

Kay, E., Cuccui, J., & Wren, B. W. (2019). Recent advances in the production of recombinant glycoconjugate vaccines. *NPJ Vaccines*, *4*(1), 16. https://doi.org/10.1038/s41541-019-0110-z, Nature Publishing Group.

Kolb, H. C., Finn, M. G., & Sharpless, K. B. (2001). Click chemistry: Diverse chemical function from a few good reactions. *Angewandte Chemie—International Edition*, *40*, 2004–2021. https://doi.org/10.1002/1521-3773(20010601)40:11<2004::AID-ANIE2004>3.0.CO;2-5.

Lairson, L. L., et al. (2008). Glycosyltransferases: Structures, functions, and mechanisms. *Annual Review of Biochemistry*, 77(1), 521–555. https://doi.org/10.1146/annurev.biochem.76.061005.092322.

Leimkuhler, C., et al. (2005). Glycosylation of glycopeptides: A comparison of chemoenzymatic and chemical methods. *Tetrahedron: Asymmetry*, *16*(2), 599–603. https://doi.org/10.1016/J.TETASY.2004.12.006, Pergamon.

Liang, H., et al. (2017). Metabolic labelling of the carbohydrate core in bacterial peptidoglycan and its applications. *Nature Communications*, *8*, 15015. The author(s). Available at https://doi.org/10.1038/ncomms15015.

Litzinger, S., & Mayer, C. (2010). In H. König, H. Claus, & A. Varma (Eds.), *The Murein Sacculus BT—Prokaryotic cell wall compounds: Structure and biochemistry* (pp. 3–52). Berlin, Heidelberg: Springer Berlin Heidelberg. https://doi.org/10.1007/978-3-642-05062-6_1.

Mayer, C. (2012). Bacterial cell wall recycling. *eLS*. https://doi.org/10.1002/9780470015902.a0021974, (Major Reference Works).

McCarthy, P. C., Sharyan, A., & Sheikhi Moghaddam, L. (2018). Meningococcal vaccines: Current status and emerging strategies. *Vaccines*, *6*(1), 12. https://doi.org/10.3390/vaccines6010012, MDPI.

Melnyk, J. E., et al. (2015). Peptidoglycan modifications tune the stability and function of the innate immune receptor Nod2. *Journal of the American Chemical Society*, *137*(22), 6987–6990. https://doi.org/10.1021/jacs.5b01607.
Mistretta, N., et al. (2019). Glycosylation of Staphylococcus aureus cell wall teichoic acid is influenced by environmental conditions. *Scientific Reports*, *9*, 3212. https://doi.org/10.1038/s41598-019-39929-1.
Ohtsubo, K., & Marth, J. D. (2006). Glycosylation in cellular mechanisms of health and disease. *Cell*, *126*(5), 855–867. https://doi.org/10.1016/j.cell.2006.08.019.
Renner-Schneck, M., et al. (2015). Crystal structure of the *N*-acetylmuramic acid α-1-phosphate (MurNAc-α1-P) uridylyltransferase MurU, a minimal sugar nucleotidyltransferase and potential drug target enzyme in Gram-negative pathogens. *Journal of Biological Chemistry*, *290*(17), 10804–10813. https://doi.org/10.1074/jbc.M114.620989.
Rostovtsev, V. V., et al. (2002). Sharpless—A stepwise Huisgen cycloaddition process.pdf. *Angewandte Chemie (International ed in English)*. https://doi.org/10.1021/jo011148j.
Sadamoto, R., et al. (2002). Cell-wall engineering of living bacteria. *American Chemical Society*, *124*, 9018–9019. https://doi.org/10.1021/JA026133X.
Schaefer, A. K., et al. (2017). Membrane association dictates ligand specificity for the innate immune receptor NOD2. *ACS Chemical Biology*, *12*, 2216–2224. https://doi.org/10.1021/acschembio.7b00469.
Scheffers, D.-J., & Pinho, M. G. (2005). Bacterial cell wall synthesis: New insights from localization studies. *Microbiology and Molecular Biology Reviews: MMBR*, *69*(4), 585–607. https://doi.org/10.1128/MMBR.69.4.585-607.2005, American Society for Microbiology.
Schleifer, K. H., & Kandler, O. (1972). Peptidoglycan types of bacterial cell walls and their taxonomic implications. *Bacteriological Reviews*, *36*(4), 407–477. Available at https://www.ncbi.nlm.nih.gov/pubmed/4568761.
Seyrek, K., Richter, M., & Lavrik, I. N. (2019). Decoding the sweet regulation of apoptosis: The role of glycosylation and galectins in apoptotic signaling pathways. *Cell Death & Differentiation*, *26*(6), 981–993. https://doi.org/10.1038/s41418-019-0317-6, Nature Publishing Group.
Silver, L. L. (2003). Novel inhibitors of bacterial cell wall synthesis. In *Current opinion in microbiology* (pp. 431–438). Elsevier Ltd. https://doi.org/10.1016/j.mib.2003.08.004.
Solá, R. J., & Griebenow, K. (2009). Effects of glycosylation on the stability of protein pharmaceuticals. *Journal of Pharmaceutical Sciences*, *98*(4), 1223–1245. https://doi.org/10.1002/jps.21504.
Strominger, J. L. (2007). Bacterial cell walls, innate immunity and immunoadjuvants. *Nature Immunology*, *8*(12), 1269–1271. https://doi.org/10.1038/ni1207-1269.
Swoboda, J. G., et al. (2010). Wall teichoic acid function, biosynthesis, and inhibition. *Chembiochem: A European Journal of Chemical Biology*, *11*(1), 35–45. https://doi.org/10.1002/cbic.200900557.
Taguchi, A., et al. (2019). FtsW is a peptidoglycan polymerase that is functional only in complex with its cognate penicillin-binding protein. *Nature Microbiology*, *4*, 587–594. https://doi.org/10.1038/s41564-018-0345-x.
Typas, A., et al. (2010). Regulation of peptidoglycan synthesis by outer-membrane proteins. *Cell*, *143*(7), 1097–1109. https://doi.org/10.1016/j.cell.2010.11.038.
Vermassen, A., et al. (2019). Cell wall hydrolases in bacteria: Insight on the diversity of cell wall amidases, glycosidases and peptidases toward peptidoglycan. *Frontiers in Microbiology*, *10*, 331. Available at: https://www.frontiersin.org/article/10.3389/fmicb.2019.00331.
Wagstaff, B. A., et al. (2015). Enzymatic synthesis of nucleobase-modified UDP-sugars: Scope and limitations. *Carbohydrate Research*, *404*, 17–25. https://doi.org/10.1016/j.carres.2014.12.005.

Walsh, C. (2000). Molecular mechanisms that confer antibacterial drug resistance. *Nature, 406*(6797), 775–781. https://doi.org/10.1038/35021219.

Watanakunakorn, C. (1984). Mode of action and in-vitro activity of vancomycin. *The Journal of Antimicrobial Chemotherapy, 14*(Suppl. D), 7–18. https://doi.org/10.1093/jac/14.suppl_d.7.

Xia, G., et al. (2010). Glycosylation of wall teichoic acid in *Staphylococcus aureus* by TarM. *Journal of Biological Chemistry, 285*, 13405–13415. https://doi.org/10.1074/jbc.M109.096172.

CHAPTER TWO

Chemical tools for selective activity profiling of bacterial penicillin-binding proteins

Shabnam Sharifzadeh[a], Nathaniel W. Brown[a,†], Joshua D. Shirley[b,†], Kevin E. Bruce[c], Malcolm E. Winkler[c], Erin E. Carlson[a,b,d,*]

[a]Department of Chemistry, University of Minnesota, Minneapolis, MN, United States
[b]Department of Medicinal Chemistry, University of Minnesota, Minneapolis, MN, United States
[c]Department of Biology, Indiana University, Bloomington, IN, United States
[d]Department of Biochemistry, Molecular Biology and Biophysics, University of Minnesota, Minneapolis, MN, United States
*Corresponding author: e-mail address: carlsone@umn.edu

Contents

† Equal contributions.

Methods in Enzymology, Volume 638
ISSN 0076-6879
https://doi.org/10.1016/bs.mie.2020.02.015

Abstract

Penicillin-binding proteins (PBPs) are membrane-associated proteins involved in the biosynthesis of peptidoglycan (PG), the main component of bacterial cell walls. These proteins were discovered and named for their affinity to bind the β-lactam antibiotic penicillin. The importance of the PBPs has long been appreciated; however, specific roles of individual family members in each bacterial strain, as well as their protein-protein interactions, are yet to be understood. The apparent functional redundancy of the 4–18 PBPs that most eubacteria possess makes determination of their individual roles difficult. Existing techniques to study PBPs are not ideal because they do not directly visualize protein activity and can suffer from artifacts and perturbations of native PBP function. Therefore, development of new methods for studying the roles of individual PBPs in cell wall synthesis is required. We recently generated a library of fluorescent chemical probes containing a β-lactone scaffold that specifically targets the PBPs, enabling the visualization of their catalytic activity. Herein, we describe a general protocol to label and detect the activity of individual PBPs in *Streptococcus pneumoniae* using our fluorescent β-lactone probes.

1. Introduction

Peptidoglycan (PG) is a heteroglycan polymer crosslinked by short peptide chains branched off of the glycan backbone that makes up the primary component of the bacterial cell wall. PG is essential for bacterial cell growth and survival, and as such presents an ideal target for antibiotics, as the cell wall and the concomitant machinery are exclusively present in prokaryotic organisms. Penicillin-binding proteins (PBPs) are a family of bacterial enzymes responsible for the synthesis of PG *via* the polymerization of the glycan chain (transglycosylation) and the cross-linking of peptide chains (transpeptidation). β-Lactam antibiotics such as penicillin inhibit the transpeptidation reaction by covalently binding to the PBPs, resulting in cell lysis, and as such are some of the most effective treatments for bacterial infection. Despite the importance of the PBPs in bacterial growth and the clinical importance of cell wall-targeting antibiotics, little is known about the functions and regulation of individual PBP isoforms within the same organism. Most bacteria contain multiple PBPs that are highly structurally homologous and functionally redundant. The inability to elucidate the catalytic activity, localization, and regulation of individual PBPs represents a critical gap in knowledge in the battle against antimicrobial resistance.

Current strategies to study bacterial cell wall synthesis and remodeling have relied extensively upon the β-lactam scaffold since it was determined that penicillin V is a global inhibitor of PBPs (Böttcher & Sieber, 2012; Falconer, Czarny, & Brown, 2011). A common strategy to study PBP catalytic activity is to use radiolabeled or fluorescently-tagged penicillin analogs, followed by gel separation and visualization of the tagged proteins (Ghuysen, 1977; Lakaye et al., 1994; Spratt & Pardee, 1975; Waxman & Strominger, 1983; Zhao, Meier, Kahl, Gee, & Blaszczak, 1999). However, this strategy cannot report on individual PBP catalytic activity, as most penicillin-based compounds inhibit all PBPs within a given organism. Fluorescent fusion proteins of the PBPs have been utilized, but these artificial fusions can interfere with native protein abundance, localization, or activity (Margolin, 2012; Swulius & Jensen, 2012). Furthermore, such fusion tags only report on protein localization, not activity.

Fluorescent small molecules offer spatiotemporal resolution while eliminating the need for genetic manipulation. For example, newly synthesized PG can be visualized with fluorescent vancomycin (Van-FL) and ramoplanin, which bind to PG precursors in Gram-positive bacteria (Daniel & Errington, 2003; Tiyanont et al., 2006). Other strategies have used fluorescent D-amino acid analogs (FDAAs) that are incorporated into the PG by PG synthesis machinery, enabling visualization of nascent peptidoglycan molecules (Fura, Kearns, & Pires, 2015; Kuru et al., 2012; Lebar et al., 2014; Pidgeon et al., 2015; Siegrist et al., 2013).

In order to tease apart the various roles of the PBPs and regulation of PG synthesis and remodeling, β-lactam-based probes have been synthesized for *in vitro* labeling, primarily targeting the PBPs in addition to other bacterial proteins (Dargis & Malouin, 1994; Staub & Sieber, 2008; Zhao et al., 1999). The Carlson lab reported that the antibiotic cephalosporin C could be modified to produce a chemical probe that selectively labeled a subset of PBPs in *S. pneumoniae* and *Bacillus subtilis* (Kocaoglu et al., 2012). In order to identify selective probes for individual PBP isoforms, a library of 20 commercially available β-lactams was screened against *S. pneumoniae*, *B. subtilis* and *Escherichia coli* (Kocaoglu & Carlson, 2015; Kocaoglu, Tsui, Winkler, & Carlson, 2015; Sharifzadeh, Dempwolff, Kearns, & Carlson, 2020). Of the screened compounds, several possessed selectivity for one or several PBPs at different concentrations of the β-lactam. However, many PBPs were not inhibited by available β-lactams, indicating a need to further explore chemical space. To that end, the Carlson lab has identified a β-lactone scaffold as a promising foundation from which selective PBP inhibitors can be developed and has applied these reagents to the selective imaging of essential PBPs, PBP2x and PBP2b, in *S. pneumoniae*.

2. Activity-based profiling of penicillin-binding proteins

Activity-based protein profiling (ABPP) methods have been utilized for the last two decades as a means to investigate complex proteomes in all three domains of life. Extensive use in eukaryotes has provided a more fundamental understanding of the biological processes involved in numerous diseases, which has played a key role in drug discovery campaigns. However, the use of ABPP in prokaryotes has lagged behind and has only gained more attention over the last decade. The alarming rise in antibiotic resistance has called for a better understanding of bacterial physiology and pathogenicity at a foundational level. ABPP provides a means by which new, clinically relevant drug targets may be identified through gaining insight into key biological processes. Several reviews are available on the use of ABPP, including their utility in prokaryotic systems (Cravatt, Wright, & Kozarich, 2008; Keller, Babin, Lakemeyer, & Bogyo, 2019; Krysiak et al., 2017; Sharifzadeh, Shirley, & Carlson et al., 2019). This chapter elaborates on the activity-based probes (ABPs) that we have applied to selectively profile individual members of the PBP enzyme family. Here, the methodology employed to design and apply a novel class of PBP ABPs containing a β-lactone scaffold is discussed in detail (Sharifzadeh et al., 2017). Our studies focused on *S. pneumoniae*, an ovoid-shaped Gram-positive bacteria that causes serious diseases such as pneumonia, bacteremia, and meningitis, and is the major cause of worldwide childhood mortality from infectious disease (Henriques-Normark & Tuomanen, 2013). In addition to being clinically significant, *S. pneumoniae* is an important model of PG biosynthesis as it possesses a relatively simple PBP profile, with three class A PBPs (PBP1a, PBP1b, and PBP2a), two class B PBPs (PBP2x and PBP2b), and one class C PBP (PBP3 or DacA; D,D-carboxypeptidase).

2.1 Activity-based probes: What they are and how they work?

ABPs represent any active-site-directed chemical probe that reacts with the enzyme in a mechanism-based manner, covalently attaches to a catalytic residue, and is retained in the active site (Heal, Dang, & Tate, 2011). As such, ABPP has emerged as a powerful approach to monitor the activity of proteins in complex proteomes. This strategy enables exclusive visualization of the catalytically active form of an enzyme. ABPs are mechanism-based irreversible inhibitors consisting of an *active warhead*, which reacts covalently with a nucleophilic active-site residue, a binding or *spacer group* that provides spacing between the active warhead and the *reporter*, which is a moiety that serves as a tag for identification and purification purposes (Fig. 1A). The active warhead is typically an electrophile that selectively reacts with an active-site nucleophile and is retained on-site due to the formation of a covalent bond. Since only catalytically competent enzymes can undergo this reaction, active isoforms are selectively visualized. Upon enzyme binding, the reporter tag of ABPs enables direct visualization of the probe (gel-based or imaging) or enrichment and protein identification (mass spectrometry) depending on the type of reporter used (Sadaghiani, Verhelst, & Bogyo, 2007). Initial ABPP studies took advantage of direct radiochemical detection through incorporation of ^{125}I, which is not applicable in large scale analysis due to limitations imposed on radioactive material (Bogyo, Verhelst, Bellingard-Dubouchaud, Toba, & Greenbaum, 2000). More recently, fluorescent and affinity tags were incorporated and have been utilized to detect enzymatic labeling of ABPs (Patricelli, Giang, Stamp, & Burbaum, 2001). Biotinylated ABPs have been developed to enable target enrichment, a more time-consuming detection strategy. Fluorescent ABPs are the most direct method for in-gel detection of activity with high sensitivity and throughput.

The irreversible, covalent nature of the ABPs' bond with their targets has made them amenable for diverse applications including live-cell imaging, target enrichment and direct biochemical analysis of the tagged proteins (Fig. 1B). Moreover, assessment of the selectivity of enzyme inhibitors is another potential application for activity-based proteomics. In bacterial systems, ABPs have been applied for various applications such as target discovery, the study of microbial pathogenesis and host-pathogen interactions, live-cell imaging, and inhibitor discovery (Böttcher & Sieber, 2008; Heal et al., 2011; Sadler & Wright, 2015; Sharifzadeh et al., 2017; Willems, Overkleeft, & van Kasteren, 2014).

2.2 PBP-selective ABPs

The PBPs are PG synthase enzymes that were discovered as the targets of the β-lactam antibiotic, penicillin. Aside from their therapeutic application, β-lactams have been used as chemical probes to study the PBPs (Chambers, Sachdeva, & Hackbarth, 1994; Curtis, Orr, Ross, & Boulton, 1979; Kocaoglu et al., 2012; Staub & Sieber, 2008; Williamson, Hakenbeck, & Tomasz, 1980; Zhao et al., 1999). β-Lactams structurally resemble the terminal D-Ala-D-Ala moiety of PG stem peptides, and hence

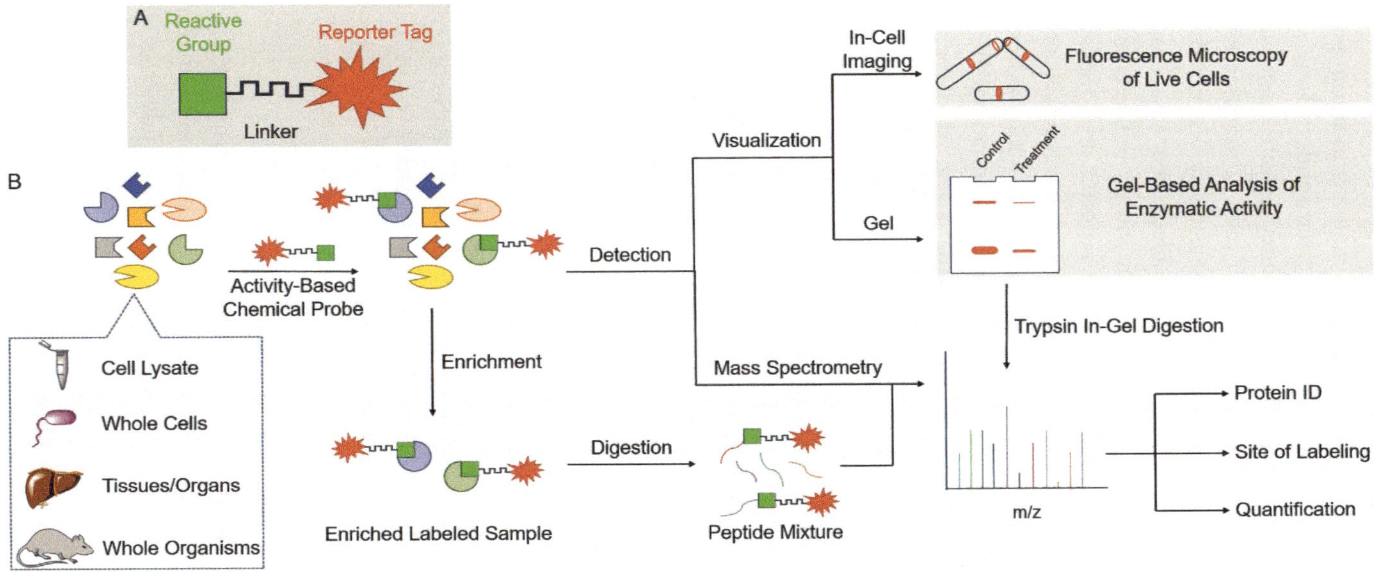

Fig. 1 (A) General structure of an ABP. An electrophilic group is used most commonly for the reactive group, which is linked to a biotin or fluorophore molecule by an unreactive linker. (B) General workflow of ABPP. An ABP is applied to a sample of interest, which can then be processed in a multitude of ways, depending on the application. The end-result is a labeled proteome that may be visualized *via* gel-based analysis or mass spectrometry (MS). Analysis of the enriched proteome could provide additional information about the enriched proteins such as identity, quantity and the site of labeling. *Adopted from Sharifzadeh, S., Shirley, J. D., & Carlson, E. E. (2019). Activity-based protein profiling methods to study bacteria: The power of small-molecule electrophiles.* Current Topics in Microbiology and Immunology, 420, *23–48. doi:10.1007/82_2018_135.*

the transpeptidation (TP) domain of the PBPs recognizes them as their natural substrate. During the TP step, a conserved serine residue in the active site attacks the β-lactam ring in a mechanism-based manner, leading to formation of an inert acyl-enzyme complex that inhibits biosynthesis of the cell wall. Given the chemical stability of this acyl-PBP intermediate, β-lactams have long been used as chemical probes to study the PBPs. While radioactive penicillins were commonly applied in this regard for several decades (Chambers et al., 1994; Preston, Wu, Blaszczak, Seitz, & Halligan, 1990; Spratt & Pardee, 1975), fluorescent analogs were more recently developed and quickly replaced their radioactive counterparts in profiling the activity of the PBPs (Kocaoglu et al., 2012; Zhao et al., 1999). Bocillin-FL (Boc-FL), comprised of penicillin V as the active warhead and BODIPY FL as a fluorescent reporter group, is commercially available. Showing global affinity for the entire PBP content in a given microorganism, Boc-FL has been exploited to monitor TP activity in pure protein samples, as well as cell lysates and whole cells in multiple microbes (Kocaoglu & Carlson, 2015; Kocaoglu et al., 2015; Stone, Butler, Phetsang, Cooper, & Blaskovich, 2018; Zhao et al., 1999). In order to obtain selectivity for an individual PBP or subset of PBPs, cephalosporin C molecules tagged with different fluorophores were developed by our lab, which targeted a subset of PBPs in *B. subtilis* and *S. pneumoniae* (Kocaoglu et al., 2012).

2.2.1 Design rationale

The β-lactone (2-oxetanone) scaffold was inspired by its presence in a number of natural products showing antibacterial activity, as well as structural mimicry of the D-Ala-D-Ala portion of the stem peptide (Fig. 2A). The strained β-lactone ring is an important pharmacophore which is present in diverse natural products and biologically active scaffolds, some of which are known to possess antibiotic activity including orlistat, obafluorin, SQ-26,517, and hymeglusin (Aldridge, Giles, & Turner, 1971; Parker, Rathnum, & Liu, 1982; Ravindran et al., 2013; Tymiak, Culver, Malley, & Gougoutas, 1985; Wells et al., 1982). β-Lactone compounds have been shown to covalently modify enzymes, including serine hydrolases (Kim et al., 2002). As such, β-lactone drugs have application in the treatment of obesity, diabetes and cancer (Basha & Sankaranarayanan, 2016; Lawasut et al., 2012; Venukadasula, Chegondi, Maitra, & Hanson, 2010). The most well-known β-lactone, tetrahydrolipstatin (THL; Orlistat®), is a long-term anti-obesity drug that functions by irreversibly inhibiting a lipase (Venukadasula et al., 2010).

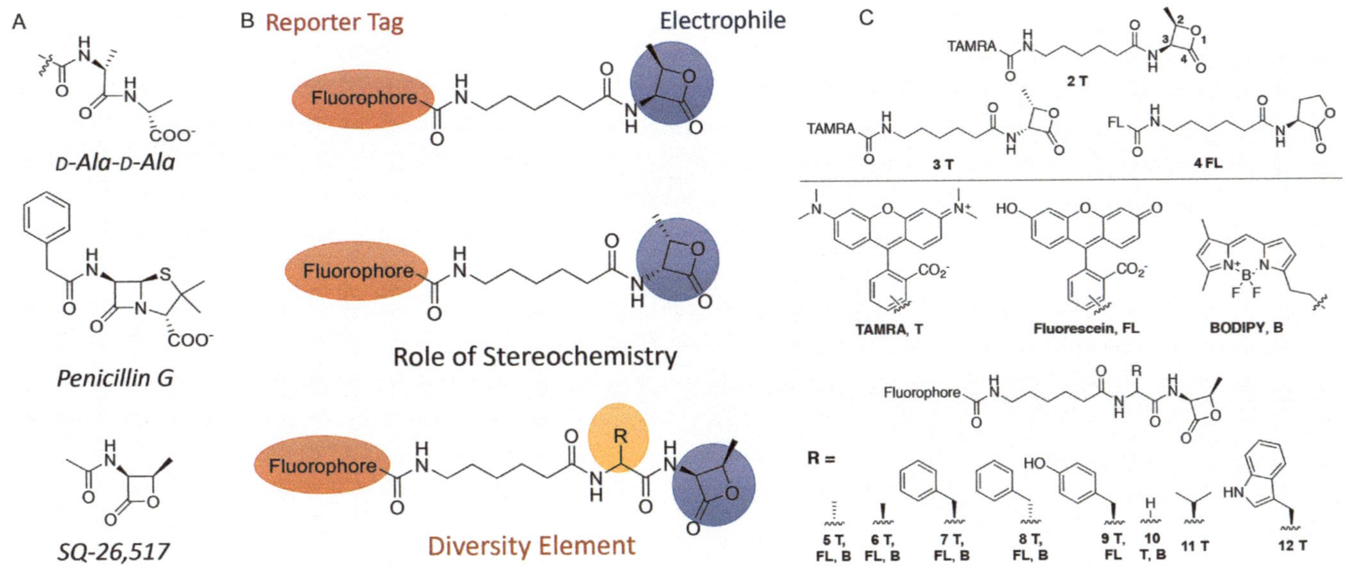

Fig. 2 β-Lactone-based probe library. (A) The design of the β-lactone probe library was inspired by the structural similarity between the terminal D-Ala-D-Ala moiety of PG as the natural substrate of PBPs, the β-lactam core in antibiotics and the β-lactone scaffold present in antibacterial natural products such as SQ-26,517. (B) A library of probes was designed containing the electrophilic threonine β-lactone ring as the warhead and different fluorophore groups were implemented as the reporter. Probes with altered stereochemistry of the β-lactone were included to study the importance of that feature. Different amino acids carrying a variety of side chains were introduced into the probe structure to expand the PBP binding spectrum. (C) Chemical structures of the synthesized probes.

Given the potential of the β-lactone scaffold to interact with active-site serine residues, including those in β-lactamases and a carboxypeptidase (Böttcher & Sieber, 2008; Compton, Schmitz, Sauer, & Sello, 2013; Wells, Trejo, Principe, & Sykes, 1984; Zeiler et al., 2011), and the structural similarity of several of these compounds to the β-lactam antibiotics, we postulated that this scaffold could be optimized to create PBP-selective probes. Additionally, our assessment of the natural product SQ-26,517 indicated that it is a weak inhibitor of PBP2a in *S. pneumoniae* (Sharifzadeh et al., 2017). We designed our molecules with an amide group proximal to the electrophilic carbonyl and the ring substituents in the *cis* conformation to mimic SQ-26,517 and the β-lactam antibiotics. Structural diversity was produced through insertion of different amino acids, with small side chains such L-alanine and D-alanine (to mimic the side chains present in the natural substrate of PBPs), as well as hydrophobic and aromatic groups such as L-valine, L- and D-phenylalanine (to mimic the side chain of penicillin G), L-tyrosine and L-tryptophan. Three fluorophores, fluorescein, BODIPY FL and tetramethylrhodamine (TAMRA) were introduced as reporter groups (Fig. 2C).

The probe library contained 24 molecules: six that possess a lactone coupled directly to a fluorophore [Fig. 2C; **2**–**4**; TAMRA (T), BODIPY FL (B) or fluorescein (FL)], six compounds with substrate-like sidechains (**5** and **6**), eight compounds with sidechains intended to mimic groups found in β-lactam antibiotics (**7**–**9**), two functionalized with glycine to assess the role of the amino acid side chain in protein labeling (**10**), and two with alternative hydrophobic side chains (L-Val and L-Trp, **11** and **12**).

2.2.2 Results

Assessment of both *cis*-functionalized analogs, (2*R*,3*S*)-β-lactone (**2**) and (2*S*,3*R*)-β-lactone (**3**), demonstrated that the stereochemistry of the substituents on the lactone ring is crucial for labeling of the PBPs (Fig. 3). The compounds displaying these groups in the same orientation as is found in β-lactams, (2*R*,3*S*)-β-lactone (**2FL** and **2T**), labeled most PBPs (PBP1a, PBP1b, PBP2x and PBP2a) in *S. pneumoniae*, while the opposite lactone configuration, (2*S*,3*R*)-β-lactone (**3T**), failed to label any PBPs (Fig. 3). We also found that lactone labeling of the PBPs is competitive with penicillin, as illustrated by the results of penicillin V pretreatment studies, where cells pretreated by penicillin V prior to labeling by β-lactone probes showed no labeling (data not shown). Moreover, treatment of β-lactone-labeled cells by penicillin V had no effect on PBP labeling (data not shown).

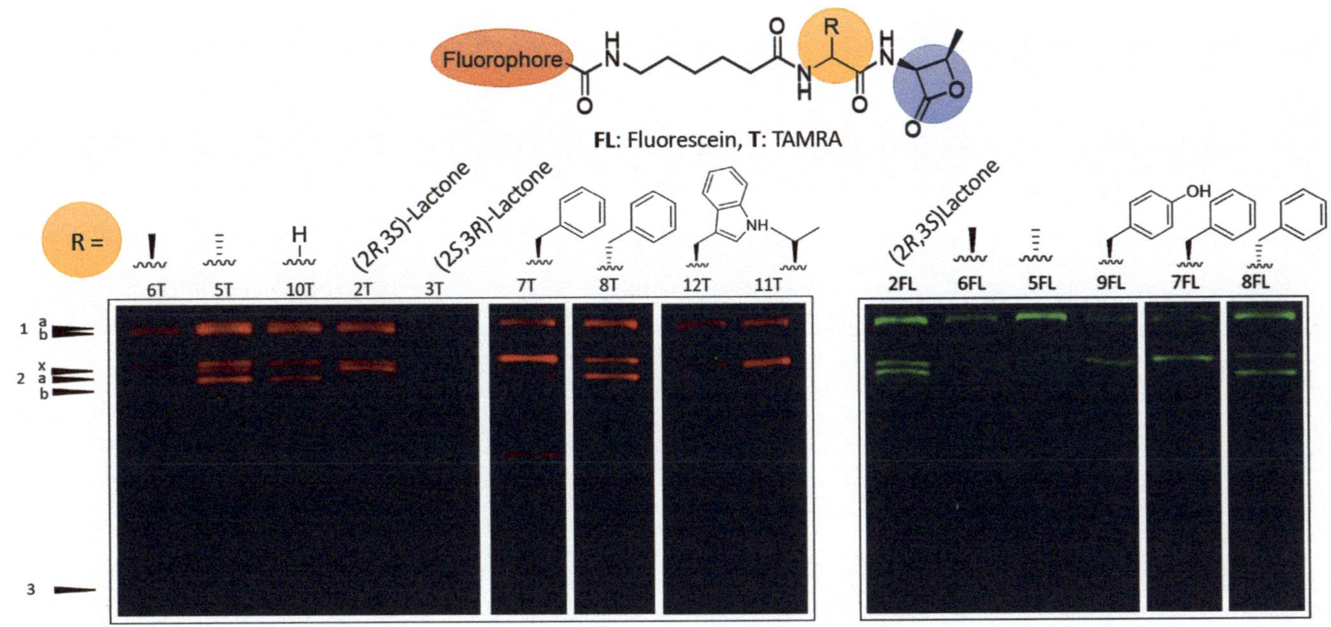

Fig. 3 Gel-based analysis of probe labeling of *S. pneumoniae* PBPs to examine ring geometry (**2** and **3**), side chains that act as substrate mimetics (**5** and **6**), side chains that mimic β-lactam antibiotics (**7–9**), and side chains of different sizes and hydrophobicity (**10**–**12**). Results of representative probes from each group are shown here, while the full labeling profile can be found in Sharifzadeh et al. (2017).

These findings suggest that the β-lactone probes are active-site directed similar to the β-lactam antibiotics, and that the resulting acylated protein complex is stable, as it is not displaced by subsequent incubation with penicillin V (Sharifzadeh et al., 2017).

Overall, we found that PBP1b is labeled by all the tested β-lactone probes, while several compounds co-selectively labeled PBP1b and PBP2x. This may indicate that these proteins are tolerant of side chains with diverse size and configuration. Intriguingly, PBP2x is a common target of the β-lactams in *S. pneumoniae* and is often co-inhibited with PBP3 (Kocaoglu et al., 2015). PBP3 was not labeled by any of the β-lactone probes in our study. In contrast, PBP1b is among the less frequently targeted PBPs by the β-lactam antibiotics (Kocaoglu et al., 2015), indicating that our probe library may be accessing a different region of binding space within the PBP active sites. PBP2b, which is the least inhibited PBP by conventional β-lactam antibiotics, was labeled by three probe scaffolds (**5T**, **5FL**, **8T**, **8FL**, **10T**). PBP1a and PBP2a labeling was only observed with compounds displaying a small side chain (Fig. 3).

PBP2x is an essential class A PBP in *S. pneumoniae* involved in septal PG synthesis (Berg, Stamsas, Straume, & Havarstein, 2013; Land et al., 2013; Tsui et al., 2014). Although we did not identify a probe that exclusively labeled PBP2x, we did uncover several compounds that co-labeled PBP2x and PBP1b, the latter of which can be deleted to yield a strain that is phenotypically indistinguishable from the wild-type organism (Land & Winkler, 2011). Thus, the Δ*pbp1b* mutant (E193) provides the ideal platform in which to image solely the activity and localization of PBP2x. We utilized the **7FL** probe in imaging studies to localize PBP2x during the course of cell division in *S. pneumoniae* E193 cells (Figs. 4–6).

PBP2b is the essential peripheral transpeptidase in *S. pneumoniae* (Tsui et al., 2016). Similar to PBP2x, it is a primary penicillin-resistance determinant (Grebe & Hakenbeck, 1996). We sought to assess the activity of PBP2b throughout the bacterial cell cycle using the lactone probe **8T**, which labels PBP1b, PBP2x and PBP2b (Figs. 4, 7 and 8). In this regard, first we treated *S. pneumoniae* E193 cells with **7FL** probe to saturate PBP2x, followed by **8T** probe (Fig. 4). Using this method, we could simultaneously visualize the activity of PBP2x and PBP2b in live pneumococcal cells (Fig. 8).

2.2.3 Live-cell imaging

Both WT and Δ*pbp1b* cells labeled with (2*R*,3*S*)-β-lactone-L-Phe-fluorescein (**7FL**) showed that PBP2x localized at the division septum

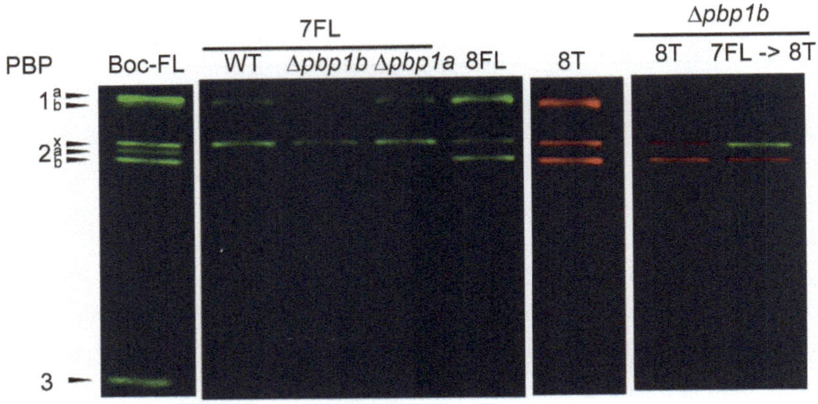

Fig. 4 PBP profiles of *S. pneumoniae* WT (IU1945), Δ*pbp1b* (E193) and Δ*pbp1a* (E177) cells that are labeled with **Boc-FL**, **7FL** and **8T** probes to resolve PBP1a and PBP1b bands. All probes were used at 5 μg/mL. Dual labeling with (2*R*,3*S*)-β-lactone-L-Phe-fluorescein (**7FL**) followed by (2*R*,3*S*)-β-lactone-D-Phe-TAMRA (**8T;** dual labeling indicated by arrow above the right-hand gel) enables separate visualization of the activity of PBP2x (green) and PBP2b (red). *Adapted from Sharifzadeh, S., Boersma, M. J., Kocaoglu, O., Shokri, A., Brown, C. L., Shirley, J. D., et al. (2017). Novel electrophilic scaffold for imaging of essential penicillin-binding proteins in Streptococcus pneumoniae.* ACS Chemical Biology, 12*(11), 2849–2857. https://doi.org/10.1021/acschembio.7b00614.*

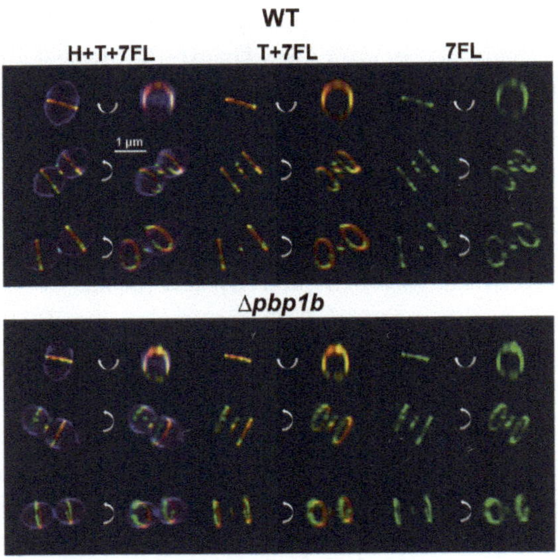

Fig. 5 Labeling of PBP2x activity (**7FL**) co-localizes with regions of new cell wall synthesis as visualized with FDAAs. Wild-type (IU1945) and Δ*pbp1b* (E193) *S. pneumoniae* cells were labeled with the hydroxycoumarin-amino-D-alanine (HADA, H, blue), followed by labeling with TAMRA-amino-D-alanine (TADA, T, red), and finally with **7FL** (green). Each row contains six different views of the same cell. The white arrows indicate rotation of the cells around the indicated axis. New cell wall synthesis, as visualized by FDAA labeling, and PBP2x activity, as visualized by **7FL** labeling, co-localize at all stages of division. For each condition, >40 cells at all stages of division from two biological replicates were imaged and a representative image is shown here. *Adapted from Sharifzadeh, S., Boersma, M. J., Kocaoglu, O., Shokri, A., Brown, C. L., Shirley, J. D., et al. (2017). Novel electrophilic scaffold for imaging of essential penicillin-binding proteins in Streptococcus pneumoniae.* ACS Chemical Biology, 12*(11), 2849–2857. https://doi.org/10.1021/acschembio.7b00614.*

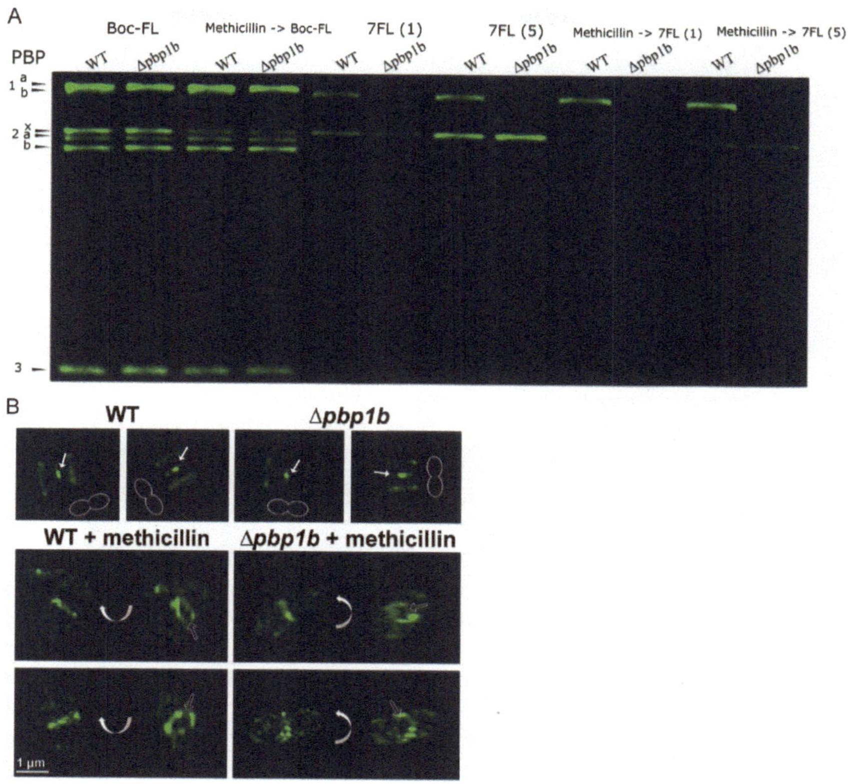

Fig. 6 Inhibition of PBP2x by methicillin. (A) Lactone labeling of PBP2x can be prevented by pretreating cells with methicillin. Treatment was carried out using 0.1 μg/mL methicillin for 30 min prior to labeling with Boc-FL or **7FL** (at 1 or 5 μg/mL, as indicated by the numbers in the parentheses). Boc-FL was used at 5 μg/mL as a control. (B) **7FL** labeling of PBP2x in wild-type *S. pneumoniae* (IU1945) and Δ*pbp1b* (E193) cells and a PBP2x-inhibiting concentration of methicillin. Cells were grown and pretreated with methicillin (0.1 μg/mL), or PBS as a control, for 20 min, followed by labeling with **7FL**. The top row consists of different cells, with the white cartoon showing the orientation of each. The bottom two rows are pairs of images of the same cell, with rotations around the indicated axis by the white arrow. The lack of PBP1b does not prevent **7FL** labeling, but methicillin-treated cells have been elongated and do not possess septal labeling, which is indicated by the empty arrows. Images represent ∼40 cells that are in the mid-to-late division stages from two biological replicates. *Adapted from Sharifzadeh, S., Boersma, M. J., Kocaoglu, O., Shokri, A., Brown, C. L., Shirley, J. D., et al. (2017). Novel electrophilic scaffold for imaging of essential penicillin-binding proteins in Streptococcus pneumoniae.* ACS Chemical Biology, 12*(11), 2849–2857. https://doi.org/10.1021/acschembio.7b00614.*

as a ring in early-to-mid divisional cells (Figs. 5 and 6). In mid-to-late divisional cells, PBP2x localized to both the constricting ring and a separate site at the center of the ring, as well as the equators of daughter cells. Finally, enzyme activity was constricted down to a single point at the division site

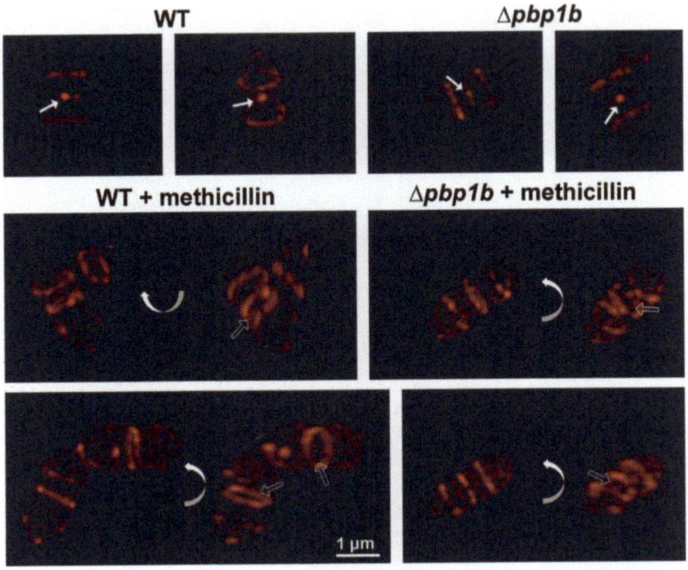

Fig. 7 **8T** labeling of *S. pneumoniae* cells with or without PBP1b and a PBP2x-inhibiting concentration of methicillin. Two cultures of WT (IU1945) and Δ*pbp1b* (E193) were pretreated with 0.1 μg/mL methicillin for 20 min, followed by labeling with **8T**. The top row contains images of different cells in each plane, while the bottom two rows contain pairs of images of the same cell, with rotations around the indicated axis with the white arrow. The absence of PBP1b does not affect the labeling of **8T**, but pretreatment of methicillin results in elongated cells and a lack of septal labeling, as shown with the empty arrows. These images are representative of >40 cells at all stages of division for each condition from two biological replicates. *Adapted from Sharifzadeh, S., Boersma, M. J., Kocaoglu, O., Shokri, A., Brown, C. L., Shirley, J. D., et al. (2017). Novel electrophilic scaffold for imaging of essential penicillin-binding proteins in Streptococcus pneumoniae.* ACS Chemical Biology, 12*(11), 2849–2857. https://doi.org/10.1021/acschembio.7b00614.*

before cell separation. This indicates that subpopulations of active PBP2x demonstrate different localization patterns during a single constriction event, which is in agreement with other reports of PBP2x localization (Land et al., 2013; Tsui et al., 2014), as well as newly synthesized PG labeled by FDAAs (Fig. 5). Pretreatment with the PBP2x-specific inhibitor methicillin (Land et al., 2013) revealed elongated cells with a labeled ring at the division site, but with minimal or no constriction evident (Fig. 6). Importantly, >98% of methicillin-treated late division cells in both the WT and Δ*pbp1b* strains displayed empty septal rings, in contrast with untreated controls from both strains that showed central septal labeling in 85% of wild-type and 73% of Δ*pbp1b* cells. This indicates a lack of active PBP2x due to methicillin

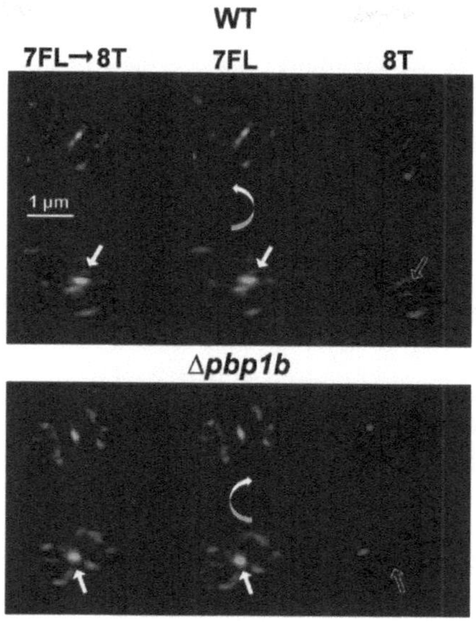

Fig. 8 Labeling with **7FL** followed by **8T** reveals distinct PBP localization patterns for PBP2x and PBP2b. Wild-type (IU1945) and Δ*pbp1b* (E193) *S. pneumoniae* were grown, labeled with **7FL**, washed, labeled with **8T**, and imaged. Each set is two rows of images of the same cell, with the bottom row rotated relative to the top row around the indicated axis. In both wild-type and PBP1b knock-out strains, **8T** labeling is restricted to peripheral rings when the cells are labeled first with **7FL**. Solid arrows highlight central septal labeling by **7FL**, and empty arrows highlight the absence of **8T** labeling from the center of the division site. For each condition, ~40 mid-to-late division cells from two biological replicates were imaged and representative results are shown. *Reproduced from Sharifzadeh, S., Boersma, M. J., Kocaoglu, O., Shokri, A., Brown, C. L., Shirley, J. D., et al. (2017). Novel electrophilic scaffold for imaging of essential penicillin-binding proteins in* Streptococcus pneumoniae. ACS Chemical Biology, *12(11), 2849–2857. doi:10.1021/ acschembio.7b00614.*

inhibition, which aligns with previous results (Tsui et al., 2013) and strongly supports the conclusion that septal **7FL** labeling is due primarily, if not exclusively, to PBP2x.

PBP2b is the essential peripheral transpeptidase in *S. pneumoniae* (Tsui et al., 2016). In order to assess the activity of PBP2b throughout the cell cycle, WT and Δ*pbp1b* cells were labeled with (2*R*,3*S*)-β-lactone-D-Phe-TAMRA (**8T**), which is specific for PBP1b, PBP2x and PBP2b (Fig. 4). Labeling with **8T** yields a nearly identical labeling pattern as **7FL** due to their shared targets (Fig. 7). Additionally, labeling with **8T** after pretreatment with methicillin also mirrors the results seen with **7FL**, reinforcing the idea that PBP2x alone is localized to the center of the constricting division site (Fig. 7).

Gel-based studies indicate that labeling Δ*pbp1b* cells with **7FL**, followed by labeling with **8T**, individually tags PBP2x (green) and PBP2b (red) (Fig. 4). Imaging of dual labeled cells revealed that **7FL** displayed a labeling pattern similar to the single labeling results (Fig. 8), as expected. However, labeling of **8T** was excluded from the center of the division site in >83% of WT and Δ*pbp1b* cells. Instead, it was restricted to the outer peripheral ring around the division site and the equatorial rings of future daughter cells, which have not yet begun to constrict. These results correlate well with previous work (Tsui et al., 2016) that demonstrated that PBP2b localization remained separate from and external to PBP2x during constriction. We have now shown that active PBP2b follows this pattern as well and ruled out the possibility that methicillin treatment disrupts PBP2b localization.

3. Gel-based analysis of PBP activity profile

3.1 Equipment

1. CO_2 incubator
2. Spectrophotometer (Genesys 20, Thermo Scientific or equivalent)
3. Plastic 17 × 100 mm culture tubes (VWR, 60818-689)
4. Micropipettors
5. Micropipettor tips
6. 1.5-mL microcentrifuge tubes
7. Bench top microcentrifuge
8. Branson Sonifier 250 OR vial tweeter UP200St (Hielscher)
9. NanoDrop 1000 Spectrophotometer (Thermo Scientific) OR NanoPhotometer P330 (IMPLEN)
10. Analog heat block (VWR)
11. XCell SureLock mini-cell gel electrophoresis system (Thermo Fisher Scientific)
12. Typhoon 9210 gel scanner (Amersham Biosciences) equipped with the following filters: 510 nm LP, 575 LP, BPB1 530DF20, and 665 LP
13. ImageJ software (NIH) for gel data analysis

3.2 Material

1. *S. pneumoniae* IU1945 (an unencapsulated derivative of serotype 2 *S. pneumoniae* strain D39), E177 (Δ*pbp1a*) and E193 (Δ*pbp1b*) cells (Lanie et al., 2007)

2. Brain heart infusion medium (powder, Bacto Brain Heart Infusion, VWR, 90003-038)
3. 1× Phosphate-buffered saline (PBS) (137 mM NaCl, 2.7 mM KCl, 10.1 mM Na_2HPO_4, 1.8 mM KH_2PO_4, pH 7.4; filter sterilized using a 0.22-μm filter)
4. **7FL** and **8T** probe (Sharifzadeh et al., 2017) stock solutions in DMSO (store at −80 °C, protected from light; stable for years)
5. Bocillin-FL (1 mg/mL, Invitrogen, cat. no. B-13233) in DMSO (store at −80 °C, protected from light; stable for months)
6. Lysozyme (10 mg/mL) in 1× PBS (from chicken egg white, Fluka)
7. 4× SDS-PAGE gel loading buffer (200 mM Tris·Cl, pH 6.8, 400 mM dithiothreitol (DTT), 8% (w/v) SDS, 0.2% (w/v) bromophenol blue, 40% (v/v) glycerol)
8. BenchMark™ Fluorescent Protein Standard (Invitrogen, LC5928)
9. 10× Tris-glycine SDS running buffer (144.0 g/L glycine, 30.0 g/L Tris, 10.0 g/L SDS; pH ~8.3)
10. Mini-polyacrylamide gels produced *in-house* [polyacrylamide gels were composed of 10% acrylamide resolving gel (10.5 mL of 1.5 M Tris-HCl buffer pH 8.8, 10.5 mL of acrylamide:bis-acrylamide 29:1 (40% solution), 21 mL of H_2O, 140 μL of aqueous 10% ammonium persulfate (APS), 15 μL of tetramethylethylenediamine (TEMED)) and 4.5% acrylamide stacking gel (2.5 mL of 0.5 M Tris-HCl buffer pH 6.8, 1.13 mL of acrylamide:bis-acrylamide 29:1 (40% solution), 6.38 mL of H_2O, 30 μL of aqueous 10% APS, 10 μL of TEMED)]
11. Coomassie brilliant blue R-250 staining solution (Bio-Rad, 1610436)
12. Additional reagents and equipment for SDS-PAGE

3.3 Protocols

3.3.1 Bacterial culturing

1. Frozen glycerol stocks of *S. pneumoniae* cells were inoculated into 5 mL BHI broth in a plastic culture tube using one loopful with a sterile inoculating loop.
2. Serial dilutions were made by mixing each tube briefly and transferring 500 μL to a tube containing 5 mL fresh BHI broth. Six dilutions were made and tubes were incubated at 37 °C in an atmosphere containing 5% CO_2 for 12–16 h.
3. Following incubation, the optical density at 620 nm (OD_{620}) of each tube was determined using a spectrophotometer.

4. A culture in exponential phase (with an OD_{620} between 0.05 and 0.4) was diluted to an OD_{620} of 0.02 in 2 mL of fresh BHI broth prewarmed to 37 °C and incubated statically without shaking at 37 °C in an atmosphere of 5% CO_2.
5. When the culture reached an OD_{620} of 0.2, cells were harvested by centrifugation.

3.3.2 β-Lactone probe labeling

1. Cell pellets from 1.5 mL of *S. pneumoniae* cultures at OD_{620} ~0.2 were harvested by centrifugation at 16,100 × *g* for 2 min at room temperature (RT).
2. Pellets were resuspended with 1.0 mL 1× PBS and centrifuged at 16,100 × *g* for 2 min at RT.
3. Cell pellets were resuspended in 50 μL of 1× PBS containing 1–30 μg/mL of β-lactone probes and incubated in the dark at RT for 20 min.

   Note: Incubation time could be optimized to enhance labeling efficiency, as needed.

   To prepare a reference sample, one washed pellet sample was resuspended in 50 μL of 1× PBS containing 5 μg/mL Boc-FL and incubated in the dark for 10 min at RT.
4. The cells were pelleted using the above-described centrifugation settings.
5. Pellets were washed with 1.0 mL 1× PBS.
6. Membrane fractions were collected for SDS-PAGE analysis (Section 3.3.4).

3.3.3 Dual labeling with 7FL and 8T

1. *S. pneumoniae* E193 (Δ*pbp1b*) cells were grown in BHI broth at 37 °C in an atmosphere of 5% CO_2 to reach an OD_{620} of 0.2–0.25, as described in Section 3.3.1.
2. Cell pellets from 1.5 mL of culture were harvested by centrifugation (16,100 × *g* for 2 min at RT).
3. Cell pellets were washed with 1 mL of 1× PBS.
4. Cell pellets were resuspended in 50 μL of 1× PBS containing 5 μg/mL **7FL** and incubated in the dark at RT for 20 min.
5. Cells were washed with 1 mL 1× PBS.
6. Cells were suspended in 50 μL of 5 μg/mL solution of **8T** in 1× PBS and incubated in the dark for 20 min at RT.

7. Cells were washed with 1 mL 1× PBS.
8. Membrane fractions were collected for SDS-PAGE analysis (Section 3.3.4).

3.3.4 Collection of membrane fraction for SDS-PAGE analysis

1. Labeled cells from Sections 3.3.2 or 3.3.3 were resuspended in 100 μL of 1× PBS containing 10 mg/mL of lysozyme and were incubated for 30 min at 37 °C.
2. The cells were lysed by a Branson Sonifier 250 (power setting 3, 30% duty cycle for 3 × 10 s intervals) or vial tweeter UP200St (70% C, 95% A, 5% adjustment snap and 1 s SD Interval/10 s for a total of six 1 min intervals with 1 min cooling time in between, 12 min total time) on ice.
3. Membrane proteome was isolated by centrifugation at 21,000 × *g* for 15 min at 4 °C.
4. Membrane proteome was resuspended in 100 μL 1× PBS and the samples were homogenized by sonication (Branson Sonifier; power setting 1, 10% duty cycle for 1 s, more cycles can be utilized if samples still appears cloudy. Alternatively, samples could be sonicated in a water bath sonifier).
5. Protein concentration was measured by NanoDrop 1000 spectrophotometer (Thermo Scientific) or NanoPhotometer P330 (IMPLEN). The protein concentration was adjusted to 5.0 mg/mL by diluting with 1× PBS. For the blank solution, 1× PBS was used. (Note: An adjusted concentration within 2.5–5.0 mg/mL would give a measurable in-gel fluorescent signal).
6. Thirty microliters of proteome sample were dispensed into a clean 1.5-mL microcentrifuge tube and 10 μL of 4× SDS-PAGE loading buffer was added to each sample.
7. The samples were heated for 5 min at 90–95 °C to denature the proteins then cooled to RT.
8. Samples were analyzed by SDS-PAGE (Section 3.3.5).

3.3.5 SDS-PAGE analysis

Ten to twelve microliters of membrane sample from Section 3.3.4 were loaded onto a 10% acrylamide gel for SDS-PAGE (acrylamide: bis-acrylamide = 29:1). The protein bands were separated by gel electrophoresis for 1.5 h, at 180 V, 60 W. The gel rig should be protected from light, to prevent photobleaching of the fluorophores. This was done by covering the gel rig with a cardboard box, but foil can also be used. The gel was rinsed with distilled water three times before fluorescence scanning (Section 3.3.6).

3.3.6 In-gel fluorescence detection of labeled proteins

After SDS-PAGE (Section 3.3.5), labeled proteins were directly visualized at 50-μm resolution in-gel using a Typhoon 9210 gel scanner with a 532 nm laser and 580-nm bandpass filter for TAMRA and a 473 nm laser and a 526-nm short-pass filter for fluorescein and BODIPY FL. All gel images were analyzed using ImageJ software (Schneider, Rasband, & Eliceiri, 2012). The background signal of the gel images was subtracted, and the brightness and contrast were manually adjusted to optimize the signal-to-noise ratio (all operations were performed over the entire gel uniformly). Following fluorescence gel analysis, gels were stained with Coomassie blue (using manufacturer's protocol) and protein levels were assessed to account for differences in sample loading onto the gel.

4. *In vivo* imaging of PBP activity

4.1 Equipment

1. CO_2 incubator
2. Spectrophotometer (Genesys 20, Thermo Scientific or equivalent)
3. Bench top centrifuges (at RT and 4 °C)
4. Bench top vortex mixer (VWR, VM-3000)
5. DeltaVision OMX 3D-SIM Super Resolution System (GE Healthcare)
6. DeltaVision OMX Software for image acquisition (GE Healthcare)
7. DeltaVision SoftWoRx Software for image processing (GE Healthcare)

4.2 Materials

1. *S. pneumoniae* cells. Cells are unencapsulated derivatives of D39, a serotype 2 strain. Wild-type (IU1945) (Lanie et al., 2007) and Δ*pbp1b* (E193) (Land & Winkler, 2011).
2. Plastic 17 × 100 mm culture tubes (VWR, 60818-689)
3. Glass 16 × 100 mm culture tubes (Fisher, 1495925B)
4. Brain heart infusion (BHI) media (BD Bacto, 237500)
5. Microcentrifuge tubes
6. Dry ice
7. Round coverslips, 12 mm diameter (Electron Microscopy Sciences, 72230-01)

8. Microscope slides (VWR, 16004-368)
9. 70% Ethanol
10. Lens paper (VWR, 52846-001)
11. Immersion oil, refractive index 1.518
12. 1× Phosphate-buffered saline (PBS) (Ambion, AM9625)
13. PBS containing 50 mM glucose
14. GTE buffer (50 mM glucose, 1 mM EDTA, 20 mM Tris-HCl, pH 7.5)
15. Vectashield Hardset Antifade mounting media (Vector Laboratories, 93952-26)
16. Methicillin (Sigma, 51454-50MG)
17. **7FL** and **8T** probe (Sharifzadeh et al., 2017) stock solutions in DMSO (stored at −80 °C, protected from light; stable for years)
18. Fluorescent D-amino acids (hydroxycoumarin-amino-D-alanine, HADA and TAMRA-amino-D-alanine, TADA) were synthesized as reported (Kuru et al., 2012), with the following change: TADA was synthesized as reported for TDL, except that Boc-D-DAP-OH (*N*-alpha-t-butyloxycarbonyl-D-2,3-diaminopropionic acid) was used in place of Boc-D-Lys-OH (*N*-alpha-t-butyloxycarbonyl-D-lysine). 500 mM stock solutions in DMSO were stored at −20 °C and shielded from light

4.3 Protocols

4.3.1 Bacterial cell culture

Exponentially growing cultures of *S. pneumoniae* at OD_{620} of 0.2 were prepared as described in Section 3.3.1.

4.3.2 β-Lactone probe labeling

1. Cells were pelleted as previously described and the supernatant discarded. Resuspend the cell pellet in 1 mL of RT PBS with brief vortexing, and centrifuge for 5 min at 16,000 × *g* at RT.
2. Discard supernatant and resuspend the cell pellet in either 50 μL PBS with 5 μg/mL **7FL** or 100 μL PBS with 5 μg/mL **8T** with brief vortexing.
3. Incubate the cells at RT for 20 min in the dark.
4. After incubation, centrifuge for 5 min at 16,000 × *g* at RT.
5. Discard supernatant and wash the cell pellet in either 1 mL PBS (**7FL**) or 1 mL PBS containing 50 mM glucose (**8T**) with brief vortexing. Centrifuge for 5 min at 16,000 × *g* at RT.
6. Discard supernatant and wash the cell pellet in either 200 μL PBS (**7FL**) or 200 μL PBS containing 50 mM glucose (**8T**) with brief vortexing. Centrifuge for 5 min at 16,000 × *g* at RT.

7. Discard supernatant and resuspend the cell pellet in 100 µL cold GTE buffer with brief vortexing. Centrifuge for 5 min at 16,000 × *g* at 4 °C.
8. Discard supernatant and resuspend the cell pellet in 15 µL Vectashield Hardset Antifade with brief vortexing. Keep cells on ice.
9. Clean coverslips and slides by applying 70% ethanol and wiping with lens paper.
10. Pipette 1.2 µL of cells onto a coverslip and carefully place a microscope slide on top. Don't move the slide, but put something over it (such as an empty tip box) to incubate in the dark for 15 min at RT.

4.3.3 3D-SIM (structured illumination microscopy) imaging

1. 3D-SIM imaging was performed using the DeltaVision OMX 3D-SIM Super Resolution System.
2. Image acquisition utilized DeltaVision OMX software.
3. The system was equipped with four Photometrics Cascade II EMCCD cameras that allow simultaneous imaging of up to four colors.
4. Laser lines used were 405 with emission filters of 419–465 (blue, for HADA), 488 with emission filters of 500–550 (green, for **7FL**), and 561 with emission filters of 609–654 (red, for **8T** or TADA). For all channels, exposure times were 5 ms and %T was 100%.
5. For all samples, immersion oil with a refractive index of 1.518 was used. For each sample tested, a stack of 9–15 *z* sections (each 0.125 µm thick) was rotated and visually checked for "halo" artifacts to ensure that the refractive index of the prepared sample matched the refractive index of the oil used.
6. Image processing utilized DeltaVision SoftWoRx software and consisted of OMX Reconstruction (Wiener filter value set at 0.001) and OMX Align.

4.3.4 Methicillin pretreatment

1. For pretreatment with methicillin followed by lactone probe labeling, make duplicate 2 mL cultures of each strain in BHI broth (Section 4.3.1, step 4) and culture as described.
2. When the culture reaches an OD_{620} of 0.12, add methicillin (final concentration of 0.1 µg/mL) to one culture of each strain, and place the strains back in the incubator.
3. After 20 min, add 1.5 mL of the culture to a microfuge tube and centrifuge at RT for 5 min at 16,000 × *g*.
4. Label and image cultures as described (starting with Section 4.3.2).

4.3.5 *Dual β-lactone labeling*

1. For dual labeling of cells with both lactone probes, the following changes were made after the 20 min incubation with **7FL** (Section 4.3.2, step 3).
2. After incubation, centrifuge for 5 min at 16,000 × *g* at RT.
3. Discard supernatant and wash cell pellet in 1 mL of PBS with brief vortexing. Centrifuge for 5 min at 16,000 × *g* at RT.
4. Discard supernatant and resuspend cell pellet in 100 μL PBS with 5 μg/mL **8T** with brief vortexing.
5. Incubate the cells at RT for 20 min in the dark.
6. Wash and image as described above for β-lactone labeling (Section 4.3.2, starting at step 4).

4.3.6 *FDAA β-lactone labeling*

1. For labeling of cells with FDAAs and lactone probes, the following changes were made after making the 2 mL culture in BHI broth (Section 4.3.1, step 4).
2. HADA stock solution (0.5 μL, stock 500 mM, final concentration 125 μM) was added to the 2 mL culture and incubated as described (Section 4.3.1, step 4).
3. When the culture reached an OD_{620} of 0.2, 0.5 mL of the culture was added to a microfuge tube and centrifuged at RT for 5 min at 16,000 × *g*.
4. Discard supernatant and resuspend cell pellet in 250 μL prewarmed BHI broth containing TADA (final concentration 500 μM) with brief vortexing.
5. Incubate cells at 37 °C for 5 min.
6. Cool cells on dry ice for exactly 20 s, then centrifuge at 4 °C for 2.5 min at 16,000 × *g*.
7. Discard supernatant, resuspend cell pellet in 200 μL ice-cold PBS with brief vortexing and centrifuge at 4 °C for 2.5 min at 16,000 × *g*.
8. Again, discard supernatant, resuspend cell pellet in 200 μL ice-cold PBS with brief vortexing and centrifuge at 4 °C for 2.5 min at 16,000 × *g*.
9. Discard supernatant, resuspend cell pellet in 50 μL RT PBS with 5 μg/mL **7FL** with brief vortexing.
10. Cells were labeled with **7FL**, washed and imaged as described (Section 4.3.2, starting at step 3).

5. Additional considerations and controls

5.1 Probe concentration

Fluorescent probe concentrations were determined by measuring UV-Vis absorption of each solution at λ_{max} of its corresponding fluorophore. Probes were stored as DMSO solutions at −80 °C. For **7FL** probe, 10× and 100× dilutions of the probe were made in 0.1M Tris-HCl (pH 8) and absorbance was read at 492nm using NanoPhotometer P330 (IMPLEN). The average of three absorbance values was used to calculate the concentration ($\varepsilon = 78{,}000\,M^{-1}\,cm^{-1}$). The absorbance values must be read between 0.1 and 0.9 arbitrary units to ensure an accurate calculation. **8T** probe was diluted in methanol, and concentration was calculated in the same way as for the **7FL** probe ($\lambda_{max} = 543\,nm$, $\varepsilon = 87{,}000\,M^{-1}\,cm^{-1}$).

5.2 Competition assays of penicillin V (Pen-V) and Boc-FL, 2F and 2T probes

In order to verify that the observed labeling is due to binding of the β-lactone probes to PBPs, we pretreated the cells with penicillin V, a well-known inhibitor of PBPs. To confirm the irreversibility of β-lactone labeling, labeled cells were treated with Pen-V after probe treatment.

In the Pen-V pretreatment assay, cells from 1.5mL of culture were harvested by centrifugation (16,100 × *g* for 2min at RT) and washed with 1mL of PBS, pH 7.4. Cell pellets were resuspended in 50μL of PBS containing 5μg/mL of Pen-V, and incubated for 30min at RT. Cells were pelleted and washed in 1mL of PBS. Next, the cells were resuspended in 50μL of PBS containing 5μg/mL Boc-FL, 5μg/mL **2F,** or 5μg/mL **2T** and incubated for 10min (Boc-FL) or 30min (**2F**, **2T**) at RT. In the Pen-V post-treatment assay, cells were labeled with Boc-FL, **2F** and **2T** as described, washed and then incubated with 5μg/mL Pen-V. Finally, the cells were resuspended in 100μL PBS containing 10mg/mL lysozyme and incubated for 30min at 37°C. Cells were lysed and sample preparation for SDS-PAGE analysis was performed as described in Sections 3.3.4 and 3.3.5.

5.3 Methicillin pretreatment controls

Cells from 1.5mL of *S. pneumoniae* IU1945 and E193 cultures at exponential phase (OD_{620} ~0.2) were harvested by centrifugation

(16,100 × *g* for 2 min at RT) and washed with 1 mL of PBS, pH 7.4. Cell pellets were resuspended in 50 μL of PBS containing 0.1 μg/mL of methicillin and incubated for 30 min at RT. Cells were pelleted and washed in 1 mL of PBS. Next, the cells were resuspended in 50 μL of PBS containing 1 or 5 μg/mL of **7FL** or 5 μg/mL Boc-FL (as control) and incubated for 20 min at RT (10 min at RT for Boc-FL). Cells were washed with 1 mL of PBS and imaged, as described. In gel-based studies, the cells were resuspended in 100 μL PBS containing 10 mg/mL lysozyme and incubated for 30 min at 37 °C. Cells were lysed and sample preparation for SDS-PAGE analysis was performed as described in Sections 3.3.4 and 3.3.5.

6. Summary

Bacterial cell wall biosynthesis remains an outstanding target for new antibiotic development. However, there are significant gaps in our knowledge about this complicated process, including the mechanisms and control of PBP activity throughout cell growth and division. Continued development of tools to target and map the activities of the individual PBPs is a critical component to our understanding of bacterial growth and may lead to the identification of new therapeutic targets. Here, we report a methodology to profile the activity of individual PBPs in *S. pneumoniae*, using chemical probes based upon a scaffold not previously known to target the PBPs, the β-lactones. These probes demonstrated exquisite specificity for PBPs in bacterial cells and distinct interaction patterns with the PBPs in comparison to β-lactam antibiotics, which enabled us to examine the activity of two essential transpeptidases, PBP2x and PBP2b, in live *S. pneumoniae*. We found that PBP2b and PBP2x co-localize as a single ring in early division, but during mid-to-late division, PBP2x activity concentrates to the central septal site, while PBP2b activity remains at the outer division ring. This is the first report of visualization of the transpeptidation activity of individual PBPs *in vivo*, which confirmed that PBP2b and PBP2x display distinct patterns of localization during constriction. This methodology can be applied to any other bacterial system, as we have already tested in two model bacteria; *B. subtilis* and *E. coli*. As such, this work highlights the importance of the development of PBP-selective probes. Such probes, in combination with powerful cell biology tools, will illuminate the localization and activation of the PBPs in myriad bacteria.

Acknowledgments

This work was supported by the National Institutes of Health (R01 GM128439-01A1 to E.E.C. and M.E.W.), a Sloan Research Fellow Award (E.E.C.), a University of Minnesota Interdisciplinary Doctoral Fellowship (S.S.), and the University of Minnesota, Department of Chemistry. N.W.B. was supported by NIH Institutional Research and Academic Career Development Award K12 GM119955. J.D.S. was supported by the National Institutes of Health's National Center for Advancing Translational Sciences, grants TL1R002493 and UL1TR002494. The content is solely the responsibility of the authors and does not necessarily represent the official views of the National Institutes of Health's National Center for Advancing Translational Sciences. The DeltaVision OMX system was obtained with equipment grant NIH 1S10OD024988-01 to the Indiana University LMIC.

References

Aldridge, D. C., Giles, D., & Turner, W. B. (1971). Antibiotic 1233A: A fungal β-lactone. *Journal of the Chemical Society Perkin Transactions 1*, *23*, 3888–3891.

Basha, R. H., & Sankaranarayanan, C. (2016). β-caryophyllene, a natural sesquiterpene lactone attenuates hyperglycemia mediated oxidative and inflammatory stress in experimental diabetic rats. *Chemico-Biological Interactions*, *245*, 50–58. https://doi.org/10.1016/j.cbi.2015.12.019.

Berg, K. H., Stamsas, G. A., Straume, D., & Havarstein, L. S. (2013). Effects of low PBP2b levels on cell morphology and peptidoglycan composition in *Streptococcus pneumoniae* R6. *Journal of Bacteriology*, *195*(19), 4342–4354. https://doi.org/10.1128/JB.00184-13.

Bogyo, M., Verhelst, S., Bellingard-Dubouchaud, V., Toba, S., & Greenbaum, D. (2000). Selective targeting of lysosomal cysteine proteases with radiolabeled electrophilic substrate analogs. *Chemistry & Biology*, 7(1), 27–38. https://doi.org/10.1016/s1074-5521(00)00061-2.

Böttcher, T., & Sieber, S. A. (2008). β-lactones as privileged structures for the active-site labeling of versatile bacterial enzyme classes. *Angewandte Chemie, International Edition*, *47*(24), 4600–4603. https://doi.org/10.1002/anie.200705768.

Böttcher, T., & Sieber, S. A. (2012). β-Lactams and β-lactones as activity-based probes in chemical biology. *Medicinal Chemistry Communications*, *3*, 408–417.

Chambers, H. F., Sachdeva, M. J., & Hackbarth, C. J. (1994). Kinetics of penicillin binding to penicillin-binding proteins of *Staphylococcus aureus*. *The Biochemical Journal*, *301*(Pt. 1), 139–144. https://doi.org/10.1042/bj3010139.

Compton, C. L., Schmitz, K. R., Sauer, R. T., & Sello, J. K. (2013). Antibacterial activity of and resistance to small molecule inhibitors of the ClpP peptidase. *ACS Chemical Biology*, *8*, 2669–2677.

Cravatt, B. F., Wright, A. T., & Kozarich, J. W. (2008). Activity-based protein profiling: From enzyme chemistry to proteomic chemistry. *Annual Review of Biochemistry*, *77*, 383–414. https://doi.org/10.1146/annurev.biochem.75.101304.124125.

Curtis, N. A., Orr, D., Ross, G. W., & Boulton, M. G. (1979). Affinities of penicillins and cephalosporins for the penicillin-binding proteins of *Escherichia coli* K-12 and their antibacterial activity. *Antimicrobial Agents and Chemotherapy*, *16*(5), 533–539. https://doi.org/10.1128/aac.16.5.533.

Daniel, R. A., & Errington, J. (2003). Control of cell morphogenesis in bacteria: Two distinct ways to make a rod-shaped cell. *Cell*, *113*(6), 767–776. https://doi.org/10.1016/s0092-8674(03)00421-5.

Dargis, M., & Malouin, F. (1994). Use of biotinylated β-lactams and chemiluminescence for study and purification of penicillin-binding proteins in bacteria. *Antimicrobial Agents and Chemotherapy*, *38*(5), 973–980.

Falconer, S. B., Czarny, T. L., & Brown, E. D. (2011). Antibiotics as probes of biological complexity. *Nature Chemical Biology*, *7*, 415–423.

Fura, J. M., Kearns, D., & Pires, M. M. (2015). D-amino acid probes for penicillin binding protein-based bacterial surface labeling. *The Journal of Biological Chemistry*, *290*(51), 30540–30550. https://doi.org/10.1074/jbc.M115.683342.

Ghuysen, J. M. (1977). *The bacterial DD-carboxypeptidase-transpeptidase enzyme system. Part III. Interactions with penicillin.* Tokyo, Japan: University of Tokyo Press.

Grebe, T., & Hakenbeck, R. (1996). Penicillin-binding proteins 2b and 2x of *Streptococcus pneumoniae* are primary resistance determinants for different classes of β-lactam antibiotics. *Antimicrobial Agents and Chemotherapy*, *40*, 829–834.

Heal, W. P., Dang, T. H., & Tate, E. W. (2011). Activity-based probes: Discovering new biology and new drug targets. *Chemical Society Reviews*, *40*(1), 246–257. https://doi.org/10.1039/c0cs00004c.

Henriques-Normark, B., & Tuomanen, E. I. (2013). The pneumococcus: Epidemiology, microbiology, and pathogenesis. *Cold Spring Harbor Perspectives in Medicine*, *3*(7), a010215. https://doi.org/10.1101/cshperspect.a010215.

Keller, L. J., Babin, B. M., Lakemeyer, M., & Bogyo, M. (2019). Applications for identification of therapeutic targets and characterization of microbial communities. *Current Opinion in Chemical Biology*, *54*, 45–53.

Kim, D. H., Park, J. I., Chung, S. J., Park, J. D., Park, N. K., & Han, J. H. (2002). Cleavage of β-lactone ring by serine protease. Mechanistic implications. *Bioorganic & Medicinal Chemistry*, *10*(8), 2553–2560.

Kocaoglu, O., Calvo, R. A., Sham, L. T., Cozy, L. M., Lanning, B. R., Francis, S., … Carlson, E. E. (2012). Selective penicillin-binding protein imaging probes reveal substructure in bacterial cell division. *ACS Chemical Biology*, 7(10), 1746–1753. https://doi.org/10.1021/cb300329r.

Kocaoglu, O., & Carlson, E. E. (2015). Profiling of β-lactam selectivity for penicillin-binding proteins in *Escherichia coli* DC2. *Antimicrobial Agents and Chemotherapy*, *59*, 2785–2790.

Kocaoglu, O., Tsui, H. C., Winkler, M. E., & Carlson, E. E. (2015). Profiling of β-lactam selectivity for penicillin-binding proteins in *Streptococcus pneumoniae* D39. *Antimicrobial Agents and Chemotherapy*, *59*(6), 3548–3555. https://doi.org/10.1128/AAC.05142-14.

Krysiak, J., Stahl, M., Vomacka, J., Fetzer, C., Lakemeyer, M., & Fux, A. (2017). Quantitative map of β-Lactone-induced virulence regulation. *Journal of Proteome Research*, *16*, 1180–1192.

Kuru, E., Hughes, H. V., Brown, P. J., Hall, E., Tekkam, S., Cava, F., … VanNieuwenhze, M. S. (2012). *In Situ* probing of newly synthesized peptidoglycan in live bacteria with fluorescent D-amino acids. *Angewandte Chemie, International Edition*, *51*(50), 12519–12523. https://doi.org/10.1002/anie.201206749.

Lakaye, B., Damblon, C., Jamin, M., Galleni, M., Lepage, S., Joris, B., … Frere, J.-M. (1994). Synthesis, purification and kinetic properties of fluorescein-labelled penicillins. *The Biochemical Journal*, *300*, 141–145.

Land, A. D., Tsui, H. C., Kocaoglu, O., Vella, S. A., Shaw, S. L., Keen, S. K., … Winkler, M. E. (2013). Requirement of essential Pbp2x and GpsB for septal ring closure in *Streptococcus pneumoniae* D39. *Molecular Microbiology*, *90*(5), 939–955. https://doi.org/10.1111/mmi.12408.

Land, A. D., & Winkler, M. E. (2011). The requirement for pneumococcal MreC and MreD is relieved by inactivation of the gene encoding PBP1a. *Journal of Bacteriology*, *193*, 4166–4179.

Lanie, J. A., Ng, W. L., Kazmierczak, K. M., Andrzejewski, T. M., Davidsen, T. M., Wayne, K. J., … Winkler, M. E. (2007). Genome sequence of Avery's virulent serotype 2 strain D39 of *Streptococcus pneumoniae* and comparison with that of unencapsulated laboratory strain R6. *Journal of Bacteriology*, *189*(1), 38–51. https://doi.org/10.1128/JB.01148-06.

Lawasut, P., Chauhan, D., Laubach, J., Hayes, C., Fabre, C., Maglio, M., … Richardson, P. G. (2012). New proteasome inhibitors in myeloma. *Current Hematologic Malignancy Reports*, 7(4), 258–266. https://doi.org/10.1007/s11899-012-0141-2.

Lebar, M. D., May, J. M., Meeske, A. J., Leiman, S. A., Lupoli, T. J., Tsukamoto, H., …Kahne, D. E. (2014). Reconstitution of peptidoglycan cross-linking leads to improved fluorescent probes of cell wall synthesis. *Journal of the American Chemical Society*, *136*, 10874–10877.

Margolin, W. (2012). The price of tags in protein localization studies. *Journal of Bacteriology*, *194*(23), 6369–6371. https://doi.org/10.1128/JB.01640-12.

Parker, W. L., Rathnum, M. L., & Liu, W. C. (1982). SQ 26,517—A βeta-lactone produced by a Bacillus species. *The Journal of Antibiotics*, *35*, 900–902.

Patricelli, M. P., Giang, D. K., Stamp, L. M., & Burbaum, J. J. (2001). Direct visualization of serine hydrolase activities in complex proteomes using fluorescent active site-directed probes. *Proteomics*, *1*(9), 1067–1071. https://doi.org/10.1002/1615-9861(200109)1:9<1067::AID-PROT1067>3.0.CO;2-4.

Pidgeon, S. E., Fura, J. M., Leon, W., Birabaharan, M., Vezenov, D., & Pires, M. M. (2015). Metabolic profiling of bacteria by unnatural C-terminated D-amino acids. *Angewandte Chemie, International Edition*, *54*, 6158–6162.

Preston, D. A., Wu, C. Y., Blaszczak, L. C., Seitz, D. E., & Halligan, N. G. (1990). Characterization of a new radioactive labeling reagent for bacterial penicillin-binding proteins. *Antimicrobial Agents and Chemotherapy*, *34*, 718–721.

Ravindran, M. S., Rao, S. P., Cheng, X., Shukla, A., Cazenave-Gassiot, A., Yao, S. Q., & Wenk, M. R. (2013). Targeting lipid esterases in mycobacteria grown under different physiological conditions using activity-based profiling with tetrahydrolipstatin (THL). *Molecular and Cellular Proteomics*, *13*, 435–448.

Sadaghiani, A. M., Verhelst, S. H., & Bogyo, M. (2007). Tagging and detection strategies for activity-based proteomics. *Current Opinion in Chemical Biology*, *11*(1), 20–28. https://doi.org/10.1016/j.cbpa.2006.11.030.

Sadler, N. C., & Wright, A. T. (2015). Activity-based protein profiling of microbes. *Current Opinion in Chemical Biology*, *24*, 139–144. https://doi.org/10.1016/j.cbpa.2014.10.022.

Schneider, C. A., Rasband, W. S., & Eliceiri, K. W. (2012). NIH Image to ImageJ: 25 years of image analysis. *Nature Methods*, *9*, 671–675.

Sharifzadeh, S., Boersma, M. J., Kocaoglu, O., Shokri, A., Brown, C. L., Shirley, J. D., …Carlson, E. E. (2017). Novel electrophilic scaffold for imaging of essential penicillin-binding proteins in *Streptococcus pneumoniae*. *ACS Chemical Biology*, *12*(11), 2849–2857. https://doi.org/10.1021/acschembio.7b00614.

Sharifzadeh, S., Dempwolff, F., Kearns, D. B., & Carlson, E. E. (2020). Harnessing β-lactam antibiotics for illumination of the activity of penicillin-binding proteins in *Bacillus subtilis*. *ACS Chemical Biology*. https://doi.org/10.1021/acschembio.9b00977.

Sharifzadeh, S., Shirley, J. D., & Carlson, E. E. (2019). Activity-based protein profiling methods to study bacteria: The power of small-molecule electrophiles. *Current Topics in Microbiology and Immunology*, *420*, 23–48. https://doi.org/10.1007/82_2018_135.

Siegrist, M. S., Whiteside, S., Jewett, J. C., Aditham, A., Cava, F., & Bertozzi, C. R. (2013). (D)-amino acid chemical reporters reveal peptidoglycan dynamics of an intracellular pathogen. *ACS Chemical Biology*, *8*(3), 500–505. https://doi.org/10.1021/cb3004995.

Spratt, B. G., & Pardee, A. B. (1975). Penicillin-binding proteins and cell shape in *E. coli*. *Nature*, *254*(5500), 516–517.

Staub, I., & Sieber, S. A. (2008). β-Lactams as selective chemical probes for the *in vivo* labeling of bacterial enzymes involved in cell wall biosynthesis, antibiotic resistance, and virulence. *Journal of the American Chemical Society*, *130*, 13400–13409.

Stone, M. R. L., Butler, M. S., Phetsang, W., Cooper, M. A., & Blaskovich, M. A. T. (2018). Fluorescent antibiotics: New research tools to fight antibiotic resistance. *Trends in Biotechnology*, *36*(5), 523–536. https://doi.org/10.1016/j.tibtech.2018.01.004.

Swulius, M. T., & Jensen, G. J. (2012). The helical MreB cytoskeleton in *Escherichia coli* MC1000/pLE7 is an artifact of the N-terminal yellow fluorescent protein tag. *Journal of Bacteriology*, *194*(23), 6382–6386. https://doi.org/10.1128/JB.00505-12.

Tiyanont, K., Doan, T., Lazarus, M. B., Fang, X., Rudner, D. Z., & Walker, S. (2006). Imaging peptidoglycan biosynthesis in *Bacillus subtilis* with fluorescent antibiotics. *Proceedings of the National Academy of Sciences of the United States of America*, *103*(29), 11033–11038. https://doi.org/10.1073/pnas.0600829103.

Tsui, H.-C. T., Land, A. D., Kocaoglu, O., Vella, S., Carlson, E. E., & Winkler, M. E. (2013). PBP2x and GpsB are essential for septal formation in *Streptococcus pneumoniae* D39. *Molecular Microbiology*, *90*, 939–955.

Tsui, H. C., Boersma, M. J., Vella, S. A., Kocaoglu, O., Kuru, E., Peceny, J. K., … Winkler, M. E. (2014). Pbp2x localizes separately from Pbp2b and other peptidoglycan synthesis proteins during later stages of cell division of *Streptococcus pneumoniae* D39. *Molecular Microbiology*, *94*(1), 21–40. https://doi.org/10.1111/mmi.12745.

Tsui, H.-C., Zheng, J. J., Magallon, A. N., Ryan, J. D., Yunck, R., Rued, B. E., … Winkler, M. E. (2016). Suppression of a deletion mutation in the gene encoding essential PBP2b reveals a new lytic transglycosylase involved in peripheral peptidoglycan synthesis in *Streptococcus pneumoniae* D39. *Molecular Microbiology*, *100*, 1039–1065.

Tymiak, A. A., Culver, C. A., Malley, M. F., & Gougoutas, J. Z. (1985). Structure of obafluorin: An antibacterial β-lactone from *Pseudomonas fluorescens*. *The Journal of Organic Chemistry*, *50*, 5491–5495.

Venukadasula, P. K. M., Chegondi, R., Maitra, S., & Hanson, P. R. (2010). A concise, phosphate-mediated approach to the total synthesis of (−)-tetrahydrolipstatin. *Organic Letters*, *12*, 1556–1559.

Waxman, D. J., & Strominger, J. L. (1983). Penicillin-binding proteins and the mechanism of action of β-lactam antibiotics. *Annual Review of Biochemistry*, *52*, 825–869. https://doi.org/10.1146/annurev.bi.52.070183.004141.

Wells, J. S., Hunter, J. C., Astle, G. L., Sherwood, J. C., Ricca, C. M., Trejo, W. H., … Sykes, R. B. (1982). Distribution of β-lactam and β-lactone producing bacteria in nature. *Journal of Antibiotics (Tokyo)*, *35*(7), 814–821.

Wells, J. S., Trejo, W. H., Principe, P. A., & Sykes, R. B. (1984). Obafluorin, a novel β-lactone produced by *Pseudomonas fluorescens*. Taxonomy, fermentation and biological properties. *The Journal of Antibiotics*, *37*, 802–803.

Willems, L. I., Overkleeft, H. S., & van Kasteren, S. I. (2014). Current developments in activity-based protein profiling. *Bioconjugate Chemistry*, *25*(7), 1181–1191. https://doi.org/10.1021/bc500208y.

Williamson, R., Hakenbeck, R., & Tomasz, A. (1980). In vivo interaction of β-lactam antibiotics with the penicillin-binding proteins of *Streptococcus pneumoniae*. *Antimicrobial Agents and Chemotherapy*, *18*(4), 629–637. https://doi.org/10.1128/aac.18.4.629.

Zeiler, E., Braun, N., Bottcher, T., Kastenmuller, A., Weinkauf, S., & Sieber, S. A. (2011). Vibralactone as a tool to study the activity and structure of the ClpP1P2 complex from *Listeria monocytogenes*. *Angewandte Chemie (International Ed in English)*, *50*(46), 11001–11004. https://doi.org/10.1002/anie.201104391.

Zhao, G., Meier, T. I., Kahl, S. D., Gee, K. R., & Blaszczak, L. C. (1999). BOCILLIN FL, a sensitive and commercially available reagent for detection of penicillin-binding proteins. *Antimicrobial Agents and Chemotherapy*, *43*(5), 1124–1128.

CHAPTER THREE

Fluorescent stem peptide mimics: In situ probes for peptidoglycan crosslinking

Samir Gautam[a,b], Taehan Kim[b], Rebecca Howell[b], David A. Spiegel[b,*]

[a]Yale University School of Medicine, Department of Internal Medicine, Section of Pulmonary, Critical Care and Sleep Medicine, New Haven, CT, United States

[b]Yale University, Department of Chemistry, New Haven, CT, United States

[*]Corresponding author: e-mail address: david.spiegel@yale.edu

Contents

Abstract

Understanding the mechanisms of bacterial cell wall synthesis is essential for microbiology and medicine alike. A key step in this process is peptidoglycan crosslinking, which confers mechanical strength to the cell wall and represents a target for numerous classes of antibiotics. However, the biology of crosslinking remains poorly understood due to a lack of tools for studying the reaction in vivo. Recently, we developed a class of synthetic probes called fluorescent stem peptide mimics (FSPMs) that meet this need, allowing quantification and localization of crosslinking activity in live bacteria. We have utilized FSPMs to describe novel aspects of peptidoglycan synthesis in the human

Methods in Enzymology, Volume 638
ISSN 0076-6879
https://doi.org/10.1016/bs.mie.2020.02.016

pathogen, *Staphylococcus aureus*. To enable wider use of our methodology, we provide detailed protocols herein for the synthesis of FSPMs, labeling of live bacteria, and evaluation of crosslinking by flow cytometry and super-resolution microscopy. We believe that FSPMs, together with complementary biosynthetic probes and traditional bacteriologic methods, will help to advance our understanding of peptidoglycan biology and accelerate the search for new antibiotics.

1. Introduction

With few exceptions, all bacteria are surrounded by a peptidoglycan cell wall—a mesh-like polymer that confers protection and shape to prokaryotic organisms (Otten, Brilli, Vollmer, Viollier, & Salje, 2018). Its ability to withstand severe environmental stresses, such as turgor pressures up to 140 kPa (Auer & Weibel, 2017), and yet undergo the intricate processes of cell growth and division has long fascinated biologists. Meanwhile, its indispensability for cell survival makes it an attractive therapeutic target for physicians treating bacterial infections. Given the spread of drug-resistant pathogens, which are projected to claim 10 million lives per year by 2050 (O'Neill, 2016), and the slowed pace of antimicrobial development in recent decades (Brown & Wright, 2016), it is imperative to elucidate the mechanisms of peptidoglycan synthesis in order to accelerate discovery of novel antibiotics.

Structurally, peptidoglycan consists of long glycan strands crosslinked together by peptide bridges (Pazos & Peters, 2019), The enzymes responsible for crosslink formation are penicillin-binding proteins (PBPs), the target of several classes of antibiotics including beta-lactams and glycopeptides (Fig. 1A) (Frère & Page, 2014; Pazos, Peters, & Vollmer, 2017; Sauvage, Kerff, Terrak, Ayala, & Charlier, 2008). The past decade has witnessed explosive progress in PBP biology, in part due to the emergence of novel chemical probes for studying PBP function in live bacteria (Gautam, Gniadek, Kim, & Spiegel, 2013; Radkov, Hsu, Booher, & VanNieuwenhze, 2018; Taguchi, Kahne, & Walker, 2019). These probes include the fluorescent stem peptide mimics (FSPMs) developed in our laboratory (Fig. 1B) (Gautam, Kim, Shoda, et al., 2015; Gautam, Kim, & Spiegel, 2015), fluorescent D-amino acids (FDAAs) introduced by VanNieuwenhze and colleagues (Kuru et al., 2012; Radkov et al., 2018), and other agents as recently reviewed (Radkov et al., 2018; Taguchi et al., 2019).

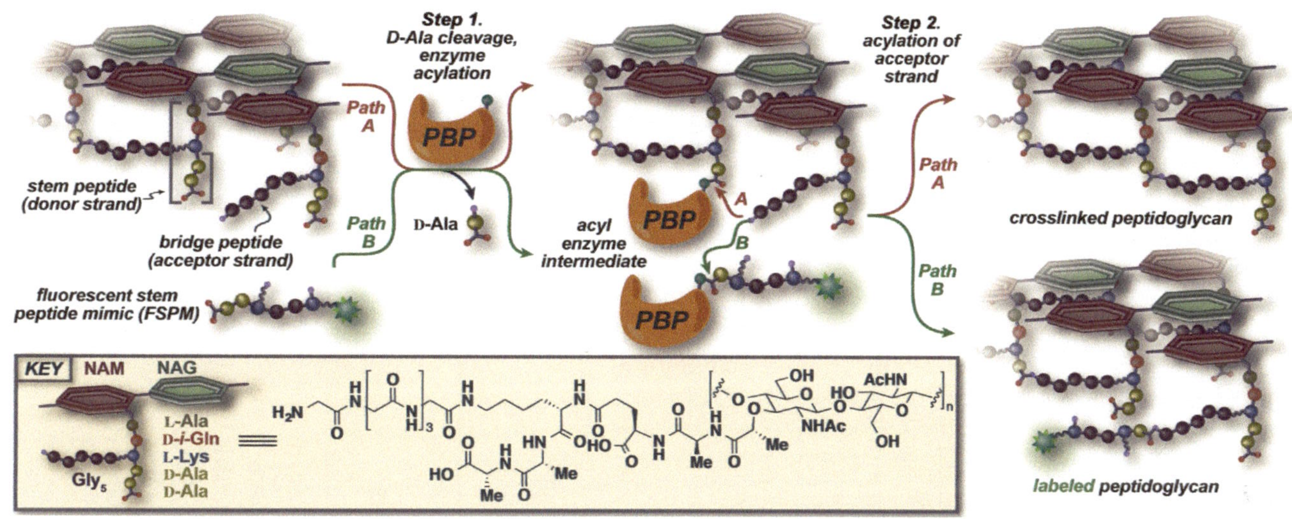

Fig. 1 Schematic of FSPM incorporation into the cell wall of *S. aureus*. Two paths are depicted: Path A (red) outlines endogenous PBP-mediated crosslinking of peptidoglycan, while Path B (green) illustrates the mechanism of FSPM incorporation. In Path A, the transpeptidase domain of a PBP recognizes the C-terminal terminal L-Lys-D-Ala-D-Ala motif within the stem peptide (donor strand) and cleaves the C-terminal D-Ala residue to form an acyl-enzyme intermediate (Step 1). This intermediate is then captured by the bridge peptide (acceptor strand) to furnish a covalent bond between the penultimate D-Ala of the stem peptide and the N-terminal glycine of the bridge peptide, i.e., a crosslink (Step 2). In Path B, FSPMs serve as mimics of the donor strand, producing the unique acyl-enzyme intermediate shown here. The acceptor strand then attacks this intermediate, leading to incorporation of a fluorescent label into peptidoglycan in lieu of a natural crosslink. Abbreviations: D-Ala, D-alanine; FSPM, fluorescent stem peptide mimic; Gly, glycine; D-Gln, D-glutamine; L-Ala, L-alanine; Lys, L-lysine; NAM, *N*-acetylmuramic acid; NAG, *N*-acetylglucosamine; PBP, penicillin-binding protein. *Figure adapted with permission from Gautam, S., Kim, T., & Spiegel, D. A. (2015). Chemical probes reveal an extraseptal mode of cross-linking in* Staphylococcus aureus. Journal of the American Chemical Society, 137*(23), 7441–7447. https://doi.org/10.1021/jacs.5b02972.*

FSPMs consist of a fluorophore conjugated to a short synthetic peptide that mimics the endogenous substrate of PBPs—the stem peptide. Covalent attachment of FSPMs to the cell wall by PBPs proceeds as described in Fig. 1B, resulting in installation of a fluorophore precisely at the site of PBP activity. Thus, FSPMs enable evaluation of crosslinking in live bacteria. This in situ labeling technique represents an important advance over conventional methods for studying crosslinking, which require lysis, enzymatic digestion, and biochemical analysis of the resulting muropeptides (Desmarais, De Pedro, Cava, & Huang, 2013).

Using the Gram-positive pathogen *Staphylococcus aureus* as a model organism, we have shown that FSPMs are exclusively incorporated by a single isoform—PBP4 (Gautam, Kim, Shoda, et al., 2015). This property enables monitoring of PBP4-specific crosslinking activity in vivo. Utilizing FSPMs we have demonstrated the PBP4 activity is restricted to the septum during division by cell surface polysaccharides called wall teichoic acids (Gautam, Kim, Shoda, et al., 2015), and occurs globally throughout the cell well between divisions (Gautam, Kim, & Spiegel, 2015). We have since extended the scope of FSPM labeling to several additional Gram-positive organisms including the rod-shaped *Bacillus subtilis* (unpublished results).

Here we provide protocols for FSPM synthesis, bacterial labeling, quantification of crosslinking by flow cytometry, and direct visualization in situ using super-resolution microscopy.

2. Probe synthesis

2.1 Peptide synthesis

Two peptides must be synthesized: a functional stem peptide mimic containing the PBP recognition motif, (L-Lys)–(D-Ala)–(D-Ala); and a diastereomeric control containing (L-Lys)–(L-Ala)–(L-Ala), which indicates non-enzymatic labeling (i.e., non-specific binding to the cell wall). Both contain a short peptide linker, (L-Lys)–(Gly)–(Gly), which allows conjugation of a fluorophore to the epsilon amine of the N-terminal lysine. The full sequences are thus (L-Lys)–(Gly)–(Gly)–(L-Lys)–(D-Ala)–(D-Ala) for the functional FSPM (e.g., D-FL), and (L-Lys)–(Gly)–(Gly)–(L-Lys)–(L-Ala)–(L-Ala) for the control (e.g., L-FL). We perform synthesis on the solid phase using a CEM Discover Liberty Microwave Peptide Synthesizer for the sake of ease, but solution phase may allow for larger scales. The amount of resin, amino acids,[a] and other reagents used in the synthesis are calculated using the manufacturer's protocol and purchased from commercial sources. It is critical to use a resin such as Wang so as to furnish a C-terminal acid upon cleavage, as the amide form is not recognized efficiently by PBPs. Upon completion of peptide synthesis, the resin is transferred to a plastic cartridge, rinsed twice with dimethylformamide (DMF) and twice with dichloromethane (DCM) before drying under vacuum. Dried peptide-resin may be stored for months at room temperature prior to conjugation.

2.2 Fluorophore conjugation[b]

The peptide-resin is transferred to a plastic cartridge sized appropriately for the scale of the reaction. A DMF solution of the fluorophore[c] and *N*,*N*-diisopropylethylamine (DIPEA) is added, and the cartridge is placed on an ATR Rotamix rotator in the dark for 24h at room temperature. Following conjugation, the DMF solution is drained under vacuum, and the resin is washed well with DMF (thrice) followed by DCM (thrice). Peptide cleavage is performed by treatment of resin with 1–3mL[d] of 95:2.5:2.5 TFA:TIPS:H_2O (trifluoroacetic acid:triisopropylsilane:water) for 90min. The resultant TFA solution is then poured into cold ether on dry ice (−78°C) to precipitate the fluorescent peptide.

2.3 FSPM purification

The crude product prepared in 2.2 must be purified by high-performance liquid chromatography (HPLC). We use a Waters HPLC system equipped with a SunFire C18 prep column (100Å, 5μm; water/acetonitrile buffered

[a] Upon completion of peptide synthesis, the epsilon amine of the N-terminal lysine must be free to enable site-specific fluorophore conjugation; the other amines must be protected. To achieve this, we use Fmoc-Lys(Boc)-OH for the linker lysine and Boc-Lys(Fmoc)-OH for the terminal lysine. Thus, the alpha amine of the N-terminal lysine and the epsilon amine of the internal lysine are Boc protected. The epsilon amine of the N-terminal lysine is protected with Fmoc, such that after treatment with piperidine (per the standard solid phase peptide synthesis protocol), the Fmoc is removed, enabling coupling to the fluorophore. In the final step, treatment with TFA/TIPS/H_2O results in removal of Boc groups and cleavage of the peptide from the resin.

[b] Other chemical moieties besides fluorophores may be attached, depending on the desired application. These include alkynes or azides to facilitate click chemistry, biotin to allow recognition or isolation of labeled peptidoglycan using streptavidin, etc.

[c] For flow cytometric analysis and basic imaging applications, inexpensive fluorophores such as fluorescein isothiocyanate (FITC) will suffice. For super-resolution imaging via three-dimensional structured illumination microscopy (3D-SIM), as described below, more photostable dyes such as Alexa fluorophores must be used.

[d] Volume depends on synthetic scale.

with 0.1% TFA). Gradient elution parameters are as follows: D-FL, 0%–5% acetonitrile over 1 min then 5%–27% acetonitrile over 29 min, D-FL at 24.2 min; L-FL, 5%–27% acetonitrile over 27 min, L-FL at 22.4 min. Fractions containing the compound of interest are pooled and solvent is evaporated using a lyophilizer. The compound is then resuspended at 100 mM in water,[e] and the concentration is confirmed via absorption on a NanoDrop spectrophotometer. The purity and identity of FSPMs are assessed via high resolution mass spectrometry (MS). For this purpose, we use a Waters Acquity UPLC system equipped with a C18 column (water/acetonitrile buffered with 0.1% formic acid) and a Xevo G2 QTof MS.

3. Bacterial growth and FSPM labeling

3.1 Maintenance of *S. aureus*

The model staphylococcal strain, Newman, is commonly used in our laboratory, but any strain expressing PBP4 can be utilized to study FSPM incorporation, including methicillin-resistant *S. aureus*. Luria-Bertani (LB) agar plates are streaked weekly from 20% glycerol stocks stored at −80 °C.

3.2 FSPM labeling

Liquid cultures of LB broth (3 mL) are inoculated with single colonies and grown overnight at 37 °C with agitation at 200 RPM to stationary phase. Cultures are then diluted 1:100 and grown 2 h to reach logarithmic phase. Probes are added at a final concentration of 1 mM, which maximizes labeling without producing toxicity. Experimental conditions should be run in technical triplicate—one set of samples with d-FL and one set with L-FL in order to control for non-specific binding of compounds. A third set treated with vehicle control (e.g., water or dimethyl sulfoxide) can be utilized as well to reveal the extent of non-specific labeling (i.e., the difference between vehicle and L-FL labeling). Volumes as low as 20 μL can be used to conserve FSPMs conjugated to more costly fluorophores (e.g., D-A568), but the atmosphere should be humidified to prevent evaporation. If labeling is performed in a 96-well plate, which is convenient for multiple samples, the plate must be shaken and the outer wells should be filled with sterile water

[e] Peptides are highly polar and so dissolve readily in water. Use of water as a diluent avoids introducing potential toxins such as dimethyl sulfoxide (DMSO).

to ensure humidification. The length of incubation with FSPMs will depend on the application, but significant incorporation is observed in as little as 5 min.

3.3 Sample preparation

After labeling, cultures are immediately diluted in ice-cold phosphate-buffered saline (PBS) to halt further probe incorporation. Cells are centrifuged at 12,000 × *g* for 2 min and then washed thrice in ice-cold PBS. Less robust bacteria (i.e., with a less-extensively cross-linked cell wall) may be spun at lower speeds to preserve cell shape and integrity, e.g., *Bacillus subtilis* at 4000 × *g*. For flow cytometry, samples are diluted in PBS to a concentration of <5000 cells/μL, placed on ice, and run immediately. Killing bacteria prior to analysis, e.g., by washing with acetone, slightly increases the fluorescein signal, likely due to abrogation of the acidic milieu maintained near the cell wall in actively respiring cells (which partly quenches fluorescein). For microscopy experiments, cells are fixed in 4% paraformaldehyde (diluted from fresh ampoules of 16% solution, Electron Microscopy Sciences), and washed twice with PBS. Cells are then resuspended in PBS, dropped onto 1.5 mm glass coverslips coated with poly-L-lysine (Santa Cruz Biotechnology), and air-dried. Coverslips are mounted in ProLong Gold antifade reagent (Invitrogen) and cured for at least 24 h before imaging. Coating with poly-L-lysine may be achieved by dispensing 1 mL of poly-L-lysine solution into wells of a 6-well plate, submerging coverslips within the wells for 30 min, and then removing coverslips to let them dry. Poly-L-lysine solution can be recovered and re-used several times for this purpose.

4. Quantification of crosslinking by flow cytometry

4.1 Flow cytometry analysis

To assess the magnitude of FSPM labeling, we use an Accuri C6 flow cytometer (Becton Dickinson) on medium speed fluidics with a minimum threshold of 40,000 FSC-H. Maximum FSC and SSC gates are set to exclude multi-cell aggregates (e.g., 50,000 FSC-A and 50,000 SSC-A for *S. aureus*). 10,000 Events per sample are analyzed. Traces are processed using FlowJo software. Representative data indicating inhibition of FSPM labeling with the PBP4-specific beta-lactam antibiotic, cefoxitin, are shown in Fig. 2A–B; Fig. 2C shows abrogation of labeling in the PBP4 deletion mutant.

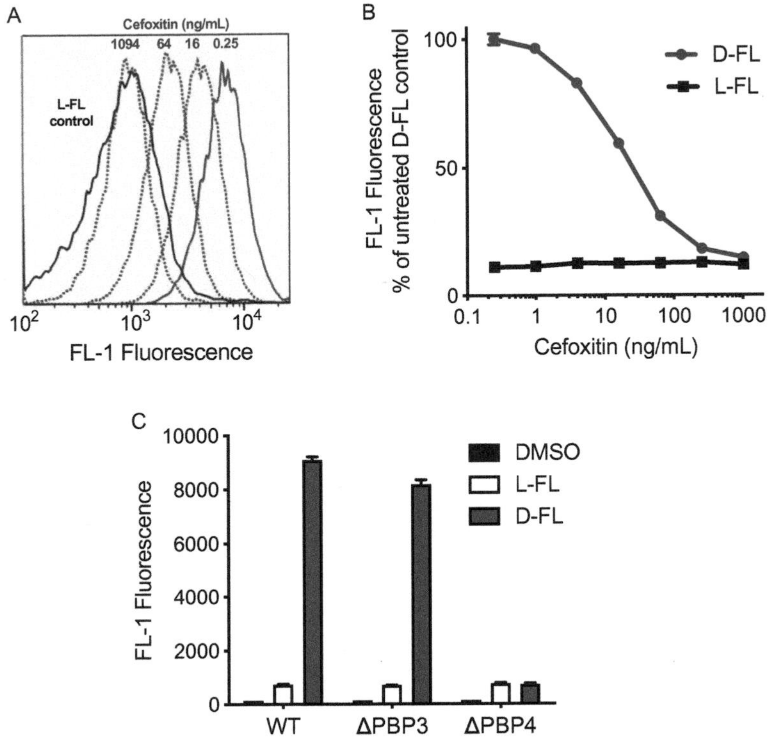

Fig. 2 Quantification of crosslinking by flow cytometry. (A) and (B) WT *S. aureus* Newman was pre-treated with cefoxitin for 1 h, labeled with probes in the presence of antibiotic for 2 h, and analyzed by flow cytometry. In (A), representative flow cytometry traces are shown. In (B), flow cytometry data are quantified and presented as percent values relative to an untreated D-FL sample. (C) *S. aureus* Newman PBP mutant strains were labeled with 1 mm D-FL, L-FL, or DMSO vehicle control for 2 h, and analyzed by flow cytometry. Representative flow cytometry experiments performed in technical triplicate are shown; error bars represent SEM. *Figure adapted with permission from Gautam, S., Kim, T., Shoda, T., Sen, S., Deep, D., Luthra, R., et al. (2015). An activity-based probe for studying crosslinking in live bacteria.* Angewandte Chemie, International Edition, *54(36), 10492–10496. https://doi.org/10.1002/anie.201503869.*

5. Localization of crosslinking by microscopy

5.1 Image acquisition

While any fluorescent microscope can be used to image FSPM incorporation, the small size of bacteria (~1 μm in diameter) often requires super-resolution imaging techniques for accurate subcellular localization of probe labeling. Here we describe our laboratory's methodology using three-dimensional structured illumination microscopy (3D-SIM). Images are acquired using a UPLANAPO 60X/1.42 PSF, oil immersion objective lens (Olympus) and CoolSNAP HQ2 CCD cameras with a pixel size of 0.080 μm (Photometrics) on the OMX version 3 system (Applied Precision) equipped with 488-, and 561-nm solid-state lasers (Coherent and MPB Communications). For 3D-SIM imaging, samples are illuminated by a coherent scrambled laser light source to create a 3D sinusoidal pattern. The pattern is shifted laterally through five phases and through three angular rotations of 60° for each *Z*-section, separated by 0.125 nm. Exposure times are between 200 and 500 ms, and the power of each laser is adjusted to achieve optimal intensities of between 2000 and 4000 counts in a raw image of 16-bit dynamic range. Raw images are processed and reconstructed to reveal structures with 100–125 nm resolution using Softworx software (Applied Precision). Representative images using unique FSPMs in the red and blue channels, along with a vancomycin-based imaging reagent in green (Van-A488) that stains the entire cell wall (Daniel & Errington, 2003), are shown in Fig. 3.

6. Concluding remarks

It is important to contrast the FSPMs described here with the other available probes for peptidoglycan synthesis—particularly FDAAs, a powerful class of reagents that label active sites of cell wall synthesis in a wide array of bacterial species (Kuru et al., 2012; Radkov et al., 2018). FDAAs are covalently incorporated into the PG stem peptide via an exchange reaction mediated by transpeptidases, wherein FDAAs serve as surrogate acceptor strands in the transpeptidation reaction (Fig. 1) (Kuru et al., 2019). Therefore, FDAA labeling represents D-amino acid exchange at the distal stem peptide. In contrast, FSPMs serve as surrogate *donor* strands in the transpeptidation reaction, and thus more specifically indicate crosslinking. Furthermore, In *S. aureus*, FSPM labeling represents PBP4-specific crosslinking (Gautam, Kim, Shoda, et al., 2015); FDAAs, meanwhile, are likely incorporated by multiple PBPs. A final difference is the ability of FDAAs to label peptidoglycan in Gram-negative organisms. FSPMs are too large and polar to cross the outer membrane, restricting their use to Gram-positive bacteria.

Guided by these considerations and others, as recently reviewed (Radkov et al., 2018; Taguchi et al., 2019), researchers must carefully select from the growing list of peptidoglycan probes to answer their particular

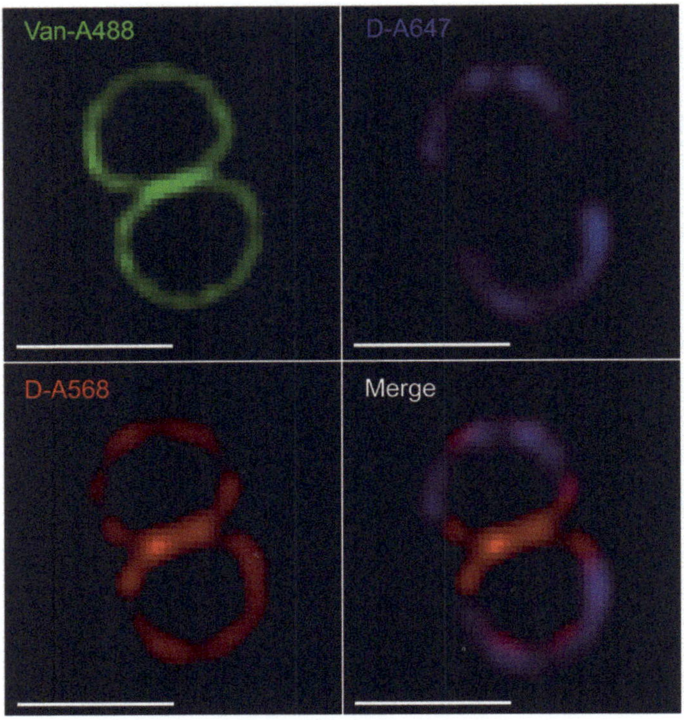

Fig. 3 Localization of crosslinking by microscopy. WT *S. aureus* Newman was labeled for 2 h with D-A647. Probe was then washed out and cells were allowed to grow for 1 h before being pulse labeled for 7.5 min with D-A568 (functional FSPM conjugated to Alexa Fluor 568) and stained with Van-A488 (vancomycin conjugated to Alexa Fluor 488). Representative SIM images of a nonseptating doublet are shown. Scale bars: 1 μm. *Figure adapted with permission from Gautam, S., Kim, T., & Spiegel, D. A. (2015). Chemical probes reveal an extraseptal mode of cross-linking in* Staphylococcus aureus. Journal of the American Chemical Society, 137*(23), 7441–7447. https://doi.org/10.1021/jacs.5b02972.*

biological question. With this caveat, such probes will continue to advance our knowledge of bacterial growth and division, providing unprecedented insights into the basic physiology of peptidoglycan synthesis, and paving the way to the much-needed development of novel antibiotics.

References

Auer, G. K., & Weibel, D. B. (2017). Bacterial cell mechanics. *Biochemistry*, *56*(29), 3710–3724. https://doi.org/10.1021/acs.biochem.7b00346.

Brown, E. D., & Wright, G. D. (2016). Antibacterial drug discovery in the resistance era. *Nature*, *529*, 336–343. https://doi.org/10.1038/nature17042.

Daniel, R. A., & Errington, J. (2003). Control of cell morphogenesis in bacteria: Two distinct ways to make a rod-shaped cell. *Cell*, *113*(6), 767–776. https://doi.org/10.1016/S0092-8674(03)00421-5.

Desmarais, S. M., De Pedro, M. A., Cava, F., & Huang, K. C. (2013). Peptidoglycan at its peaks: How chromatographic analyses can reveal bacterial cell wall structure and assembly. *Molecular Microbiology*, *89*, 1–13. https://doi.org/10.1111/mmi.12266.

Frère, J. M., & Page, M. G. P. (2014). Penicillin-binding proteins: Evergreen drug targets. *Current Opinion in Pharmacology*, *18*, 112–119. https://doi.org/10.1016/j.coph.2014.09.012.

Gautam, S., Gniadek, T. J., Kim, T., & Spiegel, D. A. (2013). Exterior design: Strategies for redecorating the bacterial surface with small molecules. *Trends in Biotechnology*, *31*, 258–267. https://doi.org/10.1016/j.tibtech.2013.01.012.

Gautam, S., Kim, T., Shoda, T., Sen, S., Deep, D., Luthra, R., et al. (2015). An activity-based probe for studying crosslinking in live bacteria. *Angewandte Chemie, International Edition*, *54*(36), 10492–10496. https://doi.org/10.1002/anie.201503869.

Gautam, S., Kim, T., & Spiegel, D. A. (2015). Chemical probes reveal an extraseptal mode of cross-linking in *Staphylococcus aureus*. *Journal of the American Chemical Society*, *137*(23), 7441–7447. https://doi.org/10.1021/jacs.5b02972.

Kuru, E., Hughes, H. V., Brown, P. J., Hall, E., Tekkam, S., Cava, F., et al. (2012). In situ probing of newly synthesized peptidoglycan in live bacteria with fluorescent D-amino acids. *Angewandte Chemie, International Edition*, *51*(50), 12519–12523. https://doi.org/10.1002/anie.201206749.

Kuru, E., Radkov, A., Meng, X., Egan, A., Alvarez, L., Dowson, A., et al. (2019). Mechanisms of incorporation for D-amino acid probes that target peptidoglycan biosynthesis. *ACS Chemical Biology*, *14*, 2745–2756. acschembio.9b00664. https://doi.org/10.1021/acschembio.9b00664.

O'Neill, J. (2016). *Tackling drug-resistant infections globally: Final report and recommendations. In The review on antimicrobial resistance*. World Health Organization, Retrieved October 9, 2019, from https://amr-review.org/sites/default/files/160518_Final paper_with cover.pdf.

Otten, C., Brilli, M., Vollmer, W., Viollier, P. H., & Salje, J. (2018). Peptidoglycan in obligate intracellular bacteria. *Molecular Microbiology*, *107*, 142–163. https://doi.org/10.1111/mmi.13880.

Pazos, M., & Peters, K. (2019). Peptidoglycan. In A. Kuhn (Ed.), *Vol. 92. Bacterial cell walls and membranes. Subcellular biochemistry*. Cham: Springer.

Pazos, M., Peters, K., & Vollmer, W. (2017). Robust peptidoglycan growth by dynamic and variable multi-protein complexes. *Current Opinion in Microbiology*, *36*, 55–61. https://doi.org/10.1016/j.mib.2017.01.006.

Radkov, A. D., Hsu, Y.-P., Booher, G., & VanNieuwenhze, M. S. (2018). Imaging bacterial cell wall biosynthesis. *Annual Review of Biochemistry*, *87*(1), 991–1014. https://doi.org/10.1146/annurev-biochem-062917-012921.

Sauvage, E., Kerff, F., Terrak, M., Ayala, J. A., & Charlier, P. (2008). The penicillin-binding proteins: Structure and role in peptidoglycan biosynthesis. *FEMS Microbiology Reviews*, *32*, 234–258. https://doi.org/10.1111/j.1574-6976.2008.00105.x.

Taguchi, A., Kahne, D., & Walker, S. (2019). Chemical tools to characterize peptidoglycan synthases. *Current Opinion in Chemical Biology*, *53*, 44–50. https://doi.org/10.1016/j.cbpa.2019.07.009.

CHAPTER FOUR

Gram-scale preparation of the antibiotic lead compound salicyl-AMS, a potent inhibitor of bacterial salicylate adenylation enzymes

Nihar Kinarivala[a], Lisa C. Standke[b], Tezcan Guney[a], Cheng Ji[a], Naoyoshi Noguchi[c,†], Yasutomi Asano[c,‡], Derek S. Tan[a,b,d,*]
[a]Chemical Biology Program, Sloan Kettering Institute, Memorial Sloan Kettering Cancer Center, New York, NY, United States
[b]Pharmacology Program, Weill Cornell Graduate School of Medical Sciences, Memorial Sloan Kettering Cancer Center, New York, NY, United States
[c]Tri-Institutional Therapeutics Discovery Institute, New York, NY, United States
[d]Tri-Institutional Research Program, Memorial Sloan Kettering Cancer Center, New York, NY, United States
*Corresponding author: e-mail address: tand@mskcc.org

Contents

† Present address: Research Division, Scohia Pharma, Inc., Shonan Health Innovation Park, 26-1, Muraoka-Higashi 2-chome, Fujisawa, Kanagawa 251-8555, Japan.
‡ Present address: Discovery Science, Axcelead Drug Discovery Partners, Inc., Shonan Health Innovation Park, 26-1, Muraoka-Higashi 2-chome, Fujisawa, Kanagawa 251-0012, Japan.

Methods in Enzymology, Volume 638
ISSN 0076-6879
https://doi.org/10.1016/bs.mie.2020.04.051

Abstract

Salicyl-AMS (**1**) is a potent inhibitor of salicylate adenylation enzymes used in bacterial siderophore biosynthesis and a promising lead compound for the treatment of tuberculosis. An optimized, multigram synthesis is presented, which provides salicyl-AMS as its sodium salt (**1·Na**) in three synthetic steps followed by a two-step salt formation process. The synthesis proceeds in 11.6% overall yield from commercially available adenosine 2′,3′-acetonide and provides highly purified material.

1. Introduction

Tuberculosis remains a global public health crisis, with an estimated 10 million new cases and 1.5 million deaths in 2018 (Furin, Cox, & Pai, 2019; World Health Organization, 2019). Current treatments are effective against drug-sensitive infections, but multi-drug resistant and extensively-drug resistant tuberculosis present major threats, and an estimated 484,000 new infections resistant to the first-line drug rifampicin were reported in 2018. Thus, there is a great need for new antibiotics to treat drug-resistant tuberculosis infections. *Mycobacterium tuberculosis*, the bacteria that causes tuberculosis, acquires the essential nutrient Fe^{3+} using iron-chelating molecules called siderophores (De Voss, Rutter, Schroeder, & Barry, 1999; De Voss, Rutter, Schroeder, Su, et al., 2000; Quadri & Ratledge, 2005; Ratledge, 2004). Siderophores bind iron with very high affinity and are used to transport iron across the cell envelope. Accordingly, inhibition of siderophore biosynthesis to limit bacterial iron uptake has been advanced as promising new antibacterial strategy (Lamb, 2015).

In collaboration with the laboratory of Luis Quadri, we have previously reported the development of the first designed siderophore biosynthesis inhibitor, salicyl-AMS (5′-*O*-[*N*-salicylsulfamoyl]adenosine) (**1**) (Fig. 1A) (Bythrow, Mohandas, Guney, Standke, et al., 2019; Ferreras, Ryu, Di Lello, Tan, et al., 2005). Salicyl-AMS is a potent inhibitor of MbtA, a salicylate adenylation enzyme that catalyzes the first step in the synthesis of the *M. tuberculosis* mycobactin siderophores **2** (McMahon, Rush, & Thomas, 2012; Quadri, Sello, Keating, Weinreb, et al., 1998). Salicyl-AMS mimics a tightly-bound reaction intermediate, salicyl-AMP (**4**), and potently inhibits MbtA biochemical activity as well as *M. tuberculosis* siderophore production and bacterial growth in cell culture. In collaboration with the laboratory of William Bishai, we demonstrated promising

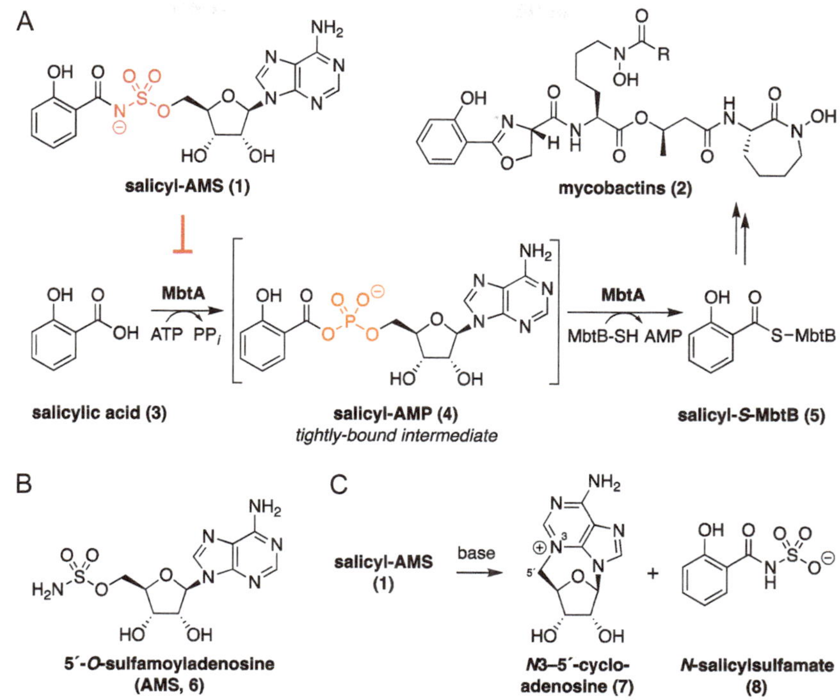

Fig. 1 (A) Salicyl-AMS (**1**) is a potent inhibitor of the salicylate adenylation enzyme MbtA, and mimics a tightly-bound salicyl-AMP intermediate (**4**), blocking production of the mycobactin siderophores (**2**). R represents various fatty acyl chains (mycobactin variants) or chains terminating in a carboxylate or methyl ester (carboxymycobactin variants). (B) Structure of highly toxic fragment 5′-*O*-sulfamoyladenosine (AMS, **6**). (C) Potential decomposition of salicyl-AMS (**1**) leading to *N*3–5′-cycloadenosine (**7**) and *N*-salicylsulfamate (**8**).

in vivo efficacy of salicyl-AMS in a mouse model of tuberculosis (Lun, Guo, Adamson, Cisar, et al., 2013). Aldrich and coworkers have also reported extensive structure–activity relationship studies of salicyl-AMS analogs (Dawadi, Boshoff, Park, Schnappinger, et al., 2018; Dawadi, Kawamura, Rubenstein, Remmel, et al., 2016; Dawadi, Viswanathan, Boshoff, Barry 3rd, et al., 2015; Duckworth, Nelson, & Aldrich, 2012; Engelhart & Aldrich, 2013; Gupte, Boshoff, Wilson, Neres, et al., 2008; Krajczyk, Zeidler, Januszczyk, Dawadi, et al., 2016; Nelson, Viswanathan, Dawadi, Duckworth, et al., 2015; Neres, Labello, Somu, Boshoff, et al., 2008; Qiao, Gupte, Boshoff, Wilson, et al., 2007; Qiao, Wilson, Bennett, & Aldrich, 2007; Somu, Boshoff, Qiao, Bennett, et al., 2006; Somu, Wilson, Bennett, Boshoff, et al., 2006; Vannada, Bennett, Wilson, Boshoff, et al., 2006).

To enable further preclinical evaluation of salicyl-AMS, we report herein an optimized, scalable protocol for synthesis, purification, and analysis of salicyl-AMS. Importantly, this protocol provides salicyl-AMS that is free of the des-salicyl fragment 5′-*O*-sulfamoyladenosine (AMS, **6**) (Fig. 1B) ($<$0.0006 mol%, below the lower limit of detection), which is known to be highly cytotoxic (NCI-60 mean $GI_{50}=1.8$ nM; NSC133114) (Bloch & Coutsogeorgopoulos, 1971; Jaffe, McCormack, & Meymerian, 1970; Rengaraju, Narayanan, Ganju, Amin, et al., 1986), as well as an *N*3–5′-cycloadenosine byproduct (**7**), which forms under basic conditions via a precedented pathway (Fig. 1C) (Duckworth, Geders, Tiwari, Boshoff, et al., 2011; Shi, Tiwari, Wilson, Seiler, et al., 2013; Shuman, Robins, & Robins, 1969). Moreover, salicyl-AMS is highly acidic at the acylsulfamate moiety ($pK_a \approx 1.5$) (Cisar & Tan,unpublished results; Somu, Boshoff, et al., 2006), and Aldrich has reported previously that conversion to the triethylammonium salt (**1·Et$_3$N**) provides increased storage stability (Somu, Boshoff, et al., 2006). Our protocol affords salicyl-AMS as the corresponding sodium salt (**1·Na**), which is more commonly used in pharmaceutical formulations (Paulekuhn, Dressman, & Saal, 2007), to enable its use in preclinical and potentially clinical studies.

2. Synthesis of salicyl-AMS, sodium salt

Salicyl-AMS (**1**) was synthesized in three steps from commercially available adenosine 2′,3′-acetonide (**13**), in up to 10-g scale and an overall yield of 14.4% (Fig. 2). The route incorporates numerous improvements to our original route (Ferreras et al., 2005), based on subsequent reports from our group and Aldrich and coworkers. Notably, only a single protecting group manipulation is required. Sulfamoyl chloride (**10**) was prepared by treatment of commercially available chlorosulfonylisocyanate (**9**) with formic acid (Fig. 2A) (Appel & Berger, 1958; Brodsky & Du Bois, 2005; Heacock, Forsyth, Shiba, & Musier-Forsyth, 1996). Salicylic acid, *N*-hydroxysuccinimide ester (**12**) was prepared by activation of commercially available salicylic acid (**11**) with dicyclohexylcarbodiimide and coupling with *N*-hydroxysuccinimide (Van Brussel & Van Sumere, 1978) (Fig. 2B).

Adenosine acetonide **13** reacted with sulfamoyl chloride (**10**) selectively at the 5′-hydroxyl in *N*,*N*-dimethylacetamide (Cisar, Ferreras, Soni, Quadri, et al., 2007; Ferreras, Stirrett, Lu, Ryu, et al., 2008; Okada, Iwashita, & Koizumi, 2000) to form 5′-*O*-sulfamoyladenosine 2′,3′-acetonide (**14**) in 62.5% yield (Fig. 2C). Acylation with salicylic acid, *N*-hydroxysuccinimide ester (**12**) using cesium carbonate base in

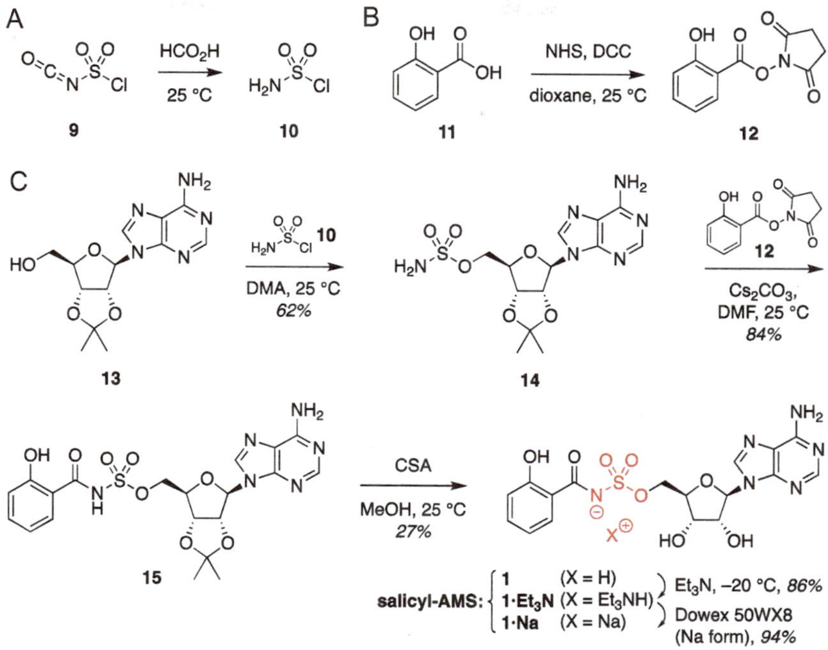

Fig. 2 (A) Preparation of sulfamoyl chloride (**10**). (B) Preparation of salicylic acid, *N*-hydroxysuccinimide ester (**12**). (C) Synthesis of salicyl-AMS (**1**) and conversion to the corresponding sodium salt (**1·Na**). CSA, camphorsulfonic acid; DCC, *N,N′*-dicyclohexylcarbodiimide; DMA, *N,N*-dimethylacetamide; DMF, *N,N*-dimethylformamide; NHS, *N*-hydroxysuccinimde.

N,N-dimethylformamide (Somu, Boshoff, et al., 2006) provided salicyl-AMS 2′,3′-acetonide (**15**) in 84.5% yield. Finally, deprotection was achieved most efficiently using camphorsulfonic acid in methanol to afford salicyl-AMS (**1**) in 27.3% yield (14.4% overall yield over three steps). This deprotection method provided higher yields compared to a variety of other conditions evaluated (5:2 trifluoroacetic acid [TFA]/H_2O; 3 equiv. TFA in 1:1 MeOH/H_2O; 3 equiv. toluenesulfonic acid in MeOH; 1 equiv. $BF_3{\cdot}Et_2O$ in CH_2Cl_2; 2 equiv. $FeCl_3$ in MeOH).

Salicyl-AMS was then converted to the corresponding sodium salt via a two-step procedure. Aldrich and coworkers have reported formation of the triethylammonium salt of salicyl-AMS directly from silica flash chromatography in the presence of 1% triethylamine (3:1 ethyl acetate/methanol) (Somu, Boshoff, et al., 2006). However, in our hands this led to incomplete conversion to the triethylammonium salt. Thus, we first attempted direct conversion of the free acid of salicyl-AMS (**1**) to the sodium salt **1·Na** using 1 equivalent of aqueous sodium hydroxide or 1 equivalent of aqueous sodium bicarbonate at various temperatures (25 °C, 0 °C, −20 °C). However, all of these reactions resulted in formation of varying amounts of an *N*3–5′-cycloadenosine byproduct (**7**), which has been reported previously for related sulfamoyladenosines (Fig. 1C) (Duckworth et al., 2011; Shi et al., 2013; Shuman et al., 1969).

Accordingly, we turned to a two-step salt formation process involving initial conversion of salicyl-AMS to the stable, isolable triethylammonium salt, followed by salt exchange to the desired sodium salt. Thus, the free acid of salicyl-AMS (**1**) was treated with excess, freshly distilled triethylamine at −20 °C to form the triethylammonium salt **1·Et$_3$N**. Purification by silica flash chromatography (4:1 ethyl acetate/methanol with 1% Et_3N) removed the cycloadenosine byproduct **7** and afforded the triethylammonium salt of salicyl-AMS in 85.8% yield. Notably, increasing the reaction temperature from −20 °C to 0 °C or 25 °C resulted in increased formation of the cycloadenosine byproduct **7**.

Next, conversion of the triethylammonium salt **1·Et$_3$N** to the sodium salt **1·Na** was achieved using cation exchange on a Dowex 50WX8 column, which was washed rigorously with water, methanol, and acetonitrile, then pretreated with aqueous sodium hydroxide to generate the sodium form. A solution of salicyl-AMS, triethylammonium salt (**1·Et$_3$N**, 0.1 M) was passed through the column and the collected solution was lyophilized to provide the salicyl-AMS, sodium salt (**1·Na**) in 94.4% yield (81.0% overall yield from free acid, 11.6% overall yield from adenosine acetonide **13**). This salt formation protocol has been applied successfully to both small-scale (100 mg) and large-scale syntheses (2.5 g). Elemental analysis indicated a sulfur/sodium ratio of 1.43:1, comparable to the theoretical value of 1.39:1.

LC-MS/MS analysis of the salicyl-AMS free acid (**1**) indicated 0.31 mol % AMS as an impurity. In contrast, analysis of the final salicyl-AMS, sodium salt (**1·Na**) indicated <0.0006 mol% AMS, below the lower limit of detection. With appropriate minor modifications, this protocol has been applied to sodium salt formation with other salicyl-AMS analogs (Bythrow et al., 2019), and may also be useful for the preparation of acyl-AMS inhibitors of other targets in the adenylate-forming enzyme superfamily (Lux, Standke, & Tan, 2019).

2.1 Equipment

- Round bottom flask
- Stir bar
- Magnetic hot plate

- Septum
- Vacuum pump
- Vacuum/argon manifold
- Thermometer
- Sonicator (water bath)
- Chromatography columns (glass, various sizes)
- Distillation apparatus
- Rotary evaporator (Buchi)
- Lyophilizer
- Automated chromatography instrument (CombiFlash; Teledyn Isco)
- NMR spectrometer (400 or 500 MHz; Bruker)
- High-performance liquid chromatograph–mass spectrometer (LC-MS; Waters)
- High-performance liquid chromatograph–triple quadrupole mass spectrometer (LC-MS/MS; Agilent)

2.2 Chemicals

Reagents were obtained from Aldrich Chemical (www.sigma-aldrich.com), Acros Organics (www.fishersci.com), Alfa Aesar (www.alfa.com), or TCI America (www.tcichemicals.com) and used without further purification unless otherwise indicated. Optima or HPLC grade solvents were obtained from Fisher Scientific (www.fishersci.com), degassed with Ar, and purified on a solvent drying system unless otherwise indicated.

- Chlorosulfonylisocyanate
- Formic acid
- Petroleum ether
- Salicylic acid
- *N*-Hydroxysuccinimide
- 1,4-Dioxane
- *N*,*N*′-Dicyclohexylcarbodiimide
- *N*,*N*-Dimethylacetamide
- Sodium bicarbonate
- Sodium sulfate
- *N*,*N*-Dimethylformamide
- Cesium carbonate
- (1*S*)-(+)-10-Camphorsulfonic acid
- Triethylamine
- Dowex 50WX8 (hydrogen form)

- 1 N Sodium hydroxide solution
- Guanosine
- Methanol
- Ethyl acetate
- Water
- Acetonitrile
- Silica gel 60 thin-layer chromatography (TLC) plates with fluorescent indicator (F254)
- Silica gel 60 (230–400 mesh)
- Celite
- Phosphate buffered saline
- Dimethylsulfoxide

3. Protocol

All reactions were performed in flame-dried glassware under positive Ar pressure with magnetic stirring unless otherwise noted. Liquid reagents and solutions were transferred thru rubber septa via syringes flushed with Ar prior to use. Cold baths were generated as follows: 0 °C, wet ice/water; −25 °C, dry ice/acetone.

TLC was performed on 0.25 mm E. Merck silica gel 60 F254 plates and visualized under UV light (254 nm) or by staining with potassium permanganate ($KMnO_4$), or cerium ammonium molybdenate (CAM). Silica flash chromatography was performed manually on E. Merck 230–400 mesh silica gel 60 or on an ISCO CombiFlash Rf+ instrument with RediSep silica gel normal phase columns or RediSep Gold silica gel normal phase columns with UV detection at 254 nm.

NMR spectra were recorded on a Bruker UltraShield Plus 400MHz Avance III NMR or UltraShield Plus 600 MHz Avance III NMR with DCH CryoProbe at 24 °C in methanol-d_4 or DMSO-d_6.

3.1 Preparation of sulfamoyl chloride (10)

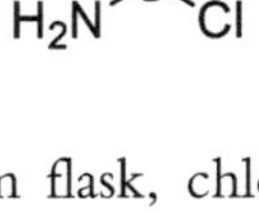

1. In a 100-mL round bottom flask, chlorosulfonylisocyanate (**9**) (20 g, 141.31 mmol, 12.27 mL, 1.0 equiv.) was cooled to 0 °C.
2. Formic acid (6.5 g, 141.31 mmol, 5.33 mL, 1.0 equiv.) was added dropwise to the reaction mixture at 0 °C.
3. The reaction mixture was allowed to warm to 25 °C and stirred for 1 h.

4. The reaction mixture was diluted with petroleum ether (50 mL), then filtered.
5. The filter cake was collected and dried under vacuum to give sulfamoyl chloride (**10**) (15 g) as a white solid, which was used without further purification.
6. *Note*: Chlorosulfonylisocyanate and sulfamoyl chloride are highly sensitive to moisture and react violently with water.

3.2 Preparation of salicylic acid, *N*-hydroxysuccinimide ester (12)

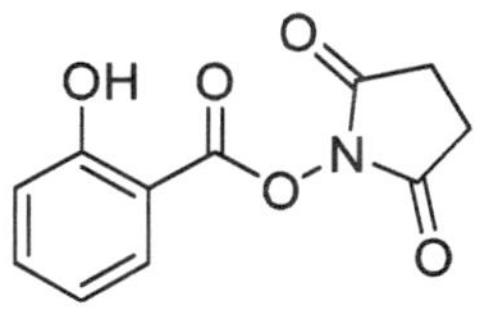

1. In a 2-L round bottom flask, salicylic acid (**11**) (42 g, 304.08 mmol, 1.0 equiv.) and *N*-hydroxysuccinimide (35 g, 304.08 mmol, 1.0 equiv.) were dissolved in 1,4-dioxane (700 mL) and cooled to 0 °C under nitrogen.
2. *N,N′*-Dicyclohexylcarbodiimide (62.74 g, 304.08 mmol, 61.51 mL, 1.0 equiv.) was dissolved in 1,4-dioxane (300 mL) and the solution was added to the reaction mixture.
3. The reaction mixture was allowed to warm to 25 °C and stirred for 12 h.
4. The reaction mixture was filtered to provide salicylic acid, *N*-hydroxysuccinimide ester (**12**) (76 g, crude) as a white solid, which was used without further purification.
5. ^{1}H NMR (DMSO-d_6, 400 MHz): δ 10.63 (br.s, 1H), 7.85 (dd, J_1= 1.6 Hz, J_2=6.4 Hz, 1H), 7.56 (td, J_1=1.6 Hz, J_2=5.6 Hz, 1H), 7.06 (d, J=8.4 Hz, 1H), 6.98–6.96 (m, 1H), 2.87 (s, 4H).

3.3 Synthesis of 5′-*O*-sulfamoyladenosine 2′,3′-acetonide (14)

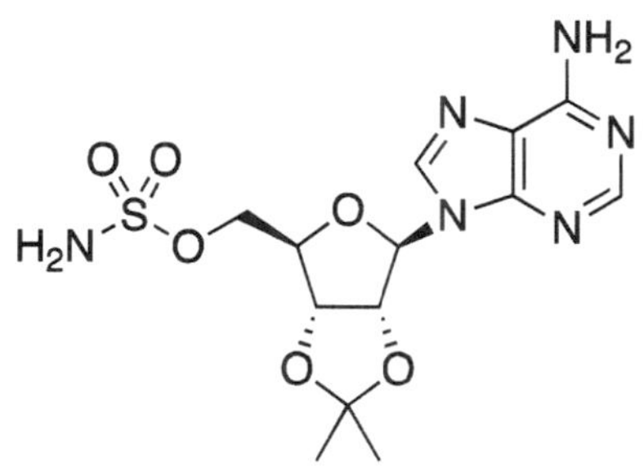

1. In a 1-L round bottom flask, adenosine 2′,3′-acetonide (**13**) (46 g, 149.69 mmol, 1.0 equiv.) was dissolved in *N,N*-dimethylacetamide (500 mL) and cooled to 0 °C under nitrogen.
2. Sulfamoyl chloride (**10**) (34.59 g, 299.38 mmol, 2.0 equiv., from Section 3.1) was added to the reaction mixture at 0 °C.
3. The reaction mixture was allowed to warm to 25 °C and stirred for 12 h with monitoring by LC-MS analysis of aliquots.
4. Upon completion, the reaction was quenched by addition of saturated aqueous sodium bicarbonate until the pH reached 7–8.
5. The mixture was extracted with ethyl acetate (5 × 500 mL), then the organic layers were combined, dried over sodium sulfate, filtered, and concentrated by rotary evaporation to provide a yellow residue.
6. The residue was purified by silica flash chromatography (ethyl acetate → 20:1 ethyl acetate/methanol) to afford 5′-*O*-sulfamoyladenosine 2′,3′-acetonide (**14**) (43 g, 84% purity: 93.48 mmol, 62.5% yield [adjusted for purity]) as a yellow oil.
7. LC-MS: $t_R = 1.092$ min, purity: 84%, m/z 387.0 $[M+H]^+$.
8. ^{1}H NMR (CD_3OD, 400 MHz): δ 8.27 (s, 1H), 8.22 (s, 1H), 6.25 (d, $J = 2.4$ Hz, 1H), 5.43 (dd, $J_1 = 2.0$ Hz, $J_2 = 6.0$ Hz, 1H), 5.14 (dd, $J_1 = 2.8$ Hz, $J_2 = 6.0$ Hz, 1H), 4.52–4.50 (m, 1H), 4.32–4.27 (m, 2H), 1.61 (s, 3H), 1.39 (s, 3H).

3.4 Synthesis of 5′-*O*-(*N*-salicylsulfamoyl)adenosine 2′,3′-acetonide (15)

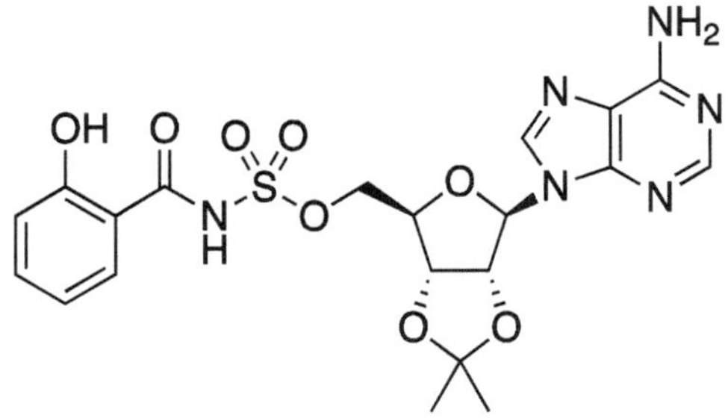

1. In a 1-L round bottom flask, 5′-*O*-sulfamoyladenosine 2′,3′-acetonide (**14**) (43 g, 84% purity: 93.48 mmol, 1.0 equiv. [adjusted for purity], from Section 3.3) was dissolved in *N,N*-dimethylformamide (450 mL) and cooled to 0 °C under nitrogen.

2. Salicylic acid, *N*-hydroxysuccinimide ester (**12**) (52.35 g, 222.58 mmol, 2.38 equiv., from Section 3.2) and cesium carbonate (108.78 g, 333.87 mmol, 3.57 equiv.) were added to the reaction mixture at 0 °C.
3. The reaction mixture was allowed to warm to 25 °C and stirred for 12 h with monitoring by LC-MS analysis of aliquots.
4. Upon completion, the reaction mixture was filtered and concentrated by rotary evaporation to provide a white residue.
5. The residue was purified by silica flash chromatography (1:1 petroleum ether/ethyl acetate → 20:1 ethyl acetate/methanol) to afford 5′-*O*-(*N*-salicylsulfamoyl)adenosine 2′,3′-acetonide (**15**) (40.2 g, 99.5% purity: 78.97 mmol, 84.5% yield [adjusted for purity]) as an off-white solid.
6. LC-MS: $t_R = 1.173$ min, purity: 99.5%, m/z 507.1 $[M+H]^+$.
7. ^{1}H NMR (CD_3OD, 400 MHz): δ 8.46 (s, 1H), 8.15 (s, 1H), 7.91 (dd, $J_1 = 1.6$ Hz, $J_2 = 7.6$ Hz, 1H), 7.33–7.29 (m, 1H), 6.82–6.79 (m, 2H), 6.25 (d, $J = 2.8$ Hz, 1H), 5.39 (dd, $J_1 = 2.8$ Hz, $J_2 = 6.0$ Hz, 1H), 5.16 (dd, $J_1 = 2.0$ Hz, $J_2 = 6.0$ Hz, 1H), 4.59–4.58 (m, 1H), 4.34 (dd, $J_1 = 2.8$ Hz, $J_2 = 4.0$ Hz, 2H), 1.60 (s, 3H), 1.27 (s, 3H).

3.5 Synthesis of salicyl-AMS (free acid) (1)

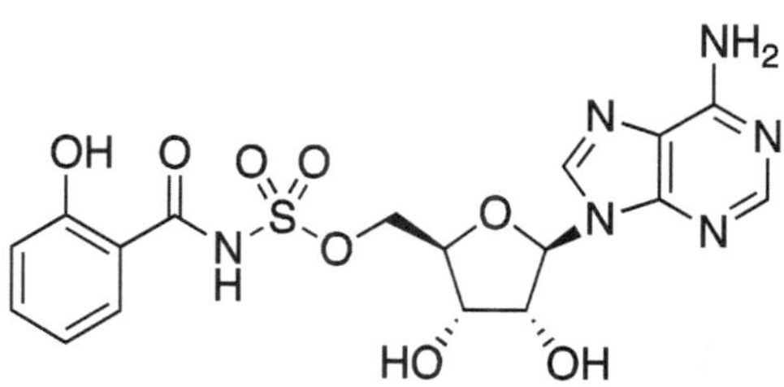

1. In a 1-L round bottom flask, 5′-*O*-(*N*-salicylsulfamoyl)adenosine 2′,3′-acetonide (**15**) (40.2 g, 99.5% purity: 78.97mmol, 1.0 equiv. [adjusted for purity], from Section 3.4) was dissolved in methanol (500 mL) and cooled to 0 °C under nitrogen.
2. Camphorsulfonic acid (55.04 g, 236.93 mmol, 3.0 equiv.) was added at 0 °C in one portion.
3. The reaction mixture was allowed to warm to 25 °C and stirred for 12 h with monitoring by LC-MS analysis of aliquots.
4. Upon completion, the reaction mixture was quenched by addition of solid sodium bicarbonate until the pH reached 7–8, then concentrated by rotary evaporation.

5. The residue was purified by silica gel chromatography (ethyl acetate → 5:1 ethyl acetate/methanol) to afford 25 g of partially purified product.
6. The partially purified product was further purified by preparative HPLC (column: Phenomenex Gemini C18 250 × 50 mm, 10 μm; mobile phase: water (0.05% v/v ammonia hydroxide) with 1–18% acetonitrile, 28 min). Lyophilization afforded salicyl-AMS (free acid) (**1**) (10.05 g, 99.9% purity: 21.53 mmol, 27.3% yield [adjusted for purity]) as a white solid.
7. ^{1}H NMR (CD_3OD, 400 MHz): δ 8.57 (s, 1H), 8.17 (s, 1H), 7.93 (dd, J_1=1.6 Hz, J_2=8.0 Hz, 1H), 7.29 (td, J_1=1.6 Hz, J_2=7.2 Hz, 1H), 6.80–6.75 (m, 2H), 6.09 (d, J=6.4 Hz, 1H), 4.72 (t, J=5.6 Hz, 1H), 4.42–4.40 (m, 2H), 4.37 (dd, J_1=3.2 Hz, J_2=11.2 Hz, 1H), 4.33–4.31 (m, 1H).

3.6 Conversion to triethylammonium salt (1·Et_3N)

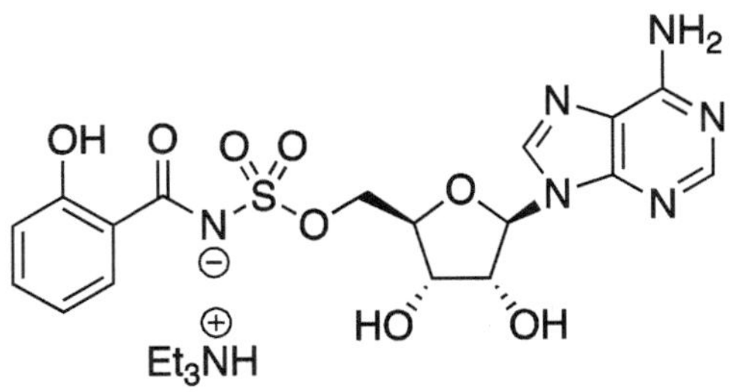

1. In a 250-mL round bottom flask, salicyl-AMS (free acid) (**1**) (2.5 g, 5.36 mmol, from Section 3.5) was dissolved in 200 mL methanol (27 mM).
2. The mixture was sonicated in a water bath for 5–10 min to increase solubilization.
3. The mixture was filtered using a 0.2 μM syringe filter.
4. The filtered solution was transferred to a 500-mL round bottom flask and cooled to −20 °C with stirring for 10 min.
5. *Note*: Salicyl-AMS may precipitate at temperatures below −25 °C.
6. Triethylamine (11.2 mL, 80.4 mmol, 15 equiv.) was added dropwise at −20 °C and the solution was stirred for 30 min at −20 °C.
7. *Note*: Freshly distilled triethylamine was used.
8. The reaction mixture was allowed to warm to room temperature and concentrated by rotary evaporation.

9. The residue was purified by silica flash chromatography (4:1 ethyl acetate/methanol + 1% triethylamine) to afford salicyl-AMS, triethylammonium salt (**1·Et$_3$N**) (2.61 g, 4.60 mmol, 85.8% yield) as a white solid.
10. *Note*: The crude product was dry-loaded onto the silica gel column by dissolving in 200 mL methanol, addition of 10 g Celite, rotary evaporation, and transfer of the Celite onto the column.
11. ^{1}H NMR (CD$_3$OD, 600 MHz): δ 8.41 (s, 1H), 8.13 (s, 1H), 7.94 (d, J=7.8 Hz, 1H), 7.28 (t, J=7.2 Hz, 1H), 6.72–6.87 (m, 2H), 6.08 (d, J=6.4 Hz, 1H), 4.69 (t, J=5.6 Hz, 1H), 4.24–4.34 (m, 4H), 3.17 (q, J=7.2 Hz, 6H), 1.33 (t, J=7.2 Hz, 9H) (Somu, Wilson, et al., 2006).

3.7 Conversion to sodium salt (1·Na)

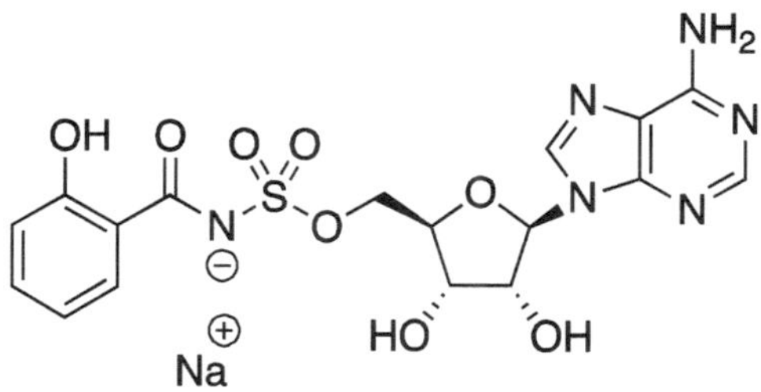

Table 1 Elemental analysis of salicyl-AMS, sodium salt (**1·Na**).

Sample	%C	%H	%N	%S	%Na
Experimental	38.65	3.89	15.54	6.39	4.47
Calculated	41.81	3.51	17.21	6.56	4.71

1. Dowex 50WX8 (hydrogen form) (1000 g) was placed in a glass column (3 in inside diameter × 12 in length, with 2000 mL reservoir).
2. The resin on the column was sequentially washed with water (5 × 1000 mL), methanol (5 × 1000 mL), water (5 × 1000 mL), acetonitrile (5 × 1000 mL), water (5 × 1000 mL) and 1 N aqueous sodium hydroxide (5 × 1000 mL).
3. The resin on the column was washed repeatedly with water until the eluent reached pH 7.

4. Salicyl-AMS, triethylammonium salt (**1·Et_3N**) (2.61 g, 4.60 mmol, from Section 3.6) was dissolved in minimal amount of 1:1 water/acetonitrile (≈50 mL).
5. The solution was loaded onto the Dowex column and incubated with the resin for 10 min.
6. The column was eluted with water (1000 mL) and 25-mL fractions were collected and analyzed by spotting on a TLC plate and visualizing under UV light (254 nm).
7. Fractions that were UV active were combined and lyophilized to afford salicyl-AMS, sodium salt (**1·Na**) (2.12 g, 4.34 mmol, 94.4% yield, 99.9% purity) as a fluffy white solid.
8. ^{1}H NMR (CD_3OD, 600 MHz): δ 8.55 (s, 1H), 8.19 (s, 1H), 7.97 (d, J=7.6 Hz, 1H), 7.31 (t, J=7.3 Hz, 1H), 8.70–6.84 (m, 2H), 6.12 (d, J=6.4 Hz, 1H), 4.73 (t, J=5.5 Hz, 1H), 4.34–4.44 (m, 4H).
9. Elemental analysis was performed to determine the percentage of carbon, hydrogen, nitrogen, sulfur, and sodium (Robertson Microlit). The sulfur:sodium ratio was calculated at 1.43:1, comparable to the theoretical ratio of 1.39:1 (1:1 molar ratio) (Table 1).

3.8 LC-MS/MS analysis of salicyl-AMS, sodium salt (1·Na) to quantify AMS impurity

3.8.1 Preparation of samples

1. Preparation of salicyl-AMS test sample solution: salicyl-AMS, sodium salt (**1·Na**) (2 mg) was dissolved in 0.4 mL of 9:1 phosphate buffered saline (PBS)/dimethylsulfoxide (DMSO) to generate a 5 mg/mL stock solution.
2. *Note*: The salicyl-AMS, sodium salt (**1·Na**) solution should be freshly prepared before use.
3. Preparation of guanosine internal standard solution: guanosine (2.5 mg) was dissolved in 0.5 mL DMSO to generate a 5 mg/mL stock solution. The stock solution was diluted 1000-fold with PBS to generate a 5 μg/mL working solution.
4. A calibration curve of AMS was prepared as follows:
   a. 5′-*O*-Sulfamoyladenosine (AMS, **6**) (Ferreras et al., 2005) (2 mg) was dissolved in 0.4 mL DMSO to generate a 5 mg/mL stock solution.
   b. The AMS stock solution was diluted 100-fold with PBS to generate a 50 μg/mL working solution.
   c. The 50 μg/mL AMS working solution was serially diluted twofold with PBS 12 times to generate calibration curve solutions at 50,

25, 12.5, 6.25, 3.13, 1.56, 0.781, 0.391, 0.195, 0.0977, 0.0488, 0.0244, 0.0122 μg/mL. A 0 μg/mL blank was also generated with PBS, for a total of 14 samples.

3.8.2 Preparation of 96-well plate

1. 100 μL of each salicyl-AMS calibration curve solution (from Section 3.8.1, step 4) was added to each well of 96-well plate, along with 100 μL of the internal standard working solution (from Section 3.8.1, step 3). The experiment was conducted in triplicates.
2. Final concentrations:
   a. Guanosine (internal standard): 2.5 μg/mL in all wells.
   b. AMS calibration curve: 0.061–25 μg/mL.
3. *Note*: When calculating the concentrations of the internal standard and of AMS in the samples, a dilution factor of 2 needs to be applied compared to the working solutions.

3.8.3 Method for LC-MS/MS analysis

LC-MS/MS analysis was carried out on an Agilent Technologies 6410 triple quadrupole LC-MS/MS system with autosampler in electrospray ionization (ESI) mode, with an Agilent Zorbax Eclipse XBD-C18 reverse phase column (50 × 4.6 mm, 5 μm) using a flow rate of 0.5 mL/min and an isocratic mobile phase of 10% acetonitrile in 0.1% aqueous formic acid over 10 min. Positive electrospray in multiple reaction monitoring (MRM) mode was employed to quantify AMS, using the ion transitions indicated below.

Compound	Precursor ion	Product ion	Fragmentor	Collision energy
AMS (**6**)	347.1	136.1	128	20
Guanosine	284.1	152.1	71	8

4. Concluding remarks

We have developed an efficient protocol for the multigram-scale preparation of the anti-tuberculosis lead compound salicyl-AMS as its sodium salt (**1·Na**). The process proceeds in three synthetic steps, followed by a two-step salt formation protocol, to afford the target compound in

11.6% overall yield from commercially available adenosine 2′,3′-acetonide (**13**). The two-step salt formation process includes an additional chromatographic purification and removes both a cycloadenosine byproduct (**7**) and a toxic trace impurity AMS (**6**), providing highly purified material suitable for preclinical evaluation. This protocol has been applied to milligram to multigram-scale syntheses, and with minor modifications to other salicyl-AMS analogs (Bythrow et al., 2019). We envision that this protocol may also be useful in the preparation of other acyl-AMS inhibitors of the adenylate-forming enzyme superfamily (Lux et al., 2019).

Acknowledgments

We thank Kazuyoshi Aso (Tri-Institutional Therapeutics Discovery Institute and Axcelead Drug Discovery Partners) and Makoto Fushimi (Tri-Institutional Therapeutics Discovery Institute and Takeda Pharmaceuticals) for helpful discussions and for coordinating large-scale synthesis of salicyl-AMS (free acid) at WuXi AppTec. Generous financial support was provided by the National Institutes of Health (R01 AI118224 to D.S.T.; F31 AI129244 and T32 GM073546-Gross to L.C.S.; and CCSG P30 CA008748 to C.B. Thompson) and the Tri-Institutional Therapeutics Discovery Institute.

Conflicts of interests

D.S.T. is an inventor on U.S. Patents 8,461,128 and 8,946,188, and D.S.T. and L.C.S. are inventors on U.S. Patent Application PCT/US2019/068107 based on this work.

References

Appel, R., & Berger, G. (1958). Hydrazinsulfonsäure-amide, I. Über das Hydrazodisulfamid. *Chemische Berichte*, *91*(6), 1339–1341.

Bloch, A., & Coutsogeorgopoulos, C. (1971). Inhibition of protein synthesis by 5′-sulfamoyladenosine. *Biochemistry*, *10*(24), 4395–4398. https://doi.org/10.1021/bi00800a007.

Brodsky, B. H., & Du Bois, J. (2005). Oxaziridine-mediated catalytic hydroxylation of unactivated 3° C-H bonds using hydrogen peroxide. *Journal of the American Chemical Society*, *127*(44), 15391–15393. https://doi.org/10.1021/ja055549i.

Bythrow, G. V., Mohandas, P., Guney, T., Standke, L. C., et al. (2019). Kinetic analyses of the siderophore biosynthesis inhibitor salicyl-AMS and analogues as MbtA inhibitors and antimycobacterial agents. *Biochemistry*, *58*(6), 833–847. https://doi.org/10.1021/acs.biochem.8b01153.

Cisar, J. S., Ferreras, J. A., Soni, R. K., Quadri, L. E. N., et al. (2007). Exploiting ligand conformation in the selective inhibition of non-ribosomal peptide synthetase amino acid adenylation with designed macrocyclic small molecules. *Journal of the American Chemical Society*, *129*, 7752–7753. https://doi.org/10.1021/ja0721521.

Cisar, J. S., & Tan, D. S., unpublished results.

Dawadi, S., Boshoff, H. I. M., Park, S. W., Schnappinger, D., et al. (2018). Conformationally constrained cinnolinone nucleoside analogues as siderophore biosynthesis inhibitors for tuberculosis. *ACS Medicinal Chemistry Letters*, *9*(4), 386–391. https://doi.org/10.1021/acsmedchemlett.8b00090.

Dawadi, S., Kawamura, S., Rubenstein, A., Remmel, R., et al. (2016). Synthesis and pharmacological evaluation of nucleoside prodrugs designed to target siderophore biosynthesis in *Mycobacterium tuberculosis*. *Bioorganic and Medicinal Chemistry*, *24*(6), 1314–1321. https://doi.org/10.1016/j.bmc.2016.02.002.

Dawadi, S., Viswanathan, K., Boshoff, H. I., Barry, C. E., 3rd, et al. (2015). Investigation and conformational analysis of fluorinated nucleoside antibiotics targeting siderophore biosynthesis. *Journal of Organic Chemistry*, *80*(10), 4835–4850. https://doi.org/10.1021/acs.joc.5b00550.

De Voss, J. J., Rutter, K., Schroeder, B. G., & Barry, C. E., 3rd (1999). Iron acquisition and metabolism by mycobacteria. *Journal of Bacteriology*, *181*(15), 4443–4451.

De Voss, J. J., Rutter, K., Schroeder, B. G., Su, H., et al. (2000). The salicylate-derived mycobactin siderophores of *Mycobacterium tuberculosis* are essential for growth in macrophages. *Proceedings of the National Academy of Sciences of the United States of America*, *97*(3), 1252–1257. https://doi.org/10.1073/pnas.97.3.1252.

Duckworth, B. P., Geders, T. W., Tiwari, D., Boshoff, H. I., et al. (2011). Bisubstrate adenylation inhibitors of biotin protein ligase from *Mycobacterium tuberculosis*. *Chemistry and Biology*, *18*(11), 1432–1441. https://doi.org/10.1016/j.chembiol.2011.08.013.

Duckworth, B. P., Nelson, K. M., & Aldrich, C. C. (2012). Adenylating enzymes in *Mycobacterium tuberculosis* as drug targets. *Current Topics in Medicinal Chemistry*, *12*(7), 766–796. https://doi.org/10.2174/156802612799984571.

Engelhart, C. A., & Aldrich, C. C. (2013). Synthesis of chromone, quinolone, and benzoxazinone sulfonamide nucleosides as conformationally constrained inhibitors of adenylating enzymes required for siderophore biosynthesis. *Journal of Organic Chemistry*, *78*(15), 7470–7481. https://doi.org/10.1021/jo400976f.

Ferreras, J. A., Ryu, J. S., Di Lello, F., Tan, D. S., et al. (2005). Small-molecule inhibition of siderophore biosynthesis in *Mycobacterium tuberculosis* and *Yersinia pestis*. *Nature Chemical Biology*, *1*(1), 29–32. https://doi.org/10.1038/nchembio706.

Ferreras, J. A., Stirrett, K. L., Lu, X., Ryu, J.-S., et al. (2008). Mycobacterial phenolic glycolipid virulence factor biosynthesis: Mechanism and small-molecule inhibition of polyketide chain initiation. *Chemistry and Biology*, *15*(1), 51–61. https://doi.org/10.1016/j.chembiol.2007.11.010.

Furin, J., Cox, H., & Pai, M. (2019). Tuberculosis. *Lancet*, *393*(10181), 1642–1656. https://doi.org/10.1016/S0140-6736(19)30308-3.

Gupte, A., Boshoff, H. I., Wilson, D. J., Neres, J., et al. (2008). Inhibition of siderophore biosynthesis by 2-triazole substituted analogues of 5′-*O*-[*N*-(salicyl)sulfamoyl]adenosine: Antibacterial nucleosides effective against *Mycobacterium tuberculosis*. *Journal of Medicinal Chemistry*, *51*(23), 7495–7507. https://doi.org/10.1021/jm8008037.

Heacock, D., Forsyth, C. J., Shiba, K., & Musier-Forsyth, K. (1996). Synthesis and aminoacyl-tRNA synthetase inhibitory activity of prolyl adenylate analogs. *Bioorganic Chemistry*, *24*(3), 273–289. https://doi.org/10.1006/bioo.1996.0025.

Jaffe, J. J., McCormack, J. J., & Meymerian, E. (1970). Trypanocidal properties of 5′-*O*-sulfamoyladenosine, a close structural analog of nucleocidin. *Experimental Parasitology*, *28*(3), 535–543. https://doi.org/10.1016/0014-4894(70)90121-9.

Krajczyk, A., Zeidler, J., Januszczyk, P., Dawadi, S., et al. (2016). 2-Aryl-8-aza-3-deazaadenosine analogues of 5′-*O*-[*N*-(salicyl)sulfamoyl]adenosine: Nucleoside antibiotics that block siderophore biosynthesis in *Mycobacterium tuberculosis*. *Bioorganic and Medicinal Chemistry*, *24*(14), 3133–3143. https://doi.org/10.1016/j.bmc.2016.05.037.

Lamb, A. L. (2015). Breaking a pathogen's iron will: Inhibiting siderophore production as an antimicrobial strategy. *Biochimica et Biophysica Acta*, *1854*(8), 1054–1070. https://doi.org/10.1016/j.bbapap.2015.05.001.

Lun, S., Guo, H., Adamson, J., Cisar, J. S., et al. (2013). Pharmacokinetic and *in vivo* efficacy studies of the mycobactin biosynthesis inhibitor salicyl-AMS in mice. *Antimicrobial Agents and Chemotherapy*, *57*(10), 5138–5140. https://doi.org/10.1128/aac.00918-13.

Lux, M. C., Standke, L. C., & Tan, D. S. (2019). Targeting adenylate-forming enzymes with designed sulfonyladenosine inhibitors. *Journal of Antibiotics (Tokyo)*, *72*(6), 325–349. https://doi.org/10.1038/s41429-019-0171-2.

McMahon, M. D., Rush, J. S., & Thomas, M. G. (2012). Analyses of MbtB, MbtE, and MbtF suggest revisions to the mycobactin biosynthesis pathway in *Mycobacterium tuberculosis*. *Journal of Bacteriology*, *194*(11), 2809–2818. https://doi.org/10.1128/JB.00088-12.

Nelson, K. M., Viswanathan, K., Dawadi, S., Duckworth, B. P., et al. (2015). Synthesis and pharmacokinetic evaluation of siderophore biosynthesis inhibitors for *Mycobacterium tuberculosis*. *Journal of Medicinal Chemistry*, *58*(14), 5459–5475. https://doi.org/10.1021/acs.jmedchem.5b00391.

Neres, J., Labello, N. P., Somu, R. V., Boshoff, H. I., et al. (2008). Inhibition of siderophore biosynthesis in *Mycobacterium tuberculosis* with nucleoside bisubstrate analogues: Structure–activity relationships of the nucleobase domain of 5′-*O*-[*N*-(salicyl)sulfamoyl]adenosine. *Journal of Medicinal Chemistry*, *51*(17), 5349–5370. https://doi.org/10.1021/jm800567v.

Okada, M., Iwashita, S., & Koizumi, N. (2000). Efficient general method for sulfamoylation of a hydroxyl group. *Tetrahedron Letters*, *41*(36), 7047–7051.

Paulekuhn, G. S., Dressman, J. B., & Saal, C. (2007). Trends in active pharmaceutical ingredient salt selection based on analysis of the Orange Book database. *Journal of Medicinal Chemistry*, *50*(26), 6665–6672. https://doi.org/10.1021/jm701032y.

Qiao, C., Gupte, A., Boshoff, H. I., Wilson, D. J., et al. (2007). 5′-*O*-[(*N*-acyl)sulfamoyl] adenosines as antitubercular agents that inhibit MbtA: An adenylation enzyme required for siderophore biosynthesis of the mycobactins. *Journal of Medicinal Chemistry*, *50*(24), 6080–6094. https://doi.org/10.1021/jm070905o.

Qiao, C., Wilson, D. J., Bennett, E. M., & Aldrich, C. C. (2007). A mechanism-based aryl carrier protein/thiolation domain affinity probe. *Journal of the American Chemical Society*, *129*(20), 6350–6351. https://doi.org/10.1021/ja069201e.

Quadri, L. E. N., & Ratledge, C. (2005). Iron metabolism in the tubercle bacillus and other mycobacteria. In S. T. Cole, K. D. Eisenach, & D. N. McMurrayet al. *Tuberculosis and the tubercle bacillus* (pp. 341–357). Washington, DC: ASM Press.

Quadri, L. E., Sello, J., Keating, T. A., Weinreb, P. H., et al. (1998). Identification of a *Mycobacterium tuberculosis* gene cluster encoding the biosynthetic enzymes for assembly of the virulence-conferring siderophore mycobactin. *Chemistry and Biology*, *5*(11), 631–645. https://doi.org/10.1016/s1074-5521(98)90291-5.

Ratledge, C. (2004). Iron, mycobacteria and tuberculosis. *Tuberculosis (Edinburgh)*, *84*(1–2), 110–130. https://doi.org/10.1016/j.tube.2003.08.012.

Rengaraju, S., Narayanan, S., Ganju, P., Amin, M., et al. (1986). 5′-*O*-Sulfamoyladenosine (defluoronucleocidin) from a *Streptomyces*. *Meiji Seika Kenkyu Nenpo*, *25*, 49–55.

Shi, C., Tiwari, D., Wilson, D. J., Seiler, C. L., et al. (2013). Bisubstrate inhibitors of biotin protein ligase in *Mycobacterium tuberculosis* resistant to cyclonucleoside formation. *ACS Medicinal Chemistry Letters*, *4*(12), 1213–1217. https://doi.org/10.1021/ml400328a.

Shuman, D., Robins, R. K., & Robins, M. J. (1969). The synthesis of adenine 5′-*O*-sulfamoyl nucleosides related to nucleocidin. *Journal of the American Chemical Society*, *91*(12), 3391–3392. https://doi.org/10.1021/ja01040a062.

Somu, R. V., Boshoff, H., Qiao, C., Bennett, E. M., et al. (2006). Rationally designed nucleoside antibiotics that inhibit siderophore biosynthesis of *Mycobacterium tuberculosis*. *Journal of Medicinal Chemistry*, *49*(1), 31–34. https://doi.org/10.1021/jm051060o.

Somu, R. V., Wilson, D. J., Bennett, E. M., Boshoff, H. I., et al. (2006). Antitubercular nucleosides that inhibit siderophore biosynthesis: SAR of the glycosyl domain. *Journal of Medicinal Chemistry*, *49*(26), 7623–7635. https://doi.org/10.1021/jm061068d.

Van Brussel, W., & Van Sumere, C. F. (1978). *N*-Acylamino acids and peptides. VI. A simple synthesis of *N*-acylglycines of the benzoyl and cinnamyl type. *Bulletin des Societes Chimiques Belges*, *87*(10), 791–797. https://doi.org/10.1002/bscb.19780871008.

Vannada, J., Bennett, E. M., Wilson, D. J., Boshoff, H. I., et al. (2006). Design, synthesis, and biological evaluation of beta-ketosulfonamide adenylation inhibitors as potential antitubercular agents. *Organic Letters*, *8*(21), 4707–4710. https://doi.org/10.1021/ol0617289.

World Health Organization. (2019). *Global tuberculosis report 2019*. https://www.who.int/tb/publications/global_report/en/.

CHAPTER FIVE

Probe-enabled approaches for function-dependent cell sorting and characterization of microbiome subpopulations

Andrea K. Steiger[a], Sarah J. Fansler[a], Christopher Whidbey[b], Carson J. Miller[a], Aaron T. Wright[a,c,*]

[a]Biological Sciences Division, Pacific Northwest National Laboratory, Richland, Washington, United States
[b]Department of Chemistry, Seattle University, Seattle, Washington, United States
[c]The Gene and Linda Voiland School of Chemical Engineering and Bioengineering, Washington State University, Pullman, Washington, United States
*Corresponding author: e-mail address: aaron.wright@pnnl.gov

Contents

Abstract

Understanding the roles that individual species or communities play within a microbiome is a significant challenge. The complexity and heterogeneity of microbiomes presents a challenge to researchers looking to unravel the function that microbiomes serve within larger environments. While identification of the species and proteins present in a

Methods in Enzymology, Volume 638
ISSN 0076-6879
https://doi.org/10.1016/bs.mie.2020.03.014

microbiome can be accomplished through genomics approaches, strategies that report on enzyme activity are limited. In this chapter, we describe the application of small molecule chemical probes in the isolation and subsequent characterization of microbiome subpopulations based on enzymatic function. We will cover protocols for labeling microbes with appropriate probes, microbiome sample preparation, and using fluorescence-activated cell sorting to isolate subpopulations based on function. We hope that the strategies outlined here will serve as a resource for researchers studying the functional role that microbiomes play in the gut and soil.

1. Introduction

Complex microbial systems, including gut, soil, aquatic and other environmental and host-associated microbiomes, are notoriously difficult to study due to their vast complexity and heterogeneity (Fierer, 2017; Gilbert et al., 2018). Early studies of complex microbiomes relied on microbe isolation followed by traditional culturing methods. However, the inability to culture a significant portion of the microbes and the resulting decrease in the diversity of the communities limited what we were able to learn from these techniques (Hiergeist, Glasner, Reischl, & Gessner, 2015). Today, questions related to microbial composition can be addressed using robust identification techniques such as sequencing and transcriptomics (Bik et al., 2006; Garoutte, Cardenas, Tiedie, & Howe, 2016; Hold, Pryde, Russell, Furrie, & Flint, 2002; Jansson & Hofmockel, 2018; Wilson & Blitchington, 1996), but it remains a challenge to identify microbial function in complex communities. Researchers are limited in the methods available to associate specific enzymatic function with the responsible microbes. While genomic techniques exist for identification of proteins and enzymes present in a microbiome of interest, confirmation of specific activity remains a challenge. Transcriptome analysis can allude to the functional state of microbial communities, but it can be a challenge to study. Microbes may be in a nutrient limited or dormant state, and mRNA is difficult to extract from heterogeneous matrices such as soil or fecal samples. Understanding of microbiome function is critical as it is now accepted that microbes are responsible for numerous critical biochemical transformations in both the gut and soil microbiomes (Whidbey & Wright, 2019).

In the gut microbiome, microbes are responsible for many biochemical transformations that the host relies upon for proper physiology. These reactions have enormous impacts on dietary metabolism, cell signaling, the effect of xenobiotic exposures, and an individual's response to therapeutics.

While it is accepted that the broad repertoire of enzymatic reactions that occur in the gut microbiome have unmatched impacts on the host, especially in the metabolism of xenobiotics, we still have a limited understanding of the exact nature of this host-microbiome-xenobiotic interaction. This is a direct result of the limited available tools as mentioned above: we are so far limited in our ability to detect and measure biochemical activity at the resolution required, such as what is offered by metagenomic analysis. (Clarke et al., 2019; Claus, Guillou, & Ellero-Simatos, 2016; Koppel, Maini Rekdal, & Balskus, 2017; Maurice, Haiser, & Turnbaugh, 2013).

Activity-based protein profiling (ABPP) pairs chemical probes with proteomic techniques to identify enzymes based on functional activity (Cravatt, Wright, & Kozarich, 2008; Fonovic & Bogyo, 2008; Heal, Dang, & Tate, 2011). This strategy has enhanced the study of complex microbiomes through the identification of functionally active enzymes that perform specific biochemical transformations. Activity-based probes (ABPs) are activated by target enzymes and rapidly label them through a covalent bond. Importantly, ABPs only label active enzymes, allowing for detection of enzyme activity rather than simply the presence of the enzyme, which may or may not be performing an activity of interest. ABPP has been used to profile bacterial pathogenesis (Krysiak & Sieber, 2017; Puri & Bogyo, 2013), xenobiotic metabolism (Bottcher & Sieber, 2010; Eirich, Orth, & Sieber, 2011; Nomura & Casida, 2011; Staub & Sieber, 2008; Whidbey et al., 2019), environmentally-relevant microbes (Chauvigne-Hines et al., 2012; Morimoto & van der Hoorn, 2016; Schroder et al., 2019; Zweerink et al., 2017), and much more (Evans & Cravatt, 2006; Sadler & Wright, 2015; Whidbey & Wright, 2019). Additionally, ABPP techniques also allow for the detection of unannotated enzymes that perform interesting biochemical activities, allowing researchers to identify the enzyme responsible for various activities in complex and poorly defined microbiomes.

By pairing ABPs capable of detecting specific enzyme activities with fluorescence-activated cell sorting (FACS), a novel strategy can be developed for the isolation of microbiome subpopulations based solely on biochemical function. This technique can be impactful for any complex microbiomes, including host-associated microbiomes such as the human gut as well as environmental and soil microbiomes. In this chapter, we will discuss approaches that exist for using carefully designed probes to first isolate, and subsequently characterize, subpopulations with specific enzymatic activities. The strategies outlined here will help to further unravel the complex interactions that exist in gut and soil microbiomes and will likely be of interest to researchers studying a variety of other complex microbiomes.

2. Probe design for FACS-enabled isolation

In order for a chemical probe to allow for function-dependent isolation of subpopulations, several requirements must be met. First, the probe must be selective for a specific enzyme. This enzyme-level selectivity is typically achieved through the use of either a known substrate or inhibitor which aids in directing the probe to the enzyme of interest. Second, the probe must respond only to active enzymes to directly and accurately report on functional activity. Probes that are modified substrates of the enzyme being labeled are selective for active enzymes only, but if inhibitors are being used, it is critical to ensure that the inhibitor structure only labels the catalytic active site to be sure that the probe will not target inactive enzymes. Third, the probe must have an intracellular response: probes targeting excreted proteins or small molecules will not provide a fluorescent cellular tag for use with flow cytometry. Finally, the probe must selectively label the cell which contains the enzyme of interest and not diffuse to non-target cells. While fluorogenic probes, which are nonfluorescent until activated by an enzyme of interest, may be used for FACS, it is important that there is some process which keeps the probe within the cell following activation to prevent off-target labeling. While many types of small molecule chemical probes exist for identifying and monitoring enzyme activities in a variety of applications, two main probe categories fulfill the above requirements and are ideal candidates for function-dependent isolation: activity-based probes and self-immobilizing probes (Fig. 1).

2.1 Activity-based probes

Activity-based protein profiling (ABPP), a chemoproteomic technique for identification and characterization of proteins, directly measures enzyme activity through the use of chemical probes that that target and covalently label active enzymes. These activity-based probes (ABPs) are unmatched in their ability to probe enzymatic activity due to their selectivity for active enzymes and their ability to rapidly and irreversibly label their target enzyme (Cravatt et al., 2008; Willems, Overkleeft, & van Kasteren, 2014). ABPs have been used extensively to identify the role of specific enzymes in various disease states (Bogyo, 2010; Nomura et al., 2010), potential drug targets and

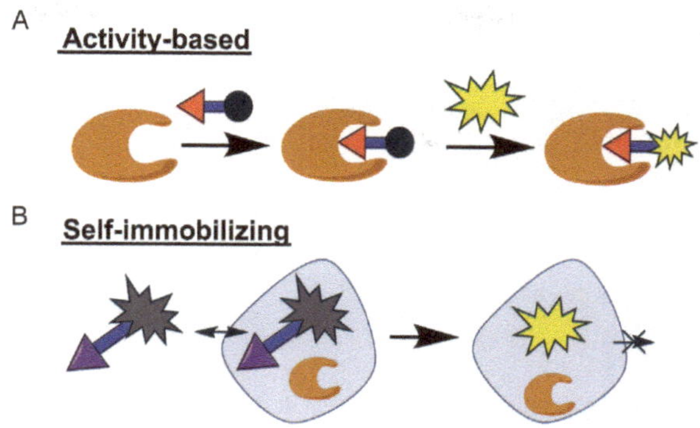

Fig. 1 Overview of the probe strategies for function-dependent isolation of subpopulations discussed here. (A) Traditional ABP design consists of enzymatic labeling followed by click chemistry to append a fluorophore; alternatively the fluorophore can be added to the probe prior to microbial community labeling. (B) Trappable, turn-on fluorescent probes are activated by the enzyme of interest and remain trapped within the cell following activation.

promising antibiotics (Fonovic & Bogyo, 2008; Heal et al., 2011; Lentz et al., 2018), and functionally study both pathogens and commensal bacteria in complex systems (Krysiak & Sieber, 2017; Sadler & Wright, 2015). The significant advantage for ABPs over other chemical probe techniques (such as fluorogenic substrates) is the inclusion of a chemical "warhead," i.e., a functional group that serves to react with and covalently modify the target enzyme. In this way, ABPs serve as a functional handle to isolate and characterize proteins of interest. Importantly, ABPs allow for labeling of active target enzymes, enrichment for LC-MS, and confirmation by proteomics (Cravatt et al., 2008; Evans & Cravatt, 2006; Whidbey & Wright, 2019). Recently, ABPs have also been used to isolate subpopulations from complex mixtures based on enzyme activity, such as β-glucuronidase (Whidbey et al., 2019). This allows for isolation of a population from a complex mixture, such as the gut microbiome, that expresses a specific biochemical activity of interest, and subsequent interrogation to carefully identify the species responsible for that activity.

A popular strategy for appending a fluorophore to an ABP is through the Cu-mediated azide-alkyne cycloaddition click reaction (CuAAC) (Haldon, Nicasio, & Perez, 2015; Martell & Weerapana, 2014; Speers, Adam, & Cravatt, 2003). While CuAAC boasts fast reaction kinetics and low background reactivity, the use of Cu(I) limits its applicability in living cells (Kennedy et al., 2011; Lallana, Riguera, & Fernandez-Megia, 2011; Li et al., 2016). While alternative bioorthogonal reactions may be selected

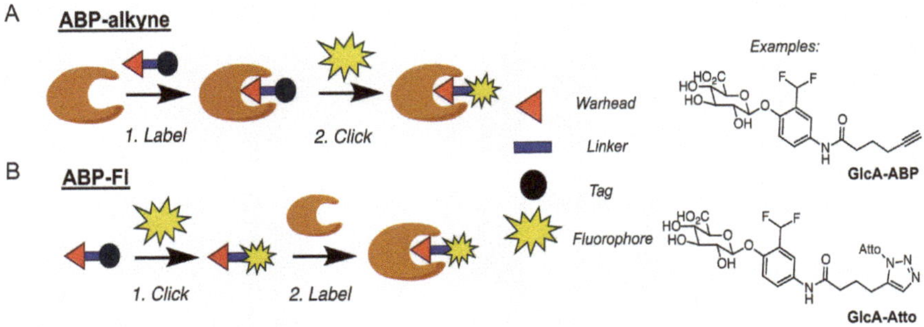

Fig. 2 Design of ABPs (A) Typical ABPs, such as GlcA-ABP, consist of an enzyme-reactive warhead, a linker, and a reactive group poised for attachment to a fluorophore. (B) Use of an ABP-Fl, such as GlcA-Atto, eliminates the need for in vitro click chemistry and contains the fluorophore already attached.

to circumvent this issue (Cañeque, Müller, & Rodriguez, 2018; Ovaa et al., 2003; Xin et al., 2019), ABPs have also been "pre-clicked" to fluorophores prior to use when cell viability must be maintained (Schroder et al., 2019; Whidbey et al., 2019). By synthesizing both the alkyne-functionalized, unreacted ABP (Fig. 2A) as well as the "pre-clicked" (ABP-Fl), fluorescent conjugate (Fig. 2B), it is possible to expand the usage of these probes to allow for both highly selective probe validation using SDS-PAGE as well as isolation of living cells for future cultivation or further analysis. Numerous prior publications have addressed the design of ABPs for targeting specific enzymes, so design and synthesis of these probes will not be addressed here (Cravatt et al., 2008; Evans & Cravatt, 2006; Fonovic & Bogyo, 2008; Heal et al., 2011; Willems et al., 2014).

One drawback of pre-clicked probes involves the potential for decreased selectivity due to the presence of a large, aromatic system appended to the ABP. Since ABPs rely on recognition by active enzymes, structural changes to the probe can decrease selectivity for the enzyme active site (Willems et al., 2014). For this reason, when careful identification of specific proteins is required (such as for proteomics), un-clicked ABPs may be preferred for some targets. However, isolation of specific populations from a complex mixture necessitates selectivity on the cellular level only, and somewhat decreased selectivity for labeling of individual enzymes may be tolerated. Therefore, pre-clicked, ABP-Fl probes may be an ideal choice for those looking to do FACS-enabled isolation of living populations, as it also eliminates the need for in vitro click chemistry.

2.1.1 Conjugation of fluorophore to ABP to give ABP-Fl

Materials

- Dry DMSO (Sigma Aldrich)
- MilliQ water
- Fluorophore-azide, i.e., Atto-633-Az (Sigma Aldrich)
- CuSO4 (Sigma Aldrich), THPTA (Click Chemistry Tools), sodium ascorbate (Sigma Aldrich)
- HPLC (Agilent 1100)

1. Dissolve ABP in DMSO to make a 50 mM stock solution.
2. Add ABP to Fl-Az (5:1) in an aqueous solution of $CuSO_4$ (4 mM), THPTA (1 mM) and ascorbic acid (10 mM).
3. Incubate at room temperature in the dark overnight while shaking.
4. Purify by HPLC using a water/acetonitrile gradient.
5. Collect the product under vacuum and obtain NMR and mass spectra.
6. Resuspend in 50 μL DMSO and determine the concentration using fluorescence measurements, using Fl-Az to generate a standard curve.

2.2 Self-immobilizing fluorescent probes

A popular strategy in the development of ABPs includes the use of quinone-methides, which, upon activation, reveal an electrophile that covalently modifies the protein of interest (Hsu et al., 2015; Nandakumar et al., 2015; Wu et al., 2019). In practice, however, these probes may suffer from a lack of selectivity: the reactive quinone-methide can diffuse throughout the cell and ultimately label off-target proteins (Wu et al., 2019). While this is a significant drawback in the use of these ABPs for proteomics, this same reactivity has also been utilized in the development of probes for applications that require selectivity at the cellular level, but not the enzyme level. In this case, quinone-methides provide signal amplification: diffusion away from the enzyme of interest and labeling of other intracellular proteins ensures that sufficient labeling occurs even when the target enzyme is present in a low abundance. Additionally, these self-immobilizing probes (Fig. 3) are less likely to covalently inhibit the enzyme of interest, allowing for isolation and cultivation of living cells that still contain active enzyme for future work (Kwan et al., 2011; Mao, Xia, Wang, & Xie, 2016).

Withers, et al., capitalized on quinone-methide diffusion through the development of a probe for glycosidase activity with a difluoromethyl-functionalized coumarin. Upon enzymatic cleavage, the probe becomes fluorescent while simultaneously self-immobilizing onto a nearby protein via a

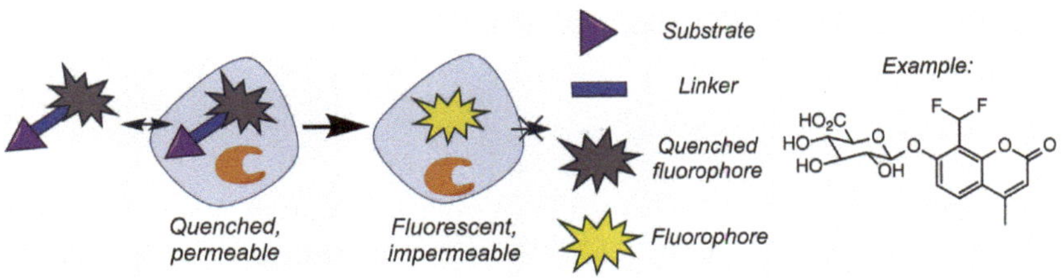

Fig. 3 Design of self-immobilizing, turn-on fluorescent probes for enzyme activity that self-immobilize inside the cell of interest following activation.

reaction with the newly revealed quinone-methide. Using this probe, glycosidase-active microbes were sorted from a complex mixture using FACS. This probe was the first example of a self-immobilizing fluorescent probe for enzyme activity, and represents a simple improvement on standard turn-on fluorescent probes to allow for imaging and FACS (Kwan et al., 2011; Mao et al., 2016). This probe is shown as an example in Fig. 3, as it is a direct comparison to the ABPs shown in Fig. 2.

3. Labeling and selectivity protocols

Following development of an appropriate chemical probe, it is crucial to validate that the probe successfully labels the target enzyme or cell of interest with a high degree of selectivity. Plate reader assays, gel electrophoresis, and flow cytometry are all techniques that can be used to confirm that a novel probe provides sufficient signal in cells of interest, and does not give false positive responses in non-target cells. Plate reader assays are an ideal option for self-immobilizing, turn-on probes, as they are nonfluorescent until activated by the enzyme of interest. By monitoring the increase in fluorescence over time at a range of concentrations, ideal labeling conditions (i.e., concentration, time, and temperature) can be established though simple assays prior to moving on to more complex experiments.

Activity-based probes, whether ABP-alkyne or ABP-Fl, are not as well suited for plate reader assays as they are fluorescent prior to labeling and rely on covalent modification of the target. For this reason, SDS-PAGE and flow cytometry are better options for validating the selectivity of these probes. SDS-PAGE allows for identification of selectivity at the protein level, while flow cytometry confirms labeling within whole cells. SDS-PAGE additionally allows for analysis of selectivity using purified proteins or cell lysates.

3.1 Plate reader assays

Materials

- Dry DMSO (Sigma Aldrich)
- Bacterial culture and appropriate media
- Fluorescence plate reader (Biotek)

1. Dissolve probe in DMSO to a final concentration of 50 mM.
2. Make stock solutions of 25, 12.5, 6.25, 3.13, 1.56, 0.79, and 0.39 mM in DMSO via serial dilution from the 50 mM stock solution. Probe stock solutions in DMSO should be stored at −80 °C and freeze/thaw cycles should be minimized.
3. Pellet overnight cultures of bacteria of interest and resuspend in fresh media to an OD of 0.2.
4. Add stock solutions of probe (1 μL) to cultures (100 μL) in 96 well plate, as well as probe in media alone and DMSO (no probe)-only controls.
5. Read fluorescence (excitation 365, emission 460 for MUB-based substrates) and OD600 on plate reader every 15 min for 24 h.
6. Repeat for target bacteria as well as non-target bacteria for comparison of fluorescence turn-on.

3.2 In vitro labeling

Materials

- Purified proteins or cell lysates
- CuAAC reagents
- Cold (−20 °C) methanol
- Tabletop centrifuge
- Nu-PAGE SDS (2 ×)
- Nu-PAGE reducing agent (10 ×)
- Nu-Page Tris-glycine mini-gels

1. Label aliquots of purified proteins (5 μM) or cell lysates (1 mg/mL) with either ABP-alkyne or ABP-Fl (0–120 μM) for 1 h at 37 °C with shaking.
2. For ABP-alkyne, perform CuAAC.
    1. Add $CuSO_4$ (1 mM final concentration), rhodamine azide (60 μM final concentration), NaAsc (5 mM final concentration), and THPTA (2 mM final concentration) to samples.
    2. Incubate with shaking in the dark at room temperature for 1 h.
    3. Add cold methanol (2 × volume) and precipitate proteins overnight at −80 °C.

   4. Centrifuge samples at 4 °C (10,000 × *g* for 5 min) and remove and discard methanol.
   5. Allow samples to evaporate in the dark open to air for 1 h.
3. Dilute samples into SDS (1 ×) with reducing agent (1 ×) and probe sonicate for 2 s on, 1 s off at medium power.
4. Heat to 95 °C for 15 min on thermal shaker.
5. Spin on tabletop centrifuge and analyze via SDS-PAGE.

3.3 Whole microbe labeling

Materials

- Bacterial cultures
- Phosphate-buffered saline (PBS)
- DMSO (Sigma Aldrich)
- Ethanol (70%)
- CuAAC reagents
- Tabletop centrifuge
- Bead-beater (Bead Ruptor Elite)
- SDS-PAGE reagents
- Flow cytometer

1. Pellet overnight cultures of bacteria of interest and resuspend in PBS (20% initial volume).
2. Label 100 μL of the culture with 0–125 μM of probe in DMSO (0.5% DMSO final) for 1 h at 37°°C with shaking.
3. Collect cells via centrifugation at 10,000 × *g* for 5 min and wash with deoxygenated PBS.
4. For ABP-alkyne, fix cells in 70% ethanol at −20°°C overnight, collect by centrifugation at 10,000 × *g* for 5 min, then perform CuAAC.
   1. Resuspend pellets in CuAAC reaction buffer (10 μM Fl-azide, 8 mM $CuSO_4$, 2 mM THPTA, 10 mM ascorbic acid in PBS:0.5% (w/v) BSA)
   2. Incubate in the dark with shaking at room temperature for 1 h
   3. Collect by centrifugation at 10,000 × *g* for 5 min.
   4. Wash 4 × by resuspension in 1 mL PBS:0.5% BSA, incubation with rotating for 5 min in the dark at room temperature, and centrifugation at 10,000 × *g* for 5 min.

3.3.1 For analysis via SDS-PAGE, lyse cells

1. Bead beat with 100 μm silica beads. Perform BCA to normalize protein concentrations prior to SDS-PAGE.

2. Dilute into SDS (1 ×) with reducing agent (1 ×) and heat to 95°C for 15 min.
3. Spin on tabletop centrifuge and load onto gel.

3.3.2 For analysis via flow cytometry

1. Load negative control (0 µM probe) onto flow cytometer and draw probe-negative gates based on fluorescence.
2. Load probe-labeled samples onto flow cytometer and confirm labeling through an increase in fluorescence in the probe-labeled population.

4. Microbiome sample preparation

Not only is the microbial makeup of microbiomes complex, but the samples from which microbial communities are collected are typically physically complex and heterogeneous in nature. Soil samples, mouse gut scrapings, and human fecal samples are all challenging to process and require detailed sample preparation protocols prior to analysis of the microbes themselves. Here we describe protocols for the processing of human fecal samples for studying the gut microbiome as well as for isolating bacteria from soil samples.

4.1 Fecal sample preparation

Note: The below protocol is for preparation of human fecal microbial communities. Protocols for preparation of mouse gut microbial communities have been published elsewhere (Maurice et al., 2013; Whidbey et al., 2019).

Materials

- Human fecal samples
- Phosphate-buffered saline (PBS)
- Glass beads (3 mm)
- Gasketed screw-cap Eppendorf tubes (2 mL)
- Silica-zirconium beads (100 µm)
- Filters (40 µm)
- Tabletop centrifuge
- Bead-beater (Bead Ruptor Elite)

1. Pre-cool a centrifuge and rotor to 4 °C and pre-chill PBS (1 ×) on ice. Prepare a biosafety cabinet (BSC) for work. The following procedure should be completed within a BSC and standard BSL2 biosafety

procedures should be followed. Any equipment that comes in contact with the fecal samples should be cleaned with 10% bleach solution for 30 min.

2. Add PBS (1 ×, chilled) to fecal sample in sample collection container to a volume of 80 mL. Thaw on ice, vortexing periodically, until sample is fully thawed.
3. Aliquot into two samples (5 g fecal matter each). Add 5 g of glass beads (3 mm), then raise the volume of each to 50 mL with PBS (1 ×).
4. Vortex each sample (2 × 30 s). Keep chilled on ice.
5. Using pre-cut pipette tips, filter the fecal content through a 40 μm filter. Transfer to new, airtight tubes. Centrifuge (8000 × *g*, 4 °C, 15 min).
6. Discard the supernatant and wash the pellet with 1 × PBS (10 mL). Centrifuge (8000 × *g*, 4 °C, 5 min) and repeat.
7. Resuspend one of the two pellets into cold PBS (1 ×, 5 mL) and add to the other pellet. Resuspend.
8. At this point, whole cells can be used for labeling experiments as needed. See Section 3.3 for whole microbe labeling protocol. If lysates are required:
    1. Add protease inhibitor (1 × final concentration)
    2. Transfer aliquots (1 mL) into 2-mL screw-top Eppendorf tubes with gaskets.
    3. Add silica-zirconium beads (200 μL, 100 μm beads) to each tube.
    4. Chill the tubes on ice for 5 min then lyse via bead beating (3 × 30 s on bacteria setting, chill on ice for 5 min in between).
    5. Chill on ice for 5 min following lysis. Centrifuge (17,500 × *g*, 30 min).
    6. Combine supernatants into 15 mL tube and vortex briefly.
    7. Determine protein concentration via the bicinchoninic acid (BCA) assay. See Section 3.2 for in vitro labeling protocol.
    8. For storage, aliquot into 1 mL tubes and quick freeze with liquid nitrogen. Limit freeze-thaw cycles for future use.

4.2 Soil sample preparation

Materials

- Soil samples
- Phosphate-buffered saline (PBS)
- Tween
- Nycodenz

1. Prior to microbe isolation, determine the moisture content of the soil and estimate the w/v ratios based on dry soil mass. Users should also consider if a biosafety exposure is of concern and work accordingly.
2. Resuspend sample in PBS (2× w/v) with 0.005% Tween-80.
3. Vortex briefly and incubate at 4 °C for 20 min with gentle agitation.
4. Allow sample to gravity settle (4 °C for 10 min) to pellet large soil particulates.
5. Aliquot Nycodenz (80%, 1 mL) into fresh 2 mL tubes, 2 tubes per sample.
6. Gently overlay supernatant into fresh 2 mL tube containing Nycodenz (80%, 1 mL).
7. Add ice-cold PBS (1 mL, 0.0005% Tween) to soil. Repeat steps 5 and 6, adding to a fresh tube containing Nycodenz (80%, 1 mL).
8. Perform a high speed Nycodenz spin at 4 °C to separate bacterial cells from soil particles (13,000 × *g*, 90 min, no breaks on accl or decel).
9. Use a pipette to recover cells from Nycodenz cushion and add to 1 mL microfuge tube with 750 μL of ice-cold PBS. Pool fractions into one tube.
10. Spin at 10,000 × *g* at 4 °C for 3 min.
11. Wash bacterial cell pellet with 1 mL of fresh, ice-cold PBS, vortex and re-centrifuge (10,000 × *g*, 4 °C for 3 min). Decant supernatant.
12. Resuspend cell pellet in 500 μL fresh ice-cold PBS. Whole cells are ready for labeling as described in Section 3.3.

5. Fluorescence-activated cell sorting (FACS) protocols

Probe-labeled subpopulations of microbiome samples can be isolated from the complex mixture using FACS. Isolation of subpopulations of interest based on enzyme activity allows for identification and further characterization of the microbes performing a given biochemical transformation. Depending on the sort setup, a user can collect a pure subpopulation containing only probe-labeled cells, a mixed population that has been enriched for a specific, probe-targeted function, or perform single-cell sorts for downstream single-cell genomics analyses.

5.1 Biosafety considerations

Fluorescence-activated cell sorters pose an aerosol hazard. Sorting of live microbial populations should be done with consideration for potential biohazards. Aerosol management systems are available for many FACS instruments and are recommended for the procedures outlined here. Additionally, if microbial samples are unknown or identified as BSL2 or higher, live samples should be sorted using a FACS that is contained within a biosafety cabinet. The protocol here assumes that the worker has prior experience with FACS and is familiar with the instrument and standard FACS operation.

5.2 FACS sorting of probe-positive subpopulations

Materials

- FACS
- Whole cell mixtures, probe labeled (Section 3.3)
- Dead cell stain (Propidium iodide (PI), ThermoFisher)
- Positive fluorescence controls (i.e., Flow Cytometry Compensation Beads, ThermoFisher)

1. Label whole cell microbe populations as described in Section 3.3.
2. Label cells with PI (use manufacturer's instructions) to eliminate dead cells.
3. Load no-probe cells onto FACS and set threshold and PMT voltages for appropriate channels. For bacteria, we recommend using side-scatter for event threshold.
4. If available, load labeled positive controls for all channels needed. Several options for positive controls are available:
    1. Probe-labeled monocultures confirmed for labeling via flow cytometry.
    2. Heat-killed monocultures labeled with PI and confirmed for labeling via flow cytometry.
    3. Pre-made compensation beads (available at ThermoFisher).
5. Follow instrument instructions for compensation setup.
6. Load individual samples onto FACS and record data. Draw gates as follows (see Fig. 4):
    1. If needed, gate based on FSC and SSC first (not shown in Fig. 4)
    2. Eliminate dead cells by gating PI negative population
    3. Gate probe-negative control so that >95% of the events in the no probe (0 μM) control sample are included
    4. Gate the remaining cells into a probe-positive population.
7. Sort probe-positive and probe-negative populations into tubes, 96-well, or 384-well plates as needed.

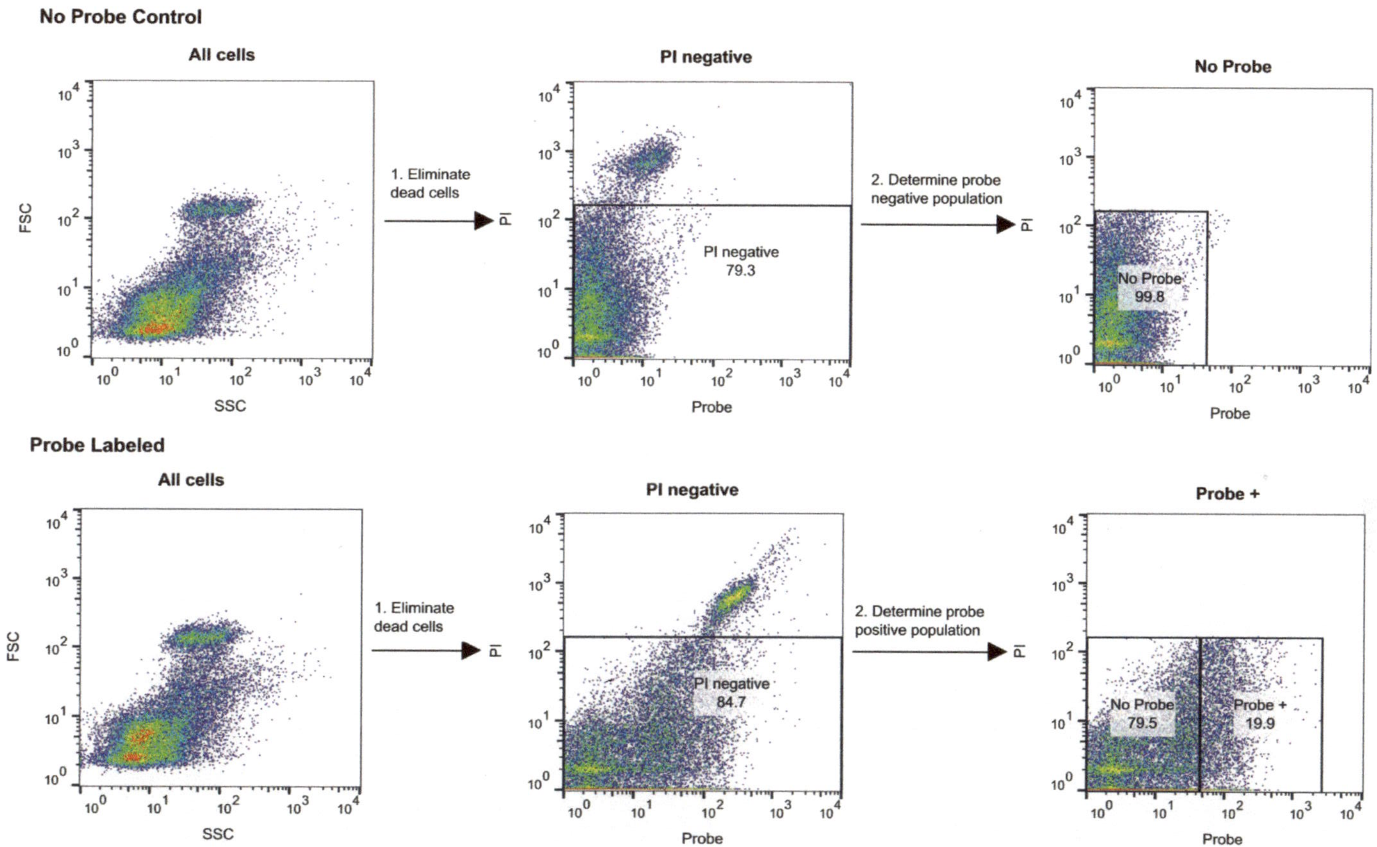

Fig. 4 Gating strategy for isolation of probe-positive populations.

5.3 Post-sort characterization

Following isolation or enrichment of subpopulations of interest, confirmation of purity should be done by loading a sorted sample onto the FACS and confirming that the sorted cells fall within the sort gates drawn. Functional characterization can then be performed. Species level taxonomy achieved from amplicon sequencing of the 16s or ITS region of the sorted population extracted DNA pool can be used to confirm sorting enrichment and to identify the species responsible for the specific activity that was probed. More comprehensive genomic conformation can be achieved by metagenome sequencing. Users can employ proteomics, metabolomics, lipidomics, and or specific enzyme assays to ensure enriched population functional capability.

6. Conclusions

The protocols described in this chapter will be of use to researchers looking to do in-depth analyses on microbiome subpopulations. ABPs and self-immobilizing probes detect active enzymes and covalently label these targets. This irreversible modification of functionally important proteins allows for detection and identification of enzymes and enables function-dependent isolation of subpopulations when used in microbiome samples and paired with FACS. Following isolation by FACS, further characterization can be completed using amplicon sequencing and omics techniques. Taken together, these techniques provide a robust method for identifying the microbial species within a complex microbiome that are responsible for important biochemical transformations.

Acknowledgments

We would like to thank Michelle Davison, William Chrisler, Carson Miller, and Agne Nixon for their contributions. This work was supported by the Defense Advanced Research Projects Agency (DARPA). PNNL is operated by Battelle for the DOE under Contract DE-AC06-76RL01830.

References

Bik, E. M., Eckburg, P. B., Gill, S. R., Nelson, K. E., Purdom, E. A., Francois, F., et al. (2006). Molecular analysis of the bacterial microbiota in the human stomach. *Proceedings of the National Academy of Sciences of the United States of America, 103*(3), 732–737. https://doi.org/10.1073/pnas.0506655103.

Bogyo, M. (2010). Finding enzymes that are actively involved in cancer. *Proceedings of the National Academy of Sciences of the United States of America, 107*(6), 2379–2380. https://doi.org/10.1073/pnas.0914955107.

Bottcher, T., & Sieber, S. A. (2010). Showdomycin as a versatile chemical tool for the detection of pathogenesis-associated enzymes in bacteria. *Journal of the American Chemical Society*, *132*(20), 6964–6972. https://doi.org/10.1021/ja909150y.

Cañeque, T., Müller, S., & Rodriguez, R. (2018). Visualizing biologically active small molecules in cells using click chemistry. *Nature Reviews. Chemistry*, *2*(9), 202–215. https://doi.org/10.1038/s41570-018-0030-x.

Chauvigne-Hines, L. M., Anderson, L. N., Weaver, H. M., Brown, J. N., Koech, P. K., Nicora, C. D., et al. (2012). Suite of activity-based probes for cellulose-degrading enzymes. *Journal of the American Chemical Society*, *134*(50), 20521–20532. https://doi.org/10.1021/ja309790w.

Clarke, G., Sandhu, K. V., Griffin, B. T., Dinan, T. G., Cryan, J. F., & Hyland, N. P. (2019). Gut reactions: Breaking down xenobiotic-microbiome interactions. *Pharmacological Reviews*, *71*(2), 198–224. https://doi.org/10.1124/pr.118.015768.

Claus, S. P., Guillou, H., & Ellero-Simatos, S. (2016). The gut microbiota: A major player in the toxicity of environmental pollutants? *NPJ Biofilms Microbiomes*, *2*, 16003. https://doi.org/10.1038/npjbiofilms.2016.3.

Cravatt, B. F., Wright, A. T., & Kozarich, J. W. (2008). Activity-based protein profiling: From enzyme chemistry to proteomic chemistry. *Annual Review of Biochemistry*, *77*, 383–414. https://doi.org/10.1146/annurev.biochem.75.101304.124125.

Eirich, J., Orth, R., & Sieber, S. A. (2011). Unraveling the protein targets of vancomycin in living S. aureus and E. faecalis cells. *Journal of the American Chemical Society*, *133*(31), 12144–12153. https://doi.org/10.1021/ja2039979.

Evans, M. J., & Cravatt, B. F. (2006). Mechanism-based profiling of enzyme families. *Chemical Reviews*, *106*(8), 3279–3301. https://doi.org/10.1021/cr050288g.

Fierer, N. (2017). Embracing the unknown: Disentangling the complexities of the soil microbiome. *Nature Reviews. Microbiology*, *15*(10), 579–590. https://doi.org/10.1038/nrmicro.2017.87.

Fonovic, M., & Bogyo, M. (2008). Activity-based probes as a tool for functional proteomic analysis of proteases. *Expert Review of Proteomics*, *5*(5), 721–730. https://doi.org/10.1586/14789450.5.5.721.

Garoutte, A., Cardenas, E., Tiedje, J., & Howe, A. (2016). Methodologies for probing the metatranscriptome of grassland soil. *Journal of Microbiological Methods*, *131*, 122–129. https://doi.org/10.1016/j.mimet.2016.10.018.

Gilbert, J. A., Blaser, M. J., Caporaso, J. G., Jansson, J. K., Lynch, S. V., & Knight, R. (2018). Current understanding of the human microbiome. *Nature Medicine*, *24*(4), 392–400. https://doi.org/10.1038/nm.4517.

Haldon, E., Nicasio, M. C., & Perez, P. J. (2015). Copper-catalysed azide-alkyne cycloadditions (CuAAC): An update. *Organic & Biomolecular Chemistry*, *13*(37), 9528–9550. https://doi.org/10.1039/c5ob01457c.

Heal, W. P., Dang, T. H., & Tate, E. W. (2011). Activity-based probes: Discovering new biology and new drug targets. *Chemical Society Reviews*, *40*(1), 246–257. https://doi.org/10.1039/c0cs00004c.

Hiergeist, A., Glasner, J., Reischl, U., & Gessner, A. (2015). Analyses of intestinal microbiota: Culture versus sequencing. *ILAR Journal*, *56*(2), 228–240. https://doi.org/10.1093/ilar/ilv017.

Hold, G. L., Pryde, S. E., Russell, V. J., Furrie, E., & Flint, H. J. (2002). Assessment of microbial diversity in human colonic samples by 16S rDNA sequence analysis. *FEMS Microbiology Ecology*, *39*(1), 33–39. https://doi.org/10.1111/j.1574-6941.2002.tb00904.x.

Hsu, Y. L., Nandakumar, M., Lai, H. Y., Chou, T. C., Chu, C. Y., Lin, C. H., et al. (2015). Development of activity-based probes for imaging human alpha-l-Fucosidases in cells. *The Journal of Organic Chemistry*, *80*(16), 8458–8463. https://doi.org/10.1021/acs.joc.5b01204.

Jansson, J. K., & Hofmockel, K. S. (2018). The soil microbiome—From metagenomics to metaphenomics. *Current Opinion in Microbiology, 43*, 162–168. https://doi.org/10.1016/j.mib.2018.01.013.

Kennedy, D. C., McKay, C. S., Legault, M. C., Danielson, D. C., Blake, J. A., Pegoraro, A. F., et al. (2011). Cellular consequences of copper complexes used to catalyze bioorthogonal click reactions. *Journal of the American Chemical Society, 133*(44), 17993–18001. https://doi.org/10.1021/ja2083027.

Koppel, N., Maini Rekdal, V., & Balskus, E. P. (2017). Chemical transformation of xenobiotics by the human gut microbiota. *Science, 356*(6344). https://doi.org/10.1126/science.aag2770.

Krysiak, J., & Sieber, S. A. (2017). Activity-based protein profiling in bacteria. *Methods in Molecular Biology, 1491*, 57–74. https://doi.org/10.1007/978-1-4939-6439-0_5.

Kwan, D. H., Chen, H. M., Ratananikom, K., Hancock, S. M., Watanabe, Y., Kongsaeree, P. T., et al. (2011). Self-immobilizing fluorogenic imaging agents of enzyme activity. *Angewandte Chemie (International Ed. in English), 50*(1), 300–303. https://doi.org/10.1002/anie.201005705.

Lallana, E., Riguera, R., & Fernandez-Megia, E. (2011). Reliable and efficient procedures for the conjugation of biomolecules through Huisgen azide-alkyne cycloadditions. *Angewandte Chemie (International Ed. in English), 50*(38), 8794–8804. https://doi.org/10.1002/anie.201101019.

Lentz, C. S., Sheldon, J. R., Crawford, L. A., Cooper, R., Garland, M., Amieva, M. R., et al. (2018). Identification of a S. aureus virulence factor by activity-based protein profiling (ABPP). *Nature Chemical Biology, 14*(6), 609–617. https://doi.org/10.1038/s41589-018-0060-1.

Li, S., Cai, H., He, J., Chen, H., Lam, S., Cai, T., et al. (2016). Extent of the oxidative side reactions to peptides and proteins during the CuAAC reaction. *Bioconjugate Chemistry, 27*(10), 2315–2322. https://doi.org/10.1021/acs.bioconjchem.6b00267.

Mao, W., Xia, L., Wang, Y., & Xie, H. (2016). A self-immobilizing and fluorogenic probe for beta-lactamase detection. *Chemistry, An Asian Journal, 11*(24), 3493–3497. https://doi.org/10.1002/asia.201601344.

Martell, J., & Weerapana, E. (2014). Applications of copper-catalyzed click chemistry in activity-based protein profiling. *Molecules, 19*(2), 1378–1393. https://doi.org/10.3390/molecules19021378.

Maurice, C. F., Haiser, H. J., & Turnbaugh, P. J. (2013). Xenobiotics shape the physiology and gene expression of the active human gut microbiome. *Cell, 152*(1–2), 39–50. https://doi.org/10.1016/j.cell.2012.10.052.

Morimoto, K., & van der Hoorn, R. A. (2016). The increasing impact of activity-based protein profiling in plant science. *Plant & Cell Physiology, 57*(3), 446–461. https://doi.org/10.1093/pcp/pcw003.

Nandakumar, M., Hsu, Y. L., Lin, J. C., Lo, C., Lo, L. C., & Lin, C. H. (2015). Detection of human alpha-L-fucosidases by a quinone methide-generating probe: Enhanced activities in response to helicobacter pylori infection. *Chembiochem, 16*(11), 1555–1559. https://doi.org/10.1002/cbic.201500178.

Nomura, D. K., & Casida, J. E. (2011). Activity-based protein profiling of organophosphorus and thiocarbamate pesticides reveals multiple serine hydrolase targets in mouse brain. *Journal of Agricultural and Food Chemistry, 59*(7), 2808–2815. https://doi.org/10.1021/jf101747r.

Nomura, D. K., Long, J. Z., Niessen, S., Hoover, H. S., Ng, S. W., & Cravatt, B. F. (2010). Monoacylglycerol lipase regulates a fatty acid network that promotes cancer pathogenesis. *Cell, 140*(1), 49–61. https://doi.org/10.1016/j.cell.2009.11.027.

Ovaa, H., van Swieten, P. F., Kessler, B. M., Leeuwenburgh, M. A., Fiebiger, E., van den Nieuwendijk, A. M., et al. (2003). Chemistry in living cells: Detection of active

proteasomes by a two-step labeling strategy. *Angewandte Chemie (International Ed. in English)*, *42*(31), 3626–3629. https://doi.org/10.1002/anie.200351314.
Puri, A. W., & Bogyo, M. (2013). Applications of small molecule probes in dissecting mechanisms of bacterial virulence and host responses. *Biochemistry*, *52*(35), 5985–5996. https://doi.org/10.1021/bi400854d.
Sadler, N. C., & Wright, A. T. (2015). Activity-based protein profiling of microbes. *Current Opinion in Chemical Biology*, *24*, 139–144. https://doi.org/10.1016/j.cbpa.2014.10.022.
Schroder, S. P., de Boer, C., McGregor, N. G. S., Rowland, R. J., Moroz, O., Blagova, E., et al. (2019). Dynamic and functional profiling of xylan-degrading enzymes in Aspergillus Secretomes using activity-based probes. *ACS Central Science*, *5*(6), 1067–1078. https://doi.org/10.1021/acscentsci.9b00221.
Speers, A. E., Adam, G. C., & Cravatt, B. F. (2003). Activity-based protein profiling in vivo using a copper(i)-catalyzed azide-alkyne [3 + 2] cycloaddition. *Journal of the American Chemical Society*, *125*(16), 4686–4687. https://doi.org/10.1021/ja034490h.
Staub, I., & Sieber, S. A. (2008). Beta-lactams as selective chemical probes for the in vivo labeling of bacterial enzymes involved in cell wall biosynthesis, antibiotic resistance, and virulence. *Journal of the American Chemical Society*, *130*(40), 13400–13409. https://doi.org/10.1021/ja803349j.
Whidbey, C., Sadler, N. C., Nair, R. N., Volk, R. F., DeLeon, A. J., Bramer, L. M., et al. (2019). A probe-enabled approach for the selective isolation and characterization of functionally active subpopulations in the gut microbiome. *Journal of the American Chemical Society*, *141*(1), 42–47. https://doi.org/10.1021/jacs.8b09668.
Whidbey, C., & Wright, A. T. (2019). Activity-based protein profiling-enabling multimodal functional studies of microbial communities. *Current Topics in Microbiology and Immunology*, *420*, 1–21. https://doi.org/10.1007/82_2018_128.
Willems, L. I., Overkleeft, H. S., & van Kasteren, S. I. (2014). Current developments in activity-based protein profiling. *Bioconjugate Chemistry*, *25*(7), 1181–1191. https://doi.org/10.1021/bc500208y.
Wilson, K. H., & Blitchington, R. B. (1996). Human colonic biota studied by ribosomal DNA sequence analysis. *Applied and Environmental Microbiology*, *62*(7), 2273–2278.
Wu, L., Armstrong, Z., Schroder, S. P., de Boer, C., Artola, M., Aerts, J. M., et al. (2019). An overview of activity-based probes for glycosidases. *Current Opinion in Chemical Biology*, *53*, 25–36. https://doi.org/10.1016/j.cbpa.2019.05.030.
Xin, B. T., Espinal, C., de Bruin, G., Filippov, D. V., van der Marel, G. A., Florea, B. I., et al. (2019). Two-step bioorthogonal activity-based protein profiling of individual human proteasome catalytic sites. *Chembiochem*, *21*, 248–255. https://doi.org/10.1002/cbic.201900551.
Zweerink, S., Kallnik, V., Ninck, S., Nickel, S., Verheyen, J., Blum, M., et al. (2017). Activity-based protein profiling as a robust method for enzyme identification and screening in extremophilic Archaea. *Nature Communications*, *8*, 15352. https://doi.org/10.1038/ncomms15352.

CHAPTER SIX

Biochemical analysis of NlpC/p60 peptidoglycan hydrolase activity

Byungchul Kim, Juliel Espinosa, Howard C. Hang*
Laboratory of Chemical Biology and Microbial Pathogenesis, The Rockefeller University, New York, NY, United States
*Corresponding author: e-mail address: hhang@rockefeller.edu

Contents

Abstract

The NlpC/p60-family of peptidoglycan hydrolases are key enzymes that facilitate bacterial cell division and also modulate microbe-host interactions. These endopeptidases utilize conserved Cys-His residues in their active site and are expressed in most bacterial species as well as some eukaryotes. Here we describe methods for biochemical analysis of *Enterococcus faecium* SagA-NlpC/p60 peptidoglycan hydrolase activity (Kim et al., 2019; Rangan et al., 2016), which includes recombinant protein preparation and biochemical analysis using both gel-based and LC-MS profiling of peptidoglycan fragments. These protocols should also facilitate the biochemical analysis of other NlpC/p60 peptidoglycan hydrolases.

Methods in Enzymology, Volume 638
ISSN 0076-6879
https://doi.org/10.1016/bs.mie.2020.02.017

1. Introduction

The cell wall that forms a protective layer around bacteria is mainly composed of the unique biopolymer, peptidoglycan (PG). PG consists of a chain of alternating *N*-acetylglucosamine (GlcNAc) and *N*-acetylmuramic acid (MurNAc) units linked by a β-1,4-glycosidic linkage between the two glycans (Vollmer, Blanot, & de Pedro, 2008). The MurNAc units of the glycan chain are linked to short peptides, and contain canonical L-amino acids and non-canonical D-amino acids. Cross-links between these peptides render the cell wall its characteristic rigidity. This highly crosslinked structure provides bacteria mechanical resistance from environmental stress and also confers their shape and morphology. PG is also dynamic, as it needs to be remodeled to allow cell growth and division (Egan & Vollmer, 2013; Wyckoff, Taylor, & Salama, 2012). For PG remodeling, a variety of enzymes need to be involved, such as PG synthetases that insert new material in the pre-existing structure and PG hydrolases which cleave the PG polymer at specific sites.

One class of PG hydrolase is the NlpC/p60 superfamily (named after new lipoprotein C from *Escherichia coli* and a 60kDa extracellular protein of *Listeria monocytogenes*) (Anantharaman & Aravind, 2003; Ishikawa, Hara, Ohnishi, & Sekiguchi, 1998). The NlpC/p60 superfamily consists of cysteine peptidases and is broadly distributed in bacteria, viruses, archaea and eukaryotes (Anantharaman & Aravind, 2003). Proteins with NlpC/p60 domains include PG endopeptidases, which are involved in growth, development, cell division, and cell wall maturation (Anantharaman & Aravind, 2003; Hess et al., 1995; Ishikawa et al., 1998). These enzymes exhibit substrate specificity for specific peptide linkages, such as between D-isoglutamine and L-lysine in gram-positive bacterial PG (Vollmer, Joris, Charlier, & Foster, 2008). Disruption of such hydrolases in various bacteria resulted in defective cell division (Hett & Rubin, 2008; Teng, Kawalec, Weinstock, Hryniewicz, & Murray, 2003). Other cell-wall-associated functions are also known to be modulated by NlpC/p60 family peptidases, such as endospore formation and resuscitation by a resuscitation-promoting factor protein SwlA from *Streptomyces coelicolor* (Haiser, Yousef, & Elliot, 2009). Collectively, these reports indicate that the NlpC/p60 family is a critical peptidase family involved in cell wall maintenance, survival, and virulence in gram-positive bacteria.

It has been shown that NlpC/p60 proteins, among other peptidoglycan hydrolases, play a modulatory role on the level of enzymatic activity and the specificity for PG substrate (Xu, Chiu, et al., 2014; Xu, Mengin-Lecreulx, et al., 2014; Xu et al., 2015). Many reports have described the characterization for NlpC/p60 activity using various assays (Bartual et al., 2014; Both, Schneider, & Schnell, 2011; Fukushima et al., 2008). The most commonly used procedure is the zymogram assay, which involves incubation of purified proteins with autoclaved bacterial cells as a substrate followed by renaturing SDS-PAGE (Claes et al., 2012). Zymography is a versatile two-stage technique involving protein separation by electrophoresis followed by detection of hydrolase activity. Protein separation by SDS-PAGE is limited by the fact that some peptidases do not renature and hence cannot be detected following treatment with SDS. Another approach is the cell wall degradation assay. It is carried out using fluorescein isothiocyanate (FITC)-labeled bacterial cells, which are incubated with purified protein, followed by measuring FITC-labeled cell wall fragments released in solution (Ruggiero et al., 2010). A major shortcoming of these assays, however, is the low detection of activity of purified NlpC/p60 domains owing to a lack of suitable synthetic substrates and low activity toward protein substrates used (Both et al., 2011; Both, Steiner, Izumi, Schneider, & Schnell, 2014; Fukushima, Uchida, Ide, Kodama, & Sekiguchi, 2018).

In this context, the characterization of activity for NlpC/p60 hydrolases, and identification of their specific substrates, is of critical importance. Biochemical analysis methods may enable assignment of observed hydrolase activity to a specific peptidoglycan fragments as substrates. This analysis requires robust methods for purification of the large quantities of this protein and substrate pool. Here, we describe a modified and highly optimized method for expression and purification of NlpC/p60 hydrolases that consistently yields large quantities of pure, functional protein using an *Escherichia coli* expression system, isolation of large quantities of bacterial peptidoglycan, and two enzymatic assays for the biochemical characterization of NlpC/p60 hydrolases. These methods will be described using the NlpC/p60 domain of *Enterococcus faecium* SagA as a representative example (Kim et al., 2019; Rangan et al., 2016).

2. Bacterial expression and purification of recombinant NlpC/p60 proteins

It is recommended to perform all protein preparation steps on ice and at 4 °C and that all buffers be pre-chilled. Here we describe a protocol to overexpress and purify NlpC/p60 hydrolases from both the cell pellet and culture supernatant of *E. coli* for subsequent biochemical characterization. We have successfully used this protocol to express and purify various constructs of the NlpC/p60 PG hydrolase SagA, which serves as the representative example. It is important to note that other methods for successful expression and purification of NlpC/p60 also exist, and each laboratory may have their own preferred method of expressing recombinant proteins.

2.1 Equipment

- 37 °C incubator with shaker
- Cell density reader CO8000 (Biochrom)
- Refrigerated centrifuge
- Swinging bucket rotor needed for Jumbosep devices (see Section 2.3.3)
- Polypropylene centrifuge tubes, conical bottom, 500 mL (Corning)
- Polyetherimide centrifuge tube cushions (Corning)
- Falcon™ conical centrifuge tubes, 15 mL (Thermo Fisher Scientific)
- Falcon™ conical centrifuge tubes, 50 mL (Thermo Fisher Scientific)
- Water bath sonicator
- Ultracentrifuge (Beckman)
- Vacuum filter/Storage bottle system, 0.22 μm pore 33.2 cm^2 PES membrane, sterile, 500 mL (Corning)
- Jumbosep™ centrifugal devices with 10 K or 30 K MWCO membrane inserts (Pall Life Sciences)
- Flat-bottomed 250 mL bottle centrifuge adapters
- Poly-Prep® chromatography columns (Bio-Rad)
- Econo-Column® chromatography columns, 1.5 × 20 cm, 35 mL max capacity (Bio-Rad)
- Econo-Column® chromatography columns, 2.5 × 20 cm, 98 mL max capacity (Bio-Rad)
- His_{60} Ni-superflow resin (Takara Bio)
- Nutating mixer
- Amicon® Ultra-4 centrifugal filter units, 10 kDa molecular weight cutoff (MWCO)
- Amicon® Ultra-15 centrifugal filter units, 10 kDa MWCO
- Slide-A-Lyzer MINI dialysis devices, 10 kDa MWCO, 0.5 mL (Thermo Fisher Scientific)
- Orbital shaker
- Superdex 75 Increase 10/300 GL column (GE Healthcare)
- Superdex 200 Increase 10/300 GL column (GE Healthcare)

2.2 Materials

- pET-21 a (+) vector (Novagen)
- 5-Alpha competent *E. coli* (NEB)
- BL21-CodonPlus® (DE3)-RIL competent cells (Stratagene)
- 100 mg/mL ampicillin stock
- 25 mg/mL chloramphenicol stock
- Luria-Bertani agar (LA) plates containing 100 μg/mL ampicillin and 25 μg/mL chloramphenicol
- Luria-Bertani (LB) broth containing 100 μg/mL ampicillin and 25 μg/mL chloramphenicol
- 1 M IPTG (isopropyl-D-thiogalactopyranoside) stock
- 1 × EDTA-free protease inhibitor cocktail tablet (Roche)
- Benzonase® Nuclease HC, purity >90%, 25 kU (EMD Millipore)
- BCA assay (Pierce Protein Biology)
- Trichloroacetic acid (TCA) for protein precipitation
- Buffers
  - Lysis: 150 mM NaCl, 50 mM Bis-Tris pH 7.5, 0.1% SDS, cOmplete™, EDTA-free protease inhibitor cocktail tablet (Roche), 0.025 U benzonase (EMD Millipore)
  - 1 × PBS (phosphate-buffered saline) (Corning)
  - Wash 1: 50 mM NaH_2PO_4, 300 mM NaCl, 20 mM imidazole, pH 8.0
  - Wash 2: 50 mM NaH_2PO_4, 300 mM NaCl, 40 mM imidazole, pH 8.0
  - Elution: 50 mM NaH_2PO_4, 300 mM NaCl, 300 mM imidazole, pH 8.0

2.3 Protocol

2.3.1 Growth and overexpression

1. Transform 50 μL *E. coli* BL21-CodonPlus® (DE3)-RIL competent cells with 1 μL of pET-21 a (+) plasmid encoding NlpC/p60 peptidoglycan hydrolase, and plate onto LA plates containing 100 μg/mL ampicillin and 25 μg/mL chloramphenicol, and incubate overnight at 37 °C.
2. Pick a single colony, inoculate 20 mL LB broth containing 100 μg/mL ampicillin and 25 μg/mL chloramphenicol, and incubate starter culture overnight at 37 °C, 220 rpm.
3. Inoculate 1 L LB broth containing 100 μg/mL ampicillin and 25 μg/mL chloramphenicol with 10 mL overnight starter culture (1:100 inoculation) and incubate at 37 °C, 220 rpm until OD_{600} ~0.5.

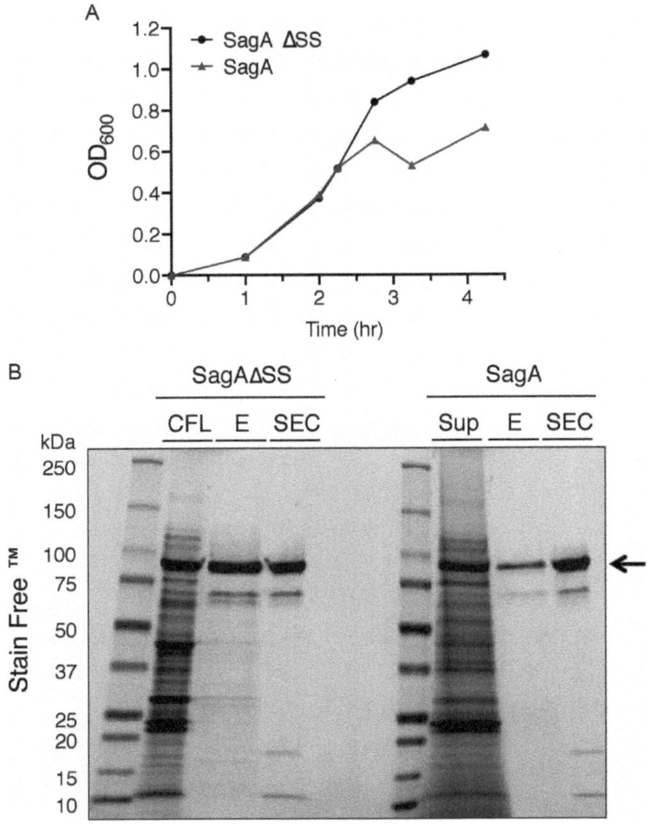

Fig. 1 Expression and purification of a SagA from *E. coli* cell pellet (SagA ΔSS)/supernatant (SagA).

4. Induce with 1 mM IPTG and incubate at 37 °C, 220 rpm for an additional 2 h. If protein is secreted, measure OD_{600} every 30 min to observe any lysis and corresponding decrease in OD_{600} (Fig. 1A).
5. Harvest cells by centrifugation at 5250 × *g* for 30 min. If protein is secreted, transfer 360 mL culture supernatant to a flask on ice and proceed as described below (Section 2.3.3).

2.3.2 Purification from bacterial cell pellet

1. Resuspend 1 L bacterial cell pellet in 20 mL lysis buffer and transfer to 50 mL Falcon™ conical centrifuge tube on ice.
2. Lyse by sonicating cells for 3 min, invert tube four times, then place on ice for 1 min five times.

3. Transfer lysate to an ultracentrifuge tube and centrifuge at 30,000 × *g* for 30 min. Disposable Poly-Prep® chromatography columns are suitable for smaller preparations.
4. Load clarified lysate to a 1.5 × 20 cm Econo-Column containing 1 mL of His_{60} Ni-superflow resin bed pre-rinsed with 10 mL 1× PBS, and incubate for 1 h at 4 °C. Longer incubation more than 1 h may yield higher amounts of protein, but with more impurities. Disposable Poly-Prep® chromatography columns are suitable for smaller preparations.
5. Collect flow-through, then in sequential order add 10 mL of each buffer (wash 1, wash 2, elution) to resin and collect in separate tubes.
6. Concentrate 10 mL protein to ≤0.5 mL by centrifugation at 4000 × *g*, 4 °C, 15 min using a 10 kDa MWCO Amicon® Ultra-15 centrifugal filter.
7. Transfer the 0.5 mL sample to a 0.5 mL 10 kDa Slide-A-Lyzer MINI dialysis device, then place in a 15 mL Falcon™ conical centrifuge tube containing approximately 13 mL 1× PBS and incubate with gentle shaking for 2 h at 4 °C.
8. Replace buffer with fresh 1× PBS and continue dialysis with gentle shaking overnight.
9. Purify the protein by loading onto either a Superdex 75 or Superdex 200 increase 10/300 GL column pre-equilibrated with 1× PBS (Fig. 1B).
10. Combine fractions containing the target protein and concentrate to >1 mg/mL by centrifugation at 4 °C, 15 min, if needed.
11. Estimate protein concentration by using BCA assay or nanodrop.
12. Flash-freeze protein in liquid nitrogen and store at −80 °C for up to 1 year.

2.3.3 Purification from culture supernatant

1. Vacuum filter the 360 mL culture supernatant containing the secreted target protein through a 0.2 μm membrane.
2. Transfer 60 mL of the filtered culture supernatant to 6 Jumbosep centrifugal devices with 10 K or 30 K MWCO membrane inserts.
3. Concentrate the 60 mL filtered culture supernatant fivefold by centrifugation at 3000 × *g*, 4 °C for 40 min.
4. Rinse each membrane with 12 mL 1× PBS and centrifuge for 10 min at 4 °C.
5. Pool concentrate and bring volume up to 72 mL with 1× PBS as a buffer-exchange step.
6. Load the concentrate to a 2.5 × 20 cm Econo-Column containing 3 mL of His_{60} Ni-superflow resin bed pre-rinsed with 30 mL 1× PBS for

overnight incubation and subsequent affinity chromatography (30 mL washes and elution), dialysis and gel filtration, as described above. Overnight incubation with the His_{60} Ni-superflow resin is recommended for culture supernatant preparations, as there are fewer impurities that can bind to the resin, and the overall yield is often lower (Fig. 1B).

2.4 Note

Many NlpC/p60 peptidoglycan hydrolases encode a signal peptide for secretion, and the protocol above describes the purification of secreted proteins from the culture media. However, secretion of NlpC/p60 peptidoglycan hydrolases may prove lytic and/or toxic to *E. coli* upon heterologous overexpression. Therefore, we recommend cloning signal peptide deletion constructs of (1) full-length NlpC/p60 PG hydrolases and (2) NlpC/p60 domain itself into the pET-21 a (+) vector as well to facilitate expression and purification of the cytoplasmically located target protein. In all steps, aliquots (10–20 μL) should be collected for SDS-PAGE analysis. For culture supernatant purifications, samples taken for SDS-PAGE must be precipitated with 10% TCA final. In case recombinant protein is not suitable, genomic insertion of an affinity tag enables the purification of endogenous secreted protein from bacterial cells using the above protocol (see Section 2.3.3).

3. Large scale isolation of peptidoglycan

Peptidoglycan isolation from diverse bacterial species has been well studied throughout the last several decades (Desmarais, De Pedro, Cava, & Huang, 2013; Glauner, 1988; Kuhner, Stahl, Demircioglu, & Bertsche, 2014; Rosenthal & Dziarski, 1994). Here we describe a modified protocol (Kuhner et al., 2014) applicable for isolating peptidoglycan from 3 L cultures of various Gram-positive bacteria, such as *E. faecium*, *E. faecalis*, or *Lactobacillus plantarum*.

3.1 Equipment

- 37 °C incubator with shaker
- Cell density reader CO8000 (Biochrom)
- Refrigerated centrifuge
- Refrigerated microcentrifuge
- Polypropylene centrifuge tubes, conical bottom, 500 mL (Corning)

- Polyetherimide centrifuge tube cushions (Corning)
- Water bath sonicator
- Water bath with heat source up to 100 °C
- Vortex mixer
- Vacuum concentrator
- Falcon™ conical centrifuge tubes, 15 mL (Thermo Fisher Scientific)
- Falcon™ conical centrifuge tubes, 50 mL (Thermo Fisher Scientific)

3.2 Materials

- Brain heart infusion (BHI) agar plates for *Enterococci*
- Brain heart infusion (BHI) broth for *Enterococci*
- Media recommended for growth of bacteria of choice
- Purified water
- Benzonase® Nuclease HC, purity >90%, 25 kU (EMD Millipore)
- 1 mg/mL Trypsin stock in purified H_2O
- Solutions A: 100 mM Bis-Tris/HCl, pH 7.5, 0.25% SDS (w/v); Solution B: 100 mM Bis-Tris/HCl, pH 7.5, 50 U benzonase; Solution C: 50 µg/mL trypsin in purified water
- 1 M HCl

3.3 Protocol

3.3.1 Isolation of peptidoglycan from gram-positive bacterial cells

1. Streak the *Enterococci* from a glycerol stock on a BHI agar plate and incubate overnight at 37 °C.
2. Pick a single colony, inoculate 60 mL BHI broth, and incubate starter culture overnight with shaking (220 rpm) at 37 °C.
3. Inoculate 3 L BHI broth with 30 mL overnight starter culture (1:100 inoculation) and incubate at 37 °C, 220 rpm until log phase (OD_{600} ~0.6).
4. Harvest cells by centrifugation $\leq 5000 \times g$ for 30 min.
5. Resuspend 3 L bacterial cell pellet in 60 mL purified water, transfer to 500 mL conical bottom tube, and harvest by centrifugation $\leq 5000 \times g$ for 30 min.
6. Set water bath to 100 °C around 1 h prior to peptidoglycan isolation. Use flask weights or 2–4 500 mL glass bottles filled with water to hold the conical bottom tubes in place.

7. Resuspend the 3 L bacterial cell pellet with 60 mL solution A, boil the suspension at 100 °C for 20 min, then cool the tube on ice for 5 min to prevent warping of the bottle during centrifugation.
   a. During all boiling steps, remove the caps and cover the tube with aluminum foil to prevent the caps from warping.
8. Pellet the cells by centrifugation $\leq 5000 \times g$ for 15 min and discard the supernatant. Supernatant is often very yellow after this step.
9. Wash the resulting insoluble cell wall preparation with purified water six times until free of SDS. Resuspend the pellet with 120 mL purified water by pipetting and vortexing, and then centrifuge $\leq 5000 \times g$ for 15 min and discard the supernatant five times. SDS has to be completely washed out in order to prevent potential inhibition of enzymatic activity.
10. Resuspend the pellet with 60 mL ultrapure water and sonicate for 30 min. Place the conical bottom bottle diagonally in the water bath sonicator to spread the suspension over a larger surface area and secure the bottle in place with tape.
11. Add 60 mL of solution B and incubate at 37 °C, 220 rpm for 1 h.
12. Add 60 mL solution C and incubate at 37 °C, 220 rpm for 1 h.
13. Boil the 180 mL suspension at 100 °C for 5 min to inactivate the enzymes and then cool the tube on ice for 5 min to prevent warping of the bottle during centrifugation.
14. Pellet the suspension by centrifugation $\leq 5000 \times g$ for 15 min and discard the supernatant.
15. Wash the pellet once with 60 mL purified water and centrifugation $\leq 5000 \times g$ for 15 min. Pellets may be delicate at this step; handle and discard supernatant carefully.
16. Resuspend the cell pellet in 60 mL 1 M HCl to release wall teichoic acids (WTAs) by incubating at 37 °C, 220 rpm for 4 h.
17. Pellet the suspension by centrifugation $\leq 5000 \times g$ for 15 min and discard the supernatant.
18. Wash the pellet with 60 mL purified water 5–6 times and centrifugation $\leq 5000 \times g$ for 10 min until pH is 4–5 by using pH strip.
19. Resuspend 60 mL suspension with 6 mL purified water and aliquot 1 mL peptidoglycan suspension into 1.5 mL Eppendorf tubes.
20. Lyophilize aliquots for 4–5 h with caps open in vacuum concentrator. Typical yield of isolated peptidoglycan from *Enterococcus faecium* culture is ~250 mg/3 L. The final peptidoglycan can be stored at −20 °C.

3.4 Note

Solution B can be replaced with mixture of DNase/RNase (Kuhner et al., 2014). Solution C can be substituted with pronase stock instead of trypsin since treatments with either trypsin or pronase are used to remove trapped high-molecular weight peptidoglycan-associated protein.

4. Biochemical analysis of NlpC/p60 peptidoglycan hydrolase activity

The purification procedure described above consistently yields fully active NlpC/p60 protein. It is imperative, however, to confirm that each preparation of the protein is indeed 100% active. Here we describe low- and high-resolution methods to evaluate the biochemical activity of NlpC/p60 peptidoglycan hydrolases *in vitro*. These methods entail the incubation of purified protein with soluble peptidoglycan fragments followed by in-gel profiling or LC-MS analysis. For the low-resolution in-gel profiling assay, muropeptide controls can be custom synthesized or obtained from *Invivogen*. For the high-resolution assay, we used an Agilent 1200 LC with a C_{18} column connected in-line with a Thermo LTQ-Orbitrap mass spectrometer equipped with an electrospray ionization source. Insoluble peptidoglycan isolated from bacteria can be treated with muramidases, such as mutanolysin, which catalyze the cleavage of the β-1,4-linkages between *N*-acetylmuramic acid and *N*-acetyl-D-glucosamine residues in peptidoglycan, thus yielding soluble non-crosslinked and crosslinked peptidoglycan fragments as potential substrates for biochemical assays. All concentrations mentioned in the following protocol are final concentrations in the mixtures.

4.1 Equipment

- 37 °C incubator with shaker
- Block heater
- Refrigerated microcentrifuge
- Vacuum concentrator
- Criterion empty cassettes, 1.0 mm thick, 18-well comb (Bio-Rad)
- Criterion vertical electrophoresis cell (Bio-Rad)
- ChemiDoc Imaging System with SYBR Safe Dye detection (Bio-Rad)
- Acclaim 120 C_{18} column (2.1 µm, 2.1 × 150 mm) (Thermo Scientific)
- Waters 2489 HPLC
- Agilent 1200 series LC/MSD TOF

4.2 Materials

- 50 mM sodium phosphate, pH 4.95
- Mutanolysin from *Streptomyces globisporus* ATCC 21553, M9901-10 kU (Sigma-Aldrich)
- ANTS (8-aminonaphthalene-1,3,6-trisulfonic acid, disodium salt) (Invitrogen)
- ANTS stock: 0.2 M ANTS (dissolve 85.5 mg ANTS in 850 μL distilled water/150 μL acetic acid), 1 M $NaBH_3CN$ (dissolve 62.84 mg in 1 mL DMSO). Sonicate reagents to dissolve and store with covered foil at −80 °C for up to 1 month
- 50% glycerol
- 2 mg/mL Orange G
- 10× Native running buffer (30 g Tris-base and 144 g glycine in 1 L distilled water)
- 50% 19:1 Acrylamide:Bis-acrylamide buffer: to a glass bottle covered with aluminum foil, add 95 g acrylamide and 5 g bis-acrylamide, and dissolve in 100 mL distilled water. When components have dissolved, increase the volume of the buffer to 200 mL with distilled water. Store at 4 °C immediately
- 1.5 M Tris-HCl, pH 8.8
- 0.5 M Tris-HCl, pH 6.8
- 10% fresh ammonium persulfate (APS) stock
- Tetramethylethylenediamine (TEMED)
- 0.5 M boric acid, pH 9.0
- Sodium borohydride pellets
- Phosphoric acid, 20%
- Buffer A: 100% HPLC grade water and 0.1% trifluoroacetic acid (TFA) final
- Buffer B: 30% HPLC grade methanol, 70% HPLC grade water and 0.1% TFA final

4.3 Protocol

4.3.1 Preparation of soluble peptidoglycan mixture and isolation of muropeptides

1. Prepare 1 mL stock of insoluble PG by carefully resuspending aliquot with purified water. Pipette gently, as the insoluble peptidoglycan can easily clog pipette tips and cause sample loss.
2. Digest 2 mg insoluble PG with 1 mg/mL mutanolysin in 50 mM sodium phosphate buffer, pH 4.9 for 16 h at 37 °C, 220 rpm. Mutanolysin digests

40–50% of insoluble peptidoglycan, therefore, twofold excess peptidoglycan is used in digests.

3. Quench the reaction by incubating at 100 °C for 5 min.
4. Centrifuge at 16,000 × *g* for 5 min in a microcentrifuge and carefully transfer the soluble muropeptide mixture to an Eppendorf tube. The soluble muropeptide mixture can be stored at −20 °C and it may be lyophilized to prepare more concentrated stocks, if needed.
5. For separating each muropeptide, inject 50 mg of mutanolysin-digested soluble peptidoglycans on a reverse-phase Luna® C_{18}, dimension of 250 × 10 mm, particle size of 5 μm, pore size of 100 Å column packed with 3 μm ODS Hypersil. Column can be run at room-temperature. The flow-rate is 2 mL/min. The linear gradient of elution started from 100% H_2O to 30% methanol over 90 min in the presence of 0.1% TFA. UV detection was performed at 210 nm. Collect each peak and lyophilize using a vacuum concentrator.

4.3.2 NlpC/p60 enzymatic assay

1. Set up reactions as follows: incubate 10 μM purified protein with 100–250 μg of the mutanolysin-digested peptidoglycan or 200 μM purified muropeptide in 50 mM Bis-Tris, pH 5.5 for 16 h at 37 °C, 220 rpm.
   a. To account for the incomplete digestion of intact peptidoglycan by mutanolysin (40–50%), the working concentration of the soluble peptidoglycan is half of the final concentration used in the mutanolysin digestion (i.e., the concentration of the soluble product is 2 mg/mL if the final concentration in the mutanolysin digestion was 4 mg/mL).
   b. We recommend using 100 μg of the soluble mutanolysin-digested peptidoglycan for in-gel profiling of peptidoglycan fragments (Section 4.3.3) and 250 μg for LC-MS analysis (Section 4.3.4).
2. Quench the reactions by incubating at 100 °C for 5 min.
3. Centrifuge the reaction at 16,000 × *g* for 5 min in a microcentrifuge and carefully transfer the muropeptide products to a new Eppendorf tube.
4. Lyophilize sample with caps open in vacuum concentrator for 2–3 h. Lyophilized muropeptide products may be stored at −20 °C.

4.3.3 In-gel profiling of peptidoglycan fragments

1. Add 5 μL of each reagent (0.2 M ANTS and 1 M $NaBH_3CN$) to the lyophilized muropeptide products, vortex to mix, and centrifuge at 16,000 × *g* for 1 min. ANTS reagents frozen at −80 °C may be thawed by sonication. Synthetic peptidoglycan fragments, such as GlcNAc,

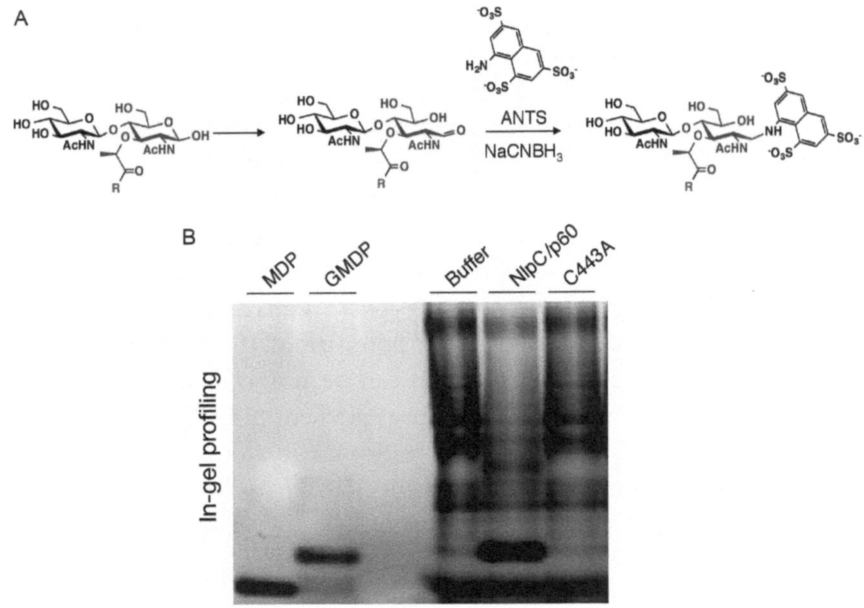

Fig. 2 In-gel profiling assay of NlpC/p60 hydrolase activity.

MurNAc, MurNAc-L-Ala, and MurNAc-L-Ala-D-isoGln may be ANTS-labeled in parallel and run for comparison (Fig. 2A).

2. Place samples in a test tube rack covered with aluminum foil and incubate overnight at 37 °C.
3. Add 10 μL 50% glycerol (v/v), vortex to mix, and centrifuge at 16,000 × *g* for 1 min.
4. Secure two criterion empty cassettes in an upright position to prepare the gels and set aside the 18-well combs. Do not remove the tape at the bottom of the cassette until prior to use of the gel. Gels can be stored in a plastic bag partially filled with 1 × native running buffer for up to 2 days at 4 °C.
5. Prepare the resolving gel of the hand-cast 37–40% Tris-glycine acrylamide gel as outlined in Table 1. Mix vigorously and immediately add 10.5 mL to each empty criterion cassette with a serological pipette followed by a top layer of 3–5 mL distilled water. Wait 1 h for the components to polymerize.
6. Remove the top layer of distilled water, prepare the stacking gel of the hand-cast 37–40% Tris-glycine acrylamide gel as outlined in Table 2. Mix vigorously, immediately add 2.5–3 mL onto top the resolving gel

Table 1 Components for resolving gel.

	Volume (mL)
50% 19:1 stock	16.28
1.5 M Tris-HCl, pH 8.8	5.5
10% APS	0.110
TEMED	0.011
Final volume	21.9

Table 2 Components for stacking gel.

	Volume (mL)
50% 19:1 stock	2.4
0.5 M Tris-HCl, pH 6.8	1.5
10% APS	0.030
TEMED	0.006
Distilled water	2
Final volume	5.9

in each criterion cassette with a serological pipette, and then carefully re-insert the 18-well comb. Wait 30 min for the components to polymerize.

7. Remove the tape from the bottom of the cassette, place the cassette in a criterion vertical electrophoresis cell, and fill the top of the cassette and the cell with 1 × native running buffer before removing the 18-well comb. Thoroughly rinse the wells with buffer using syringe and needle before use, if required.
8. Add 4 μL of the sample to the wells, 1 μL of 2 mg/mL Orange G dye to empty wells as a tracking dye, and run at 80 V for 1 h to allow for migration through the stacking gel. We recommend adding samples closer to the middle of the gel for the best resolution.
9. Increase the voltage to 100–130 V and run for an additional 2–3 h.
10. Remove the cassette from the cell, wipe down with ethanol and a Kim wipe, and image in a ChemiDoc system using the Sybr-safe UV imaging setting and 2–4 s exposure (Fig. 2B).
11. For best signal, carefully open the cassette and directly image the gel using the settings described in the previous step. The gels are very brittle

and must be handled with care to prevent any rips. We recommend handling the gels as followed: with the gel still attached to one side of the cassette, cut-off the left and right sides and directly above the ANTS dye-front (faint yellow line). Then, transfer the gel to a plastic box filled with distilled water for several seconds before imagining. More water can be added directly to the gel once it is in the ChemiDoc system to prevent it from drying.

4.3.4 LC-MS/MS analysis

1. Add 20 μL of a 1:1 mix of ultrapure water and 0.5 M boric acid, pH 9.0 to the lyophilized muropeptide products, vortex to mix, and centrifuge at 16,000 × *g* for 1 min.
2. To each sample add 1–2 grains of sodium borohydride to reduce sample. Carefully flick to dissolve, and leave the tubes on a rack with open caps for 1 h. This reduction reaction is highly volatile and will generate bubbles.
3. Adjust samples to pH 3–4 using pH indicator paper by adding 10 μL increments 20% phosphoric acid and wait for 1 h to quench the reaction completely. Keep the total sample volume minimal between 30 and 60 μL.
4. Centrifuge reduced samples at 16,000 × *g* for 1 min and carefully transfer supernatant to new Eppendorf tube. Filter the samples through 0.22 μm before injection, if needed.
5. Inject 10–15 μL sample to an Acclaim 120 reversed-phase C_{18} column (2.1 μm, 2.1 × 150 mm) column operating at 52 °C and run at 0.2 mL/min in a mobile phase (A) and an eluent (B) using linear gradients of 0–100% (B) over 60 min. Products can be analyzed with an Agilent 1200 series LC/MS-TOF using electrospray ionization in positive mode, acquiring the mass range of 50–2000 *m*/*z* (Fig. 3).

4.4 Note

The NlpC/p60 domain of SagA was shown to be more active in a low-pH buffer. An advantage of utilizing the in-gel profiling assay for evaluating enzymatic activity is that it allows for more rapid analysis of various protein constructs when compared with the LC-MS/MS system. After initial activity screening using the in-gel profiling assay, LC-MS/MS can be used for higher sensitivity and resolution, as well as direct mass measurements of the products. When preparing samples for LC-MS injection, it is critical to adjust the pH correctly to the isoelectric point of the muropeptides (~3.5) after the reduction step. Adding excess amount of acid to sample can lead to the precipitation due to low-pH than 3.

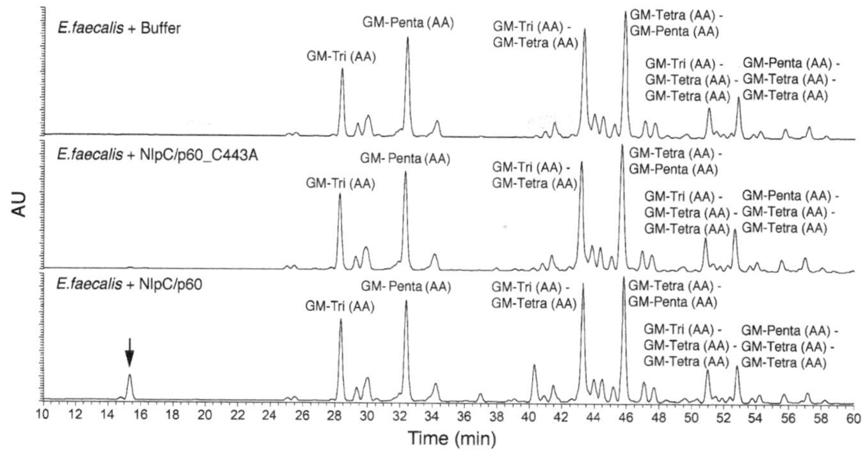

Fig. 3 LC-MS analysis of PG mixture from *E. faecalis* digested by mutanolysin followed by each construct.

5. Summary and conclusions

The methodologies that we have described should be generally applicable for the preparation of NlpC/p60 proteins from other bacterial species and the biochemical characterization of their enzymatic activities. The purified, recombinant proteins that can now be obtained for many isoforms and constructs can be used for many other *in vitro* applications. *In vitro* assays offer powerful tools to study peptidoglycan hydrolases. They can be used to identify and analyze substrates, and determine their specific activities. The present chapter should be a reference for any such studies employing recombinant NlpC/p60 proteins for studying function and regulation of this unique family in peptidoglycan remodeling.

Acknowledgments

We would like to acknowledge all past and present members of Hang Laboratory who have contributed to the studies discussed in this chapter. This research is supported by NIH grant R01GM103593 to H.C.H. This work was also partially funded by NRSA Training Grant #A1070084.

Conflict of interest statement

The authors declared that no conflict of interest exists.

References

Anantharaman, V., & Aravind, L. (2003). Evolutionary history, structural features and biochemical diversity of the NlpC/p60 superfamily of enzymes. *Genome Biology*, *4*, R11.

Bartual, S. G., Straume, D., Stamsas, G. A., Munoz, I. G., Alfonso, C., Martinez-Ripoll, M., et al. (2014). Structural basis of PcsB-mediated cell separation in Streptococcus pneumoniae. *Nature Communications*, *5*, 3842.

Both, D., Schneider, G., & Schnell, R. (2011). Peptidoglycan remodeling in Mycobacterium tuberculosis: Comparison of structures and catalytic activities of RipA and RipB. *Journal of Molecular Biology*, *413*, 247–260.

Both, D., Steiner, E. M., Izumi, A., Schneider, G., & Schnell, R. (2014). RipD (Rv1566c) from Mycobacterium tuberculosis: Adaptation of an NlpC/p60 domain to a non-catalytic peptidoglycan-binding function. *Biochemical Journal*, *457*, 33–41.

Claes, I. J., Schoofs, G., Regulski, K., Courtin, P., Chapot-Chartier, M. P., Rolain, T., et al. (2012). Genetic and biochemical characterization of the cell wall hydrolase activity of the major secreted protein of Lactobacillus rhamnosus GG. *PLoS One*, *7*, e31588.

Desmarais, S. M., De Pedro, M. A., Cava, F., & Huang, K. C. (2013). Peptidoglycan at its peaks: How chromatographic analyses can reveal bacterial cell wall structure and assembly. *Molecular Microbiology*, *89*, 1–13.

Egan, A. J., & Vollmer, W. (2013). The physiology of bacterial cell division. *Annals of the New York Academy of Sciences*, *1277*, 8–28.

Fukushima, T., Kitajima, T., Yamaguchi, H., Ouyang, Q., Furuhata, K., Yamamoto, H., et al. (2008). Identification and characterization of novel cell wall hydrolase CwlT: A two-domain autolysin exhibiting n-acetylmuramidase and DL-endopeptidase activities. *Journal of Biological Chemistry*, *283*, 11117–11125.

Fukushima, T., Uchida, N., Ide, M., Kodama, T., & Sekiguchi, J. (2018). DL-endopeptidases function as both cell wall hydrolases and poly-gamma-glutamic acid hydrolases. *Microbiology*, *164*, 277–286.

Glauner, B. (1988). Separation and quantification of muropeptides with high-performance liquid chromatography. *Analytical Biochemistry*, *172*, 451–464.

Haiser, H. J., Yousef, M. R., & Elliot, M. A. (2009). Cell wall hydrolases affect germination, vegetative growth, and sporulation in Streptomyces coelicolor. *Journal of Bacteriology*, *191*, 6501–6512.

Hess, J., Gentschev, I., Szalay, G., Ladel, C., Bubert, A., Goebel, W., et al. (1995). Listeria monocytogenes p60 supports host cell invasion by and in vivo survival of attenuated Salmonella typhimurium. *Infection and Immunity*, *63*, 2047–2053.

Hett, E. C., & Rubin, E. J. (2008). Bacterial growth and cell division: A mycobacterial perspective. *Microbiology and Molecular Biology Reviews*, *72*, 126–156, Table of contents.

Ishikawa, S., Hara, Y., Ohnishi, R., & Sekiguchi, J. (1998). Regulation of a new cell wall hydrolase gene, cwlF, which affects cell separation in Bacillus subtilis. *Journal of Bacteriology*, *180*, 2549–2555.

Kim, B., Wang, Y. C., Hespen, C. W., Espinosa, J., Salje, J., Rangan, K. J., et al. (2019). Enterococcus faecium secreted antigen A generates muropeptides to enhance host immunity and limit bacterial pathogenesis. *eLife*, *8*, e45343.

Kuhner, D., Stahl, M., Demircioglu, D. D., & Bertsche, U. (2014). From cells to muropeptide structures in 24 h: Peptidoglycan mapping by UPLC-MS. *Scientific Reports*, *4*, 7494.

Rangan, K. J., Pedicord, V. A., Wang, Y. C., Kim, B., Lu, Y., Shaham, S., et al. (2016). A secreted bacterial peptidoglycan hydrolase enhances tolerance to enteric pathogens. *Science*, *353*, 1434–1437.

Rosenthal, R. S., & Dziarski, R. (1994). Isolation of peptidoglycan and soluble peptidoglycan fragments. *Methods in Enzymology*, *235*, 253–285.

Ruggiero, A., Marasco, D., Squeglia, F., Soldini, S., Pedone, E., Pedone, C., et al. (2010). Structure and functional regulation of RipA, a mycobacterial enzyme essential for daughter cell separation. *Structure*, *18*, 1184–1190.

Teng, F., Kawalec, M., Weinstock, G. M., Hryniewicz, W., & Murray, B. E. (2003). An Enterococcus faecium secreted antigen, SagA, exhibits broad-spectrum binding to extracellular matrix proteins and appears essential for E. faecium growth. *Infection and Immunity*, *71*, 5033–5041.

Vollmer, W., Blanot, D., & de Pedro, M. A. (2008). Peptidoglycan structure and architecture. *FEMS Microbiology Reviews*, *32*, 149–167.

Vollmer, W., Joris, B., Charlier, P., & Foster, S. (2008). Bacterial peptidoglycan (murein) hydrolases. *FEMS Microbiology Reviews*, *32*, 259–286.

Wyckoff, T. J., Taylor, J. A., & Salama, N. R. (2012). Beyond growth: Novel functions for bacterial cell wall hydrolases. *Trends in Microbiology*, *20*, 540–547.

Xu, Q., Chiu, H. J., Farr, C. L., Jaroszewski, L., Knuth, M. W., Miller, M. D., et al. (2014). Structures of a bifunctional cell wall hydrolase CwlT containing a novel bacterial lysozyme and an NlpC/P60 DL-endopeptidase. *Journal of Molecular Biology*, *426*, 169–184.

Xu, Q., Mengin-Lecreulx, D., Liu, X. W., Patin, D., Farr, C. L., Grant, J. C., et al. (2015). Insights into substrate specificity of NlpC/P60 cell wall hydrolases containing bacterial SH3 domains. *mBio*, *6*, e02327-02314.

Xu, Q., Mengin-Lecreulx, D., Patin, D., Grant, J. C., Chiu, H. J., Jaroszewski, L., et al. (2014). Structure-guided functional characterization of DUF1460 reveals a highly specific NlpC/P60 amidase family. *Structure*, *22*, 1799–1809.

CHAPTER SEVEN

Controlled release of bioactive signaling molecules

Maulik S. Jani[a,b], Aneesh T. Veetil[a,b], Yamuna Krishnan[a,b,*]
[a]Department of Chemistry, University of Chicago, Chicago, IL, United States
[b]Grossman Institute of Neuroscience, Quantitative Biology and Human Behavior, University of Chicago, Chicago, IL, United States
[*]Corresponding author: e-mail address: yamuna@uchicago.edu

Contents

Abstract

Membrane-initiated steroid signaling (MISS) involves rapid second messenger based intracellular signaling without coupling to transcription or translation. MISS activates important cellular signaling cascades such as mitogen-activated protein kinase (MAPK) or adenylate cyclase pathways. Despite its vital role in signaling, the downstream second messengers involved in MISS and their temporal dynamics remain elusive. A technology which can offer pristine spatiotemporal control over the release of the steroid initiator could pave the way to understand these rapid and ultrasensitive signaling processes. Toward this, we describe a DNA-nanocapsule based technology to chemically release steroids and study MISS in endothelial cells. Here we discuss the synthesis and cellular protocols for investigators who seek to utilize DNA-nanocapsules for the chemically triggered release of small molecules.

Methods in Enzymology, Volume 638
ISSN 0076-6879
https://doi.org/10.1016/bs.mie.2020.02.018

1. Introduction

The self-assembly of DNA molecules by Watson-Crick base pairing has produced diverse 3D-polyhedra such as the DNA-tetrahedron (Goodman et al., 2005), octahedron (Shih, Quispe, & Joyce, 2004), cube (Chen & Seeman, 1991), and icosahedron (Bhatia et al., 2009). Among these, the DNA-icosahedron offers the advantage of maximum encapsulation volume coupled with a uniformly small pore size. This unique property has been exploited to encapsulate diverse cargo molecules such as gold nanoparticles, quantum dots, and fluorescently labeled dextran polymers (Bhatia et al., 2009, 2016; Surana, Bhatia, & Krishnan, 2013). Such cargo-loaded DNA-icosahedra have been employed for bioimaging and single molecule tracking applications in cells and in living systems (Bhatia et al., 2016). The DNA-icosahedron is designed to have several isotropic pores of uniform size and hence it is ideal for controlled release of bioactive small-molecules. Additionally, encapsulating bioactive small-molecules within in the DNA-icosahedron offers unique advantages in studying signaling. Encapsulation prevents the premature activation of receptors by sequestering bioactive molecule within the cavity of the DNA-icosahedron. Release of the molecule from the capsule would be a pristine way to initiate ligand-binding and following rapid signaling as seen with neurotransmitters or membrane-initiated steroid signaling (MISS). We recently applied cargo-loaded, icosahedral DNA-nanocapsules for rapid and controlled delivery of steroids to study cell signaling in real time (Veetil et al., 2017; Veetil, Jani, & Krishnan, 2018).

We described how to apply this technology to study rapid cell signaling taking MISS as a test case. MISS and nuclear-initiated steroid signaling (NISS) are the two ways by which steroids regulate diverse cellular processes (Cato, Nestl, & Mink, 2002; Chow, Handelsman, & Ng, 2010; Daufeldt, Lanz, & Alléra, 2003). NISS involves intracellular receptors, which upon binding the steroid hormone receptor translocates to the nucleus to modulate gene expression. This mode of steroid action occurs on a time scale of hours to days, involving translocation, transcription and translation (Pedram, Razandi, Aitkenhead, Hughes, & Levin, 2002). In contrast, MISS happens on time scales of seconds to minutes, and are too rapid for genomic changes. It involves the recognition of specific steroids at the plasma membrane. MISS regulates the rapid activation of mitogen-activated protein kinases (MAPKs), protein kinase C (PKC) and adenylate cyclase (Cato et al., 2002). Despite the emerging role of MISS in cell signaling, tools to manipulate these complex and rapid signaling processes are limited. Technologies that can enable the controlled release of signaling molecules such as steroids, in this case, in pristine ways could help study and manipulate MISS processes. Here we describe the application of DNA-icosahedron for investigating MISS induced by 17-β-estradiol in endothelial cells (Veetil et al., 2018).

This chapter has been divided into following parts: An introduction to MISS, DNA icosahedron, synthesis of stimuli responsive biopolymers for encapsulation and in-cellulo assay protocols to modulate MISS in mammalian endothelial cells through controlled small-molecule delivery are presented in Sections 2 and 3. We also describe assay analysis and troubleshooting to common glitches that experimenter might face while running these assays.

2. Protocol

We have synthesized three different chemically caged small-molecule–dextran conjugates for encapsulation using DNA icosahedra. Dextran (10kDa, Sigma) was chemically linked to 4-methyl-umbelliferone (4-MU, Cat #M1381, Sigma-Aldrich) or methoxyfluorescein through a disulfide linkage to display either fluoresceins (FLD_{10}) or coumarins (CD_{10}), respectively, on the dextran scaffold (Fig. 1). We have also synthesized biologically inactive estradiol-dextran conjugate (ED_{10}). Exogenous addition of a small-molecule thiol such as 1,4-dithioerythritol (DTT, Cat #D8255, Sigma-Aldrich) or glutathione to dextran-conjugates is expected to result in the release of fluorophores or estradiol through rapid disulfide-exchange chemistry.

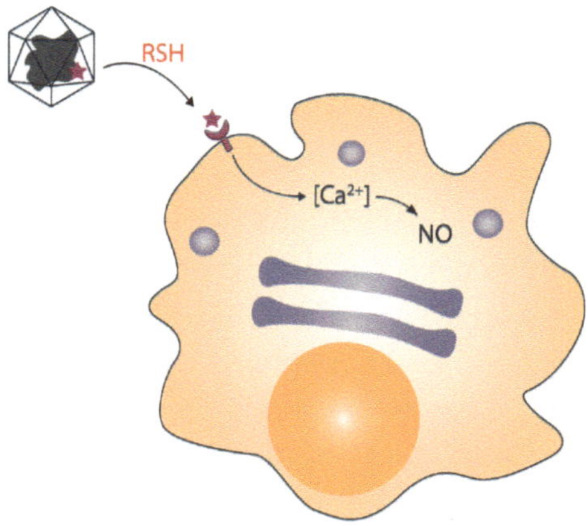

Fig. 1 Schematic showing triggered release of encapsulate steroid (purple star) from DNA-icosahedron and subsequent MISS.

Protocols for the synthesis of three different dextran-small molecule conjugates, its in vitro characterization, encapsulation and triggered release from DNA-icosahedron are detailed in subsequent sections.

2.1 Synthesis of stimuli responsive dextran-fluorophore conjugates

1. Fluorescence quenched (caged) CD_{10} and FLD_{10} were synthesized from corresponding chemically caged and bio-conjugatable precursor molecules (Fig. 1).
2. Chemical caging of 4-MU was achieved using a disulfide linker in a one-pot protocol. Dry triethyl amine (TEA, 57 μL, 0.51 mmol) was added to 4-MU (70 mg, 0.397 mmol) dissolved in extra dry acetonitrile (MeCN, 10 mL). The reaction mixture was subsequently stirred under nitrogen atmosphere at room temperature. The reaction mixture was subsequently cooled to 0 °C using an ice bath and compound 2-hydroxyethyl disulfide (**1b**, 127 mg, 0.397 mmol, Cat #380474, Sigma-Aldrich) solution in dry acetonitrile was added over a period of 15 min and the mixture was allowed to stir at RT for 12 h. The progress of the reaction was monitored by using thin layer chromatography (TLC, 1:1, ethyl acetate: hexane). Upon completion of the reaction, the reaction mixture was acidified with 10% HCl and extracted to ethyl acetate, washed with brine and dried over anhydrous sodium sulfate. The filtrate was evaporated and the residue was subjected to column chromatography using 20%–50% gradient of ethyl acetate in hexane to obtain a viscous liquid as the product (**1c**, Fig. 2) in 32% yield. *Note*: Presence of water molecule in acetonitrile will lead to hydrolysis of the product and so it is highly recommended to use dry acetonitrile and TEA in this reaction.
3. Chemical synthesis of activated ester of disulfide-caged 4-MU is achieved using 1,1′-carbonyldiimidazole (CDI, Cat #705284, Sigma-Aldrich).
4. To compound **1c** (20 mg, 0.056 mmol) dissolved in dry dichloromethane (DCM, Thermo) and kept under nitrogen atmosphere, CDI (95 mg, 0.084 mmol) was added and the mixture was stirred for 1 h at RT. Completion of the reaction is verified using TLC (1:1, ethyl acetate: hexane). DCM was evaporated from the reaction mixture to obtain crude product **1d** (Fig. 2) as a white solid and it was directly used for conjugation with 10 kDa dextran to synthesize CD_{10}. *Note*: It is recommended that product **1d** to be used for the next-step on the same day of the reaction. Long-term storage might induce hydrolysis of the activated ester.

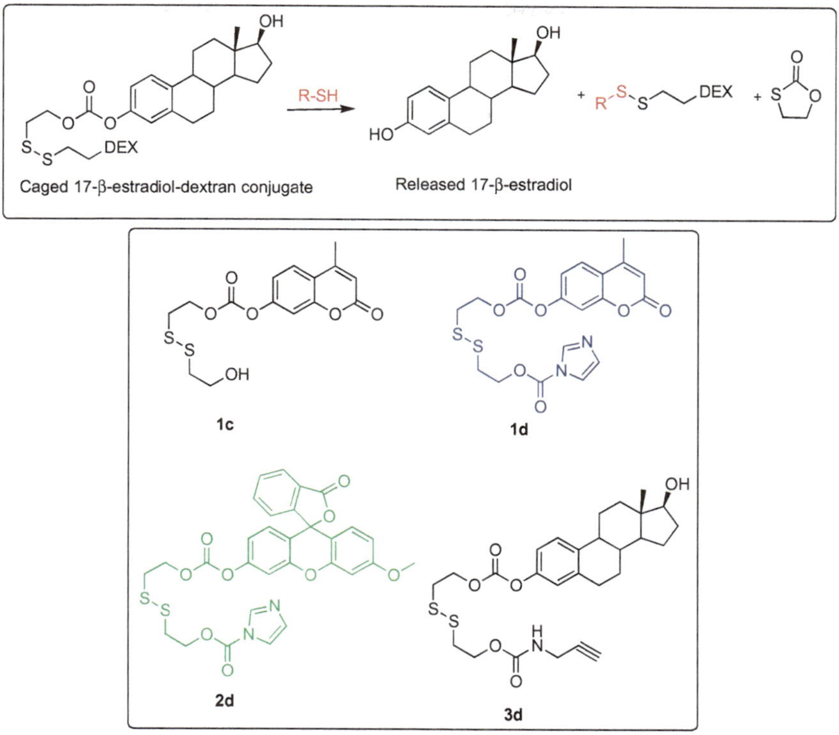

Fig. 2 Reaction scheme of thiol-triggered uncaging of 17-β-estradiol from dextran, synthetic intermediates used in the study are shown below.

5. To synthesize CD_{10}, dextran (1 g, 0.1 mmol, 10 kDa, Sigma) was dissolved in dry DMSO (10 mL). To this solution a dry DMSO solution of **1d** (1 mL, 0.02 mmol) and 4-dimethyl aminopyridine (DMAP, 0.04 mmol, Cat #851055, Sigma-Aldrich) was added.
6. The reaction mixture was stirred under nitrogen atmosphere for 12 h.
7. Dextran-4-MU conjugate (CD_{10}) was precipitated as white solid from DMSO by adding ice-cold acetone (10 mL, Sigma).
8. Precipitated CD_{10} was washed three times with diethyl ether (Sigma) and further dried in vacuum to obtain a dry white powder of CD_{10}.
9. CD_{10} (1 g) was subsequently dissolved in Milli-Q water to obtain a clear solution and dialysis of this solution was performed using a 3 kDa filter (MWCO, Millipore-Sigma, C82301) for 1 day at 4 °C to remove excess of unreacted **1c**.
10. The dialyzed solution was lyophilized to obtain pure CD_{10} as a white powder and this compound was stored at −80 °C for long-term use.

11. To synthesize FLD_{10} a similar protocol employed for CD_{10} was followed with minor modifications. In step 6, instead of compound **1d**, activated fluorescein derivative **2d** (Fig. 2) was used. Steps 7–11 was followed exactly as in the case of CD_{10} to obtain FLD_{10} as a white solid.

2.2 Synthesis of stimuli responsive dextran-estradiol conjugate

1. Thiol-activatable caged estradiol molecule (**3d**, Fig. 2) was synthesized using previously reported protocol (Veetil et al., 2018).
2. Triethylamine (35 μL, 0.25 mmol) was added to activated disulfide linker **3c** (see supporting information for chemical structure, Veetil et al., 2018) dissolved in dry dichloromethane.
3. The mixture was stirred for 10 min at room temperature and 17-β-estradiol (35 mg, 0.125 mmol, Cat #10006315, Cayman Chemicals) in CH_2Cl_2 was added dropwise to the reaction mixture.
4. The mixture was stirred at RT for 3 days under N_2 atm. to obtain the crude product, **3d** (Fig. 2).
5. Progress of the reaction over 3 days was monitored using TLC (hexane: ethylacetate, 1:1).
6. The crude reaction mixture after vacuum evaporation of DCM was purified using preparative silica TLC (hexane/ethyl acetate, 1:1) to obtain pure **3d**.
7. To synthesize ED_{10}, commercially available amino-dextran (10 mg, 10 kDa, Sigma) was first converted to corresponding azido-dextran (DEX-N3). To achieve this, amino-dextran (10 mg, 1 equivalent), dissolved in dry DMSO (Sigma xx, 100 μL) was reacted with NHS-PEG_4-azide (Thermo, Cat # 26130, 4 mg, 10 eq.) in DMSO for 2 h.
8. Excess of NHS-PEG_4-azide was removed using ultra centrifugation using a 3 kDa filter (MWCO, Millipore Sigma).
9. The resulting solution was lyophilized to obtain a white fluffy powder of DEX-N3. DEX-N3 was converted to ED_{10} using copper click-chemistry according to previously published protocol (Van Thienen et al., 2005).

2.3 Cargo encapsulation and characterization of DNA-icosahedron

1. Half icosahedron (VU_5 and VL_5) were synthesized using previously well-established protocols (Bhatia et al., 2009).

2. To achieve encapsulation of the cargo, VU_5 and VL_5 (3 μM each, 30 μL, in 50 mM phosphate buffer, pH 6.0) containing 1 mM solution of the respective cargo (CD_{10}, FLD_{10} or ED_{10}) was mixed by vertexing and incubated at 40 °C for 15 min. Incubation temperature was brought down to 20 °C with a rate of cooling 1 °C/min. The solution was subsequently transferred to 4 °C for further incubation up to 48 h.
3. Progress of the cargo encapsulation was monitored by withdrawing aliquots of samples from the mixture at definite intervals of time (10h) and to characterize using 0.8% agarose gel using EtBr staining.
4. Cargo encapsulated DNA-icosahedron ($\mathbf{I_{CD10}}$) was purified using centrifugation at 10,000 rpm at 4 °C using a 30 kDa Amicon filter (MWCO, Millipore) to remove excess of non-encapsulated dextran cargo.
5. Centrifugation in step 4 was repeated several times until no trace of dextran (e.g., CD_{10} and FLD_{10}) was observed using fluorimeter (Horiba Scientific, USA) or by using UV-VIS spectrophotometer (Shimadzu, Japan) for ED_{10} detection, 280 nm.
6. The centrifuged Dextran-small molecule conjugate was further purified by using size exclusion chromatography (SEC) using Shimadzu HPLC system equipped with size exclusion column (SEC-2000, Phenomenex, pore size 14.5 nm).
7. In a typical experiment, 1 mg of Dextran-small molecule conjugate (e.g., CD_{10}), dissolved in Milli-Q water was injected to a SEC-2000 column with a flow rate of 0.5 mL/min using 0.1% trifluoroacetic acid (TFA) in Milli-Q water as an eluent. Pure fractions of the dextran-small molecule conjugates were collected according the UV-activity of the HPLC-peaks at 260 nm (for DNA) and at 310 nm (for CD_{10}), respectively.
8. Further characterization of cargo encapsulated, HPLC-purified fractions of DNA-icosahedron was performed using Dynamic Light Scattering instrument (DynaPro NanoStar, Wyatt). HPLC-purified fractions were passed through a 0.22 μM filter (Millipore) to remove any dust particles. Filtered sample solutions were centrifuges at 10,000 rpm for 2 min and about 2 μL of the supernatant was used for DLS experiments. Samples were illuminated with laser wavelength at 658 nm, at sensitivity of 80% and a scattering angle set at 90 °C with 10 s acquisition time for data collection. Dynamics (Wyatt) software was used to plot the Gaussian distributions of hydrodynamic radii (RH). Percentage intensity observed for each sample was plotted against respective diameter values.

2.4 In-cell assay design

2.4.1 Time lapse imaging of calcium

1. Human umbilical vein endothelial cells (HUVECs) were purchased from ATCC (CRL-1730™) and were grown in 3.5 cm glass bottom dishes and maintained in Ham's F-12 medium (Gibco, Life technologies) supplemented with 100 units/mL of Penicillin, 100 μg/mL of streptomycin (Life technologies), 10% final concentration of heat inactivated fetal bovine serum (Gibco, Life technologies), 0.1 mg/mL of Heparin (Sigma) and 0.04 mg/mL of Endothelial cell growth supplement (Sigma). Cells were maintained at 37 °C in a humidified chamber at 5% CO_2 concentration.
2. Cells were washed with HBSS (HyClone, GE) followed by starvation in serum free Ham's F-12 (Gibco, Life technologies) medium containing 100 units/mL of Penicillin, 100 μg/mL of streptomycin for 24 h.
3. Cells were washed with HBSS followed 40 min incubation with 10 μM calcium sensitive fluorophore Fluo-4, AM (Thermo Fisher Scientific, #F-14201, 1 mM stock in DMSO). Subsequently, cells were washed three times with HBSS to remove excess Fluo-4, AM.
4. Cells were then treated with HBSS containing either E_2 (100 nM) or ED_{10} (100 nM) or I_{ED10} (1 μM) in the presence or absence of L-Cysteine (1 mM, Cat #C7352).
5. Care was taken not to exceed the final concentration of DMSO >1% at any time of the experiment. Additionally, 1% DMSO was used as a negative control in untreated cells.
6. Time-lapse fluorescence images were acquired on IX81 wide-field microscope (Olympus) using 40 × oil objective with 450–490 nm excitation filter and 515–586 nm emission filter (FITC/GFP filter cube). Images were recorded for 1 min with 100 ms of time delay between frames.

2.4.2 Time lapse imaging of NO

1. Repeat steps 1 and 2 from Section 2.4.1.
2. Cells were washed with HBSS followed loading of nitric Oxide (NO) sensitive fluorophore DAF-2 diacetate (Cayman, Cat # 85165, 1 mM DMSO stock). For DAF-2 loading in HUVECs, 10 μM of the probe was treated to the cells for 1 h at 37 °C in HBSS. Cells were then washed three times with HBSS to remove unloaded DAF-2.
3. Cells were then treated with HBSS containing either E_2 (100 nM) or ED_{10} (100 nM) or I_{ED10} (1 μM) in the presence or absence of Cysteine (1 mM).

4. Care was taken not to exceed the final concentration of DMSO >1% at any time of the experiment. Additionally, 1% DMSO was used as a negative control in untreated cells.
5. Confocal images were recorded on a SP8 microscope (Leica, 63×, 1.3 NA, white light laser) using 488 nm excitation and 510 nm emission with acousto-optical tunable filters. Images were recorded every 3 min for 18 min.

2.5 Imaging analysis

1. Images obtained from both IX81 and SP8 microscopes were analyzed using Fiji, Fiji is a distribution of NIH open source software ImageJ.
2. Background intensities for DAF-2 and Fluo-4 were recorded from control dishes (1% DMSO) and it is subtracted from all the images before analysis.
3. Cell outlines were drawn on all images keeping bright field images as reference, these cell outlines were later taken as ROIs to calculate mean intensities in FITC/GFP channels (Fluo-4 or DAF-2) from individual cells.
4. Normalized mean intensities were plotted against time for time lapse images or against different treatment groups for stead state imaging in Origin 2018 (origin Lab).

3. Summary

In summary, we described a detailed protocol using DNA-icosahedron to study steroid-induced rapid singling events in cells. We have described synthesis of stimuli responsive bioactive cargos, their encapsulation within DNA-icosahedron, in vitro and in-cell assays. Encapsulation and release of 17-β-estradiol from DNA-icosahedron offers a pristine technology to investigate steroid signaling without causing premature activation of the cognate receptors. This technology also has the advantage of controlled release, which enable the user to spatiotemporally investigate signaling events. This technology can be generalized to a wide-variety of small molecules (e.g., neurotransmitters, agonists, antagonists, lipid molecules, etc.) to investigate various rapid singling events.

Acknowledgments

We thank the Integrated Light Microscopy facility and BioPhysics core facility at the University of Chicago. This work was supported by a research grant from the University of Chicago Women's Board; a Pilot and Feasibility award from an NIDDK center grant no. P30DK42086 to the University of Chicago Digestive Diseases Research Core Center; the Chicago Biomedical Consortium, with support from the Searle Funds at The Chicago Community Trust, C-084; and University of Chicago start-up funds to Y.K.

References

Bhatia, D., Arumugam, S., Nasilowski, M., Joshi, H., Wunder, C., Chambon, V., et al. (2016). Quantum dot-loaded monofunctionalized DNA icosahedra for single-particle tracking of endocytic pathways. *Nature Nanotechnology*, *11*, 1112–1119.

Bhatia, D., Mehtab, S., Krishnan, R., Indi, S. S., Basu, A., & Krishnan, Y. (2009). Icosahedral DNA nanocapsules by modular assembly. *Angewandte Chemie (International Ed. in English)*, *48*, 4134–4137.

Cato, A. C. B., Nestl, A., & Mink, S. (2002). Rapid actions of steroid receptors in cellular signaling pathways. *Science's STKE: Signal Transduction Knowledge Environment*, *2002*, re9.

Chen, J. H., & Seeman, N. C. (1991). Synthesis from DNA of a molecule with the connectivity of a cube. *Nature*, *350*, 631–633.

Chow, R. W. Y., Handelsman, D. J., & Ng, M. K. C. (2010). Minireview: Rapid actions of sex steroids in the endothelium. *Endocrinology*, *151*, 2411–2422.

Daufeldt, S., Lanz, R., & Alléra, A. (2003). Membrane-initiated steroid signaling (MISS): Genomic steroid action starts at the plasma membrane. *The Journal of Steroid Biochemistry and Molecular Biology*, *85*, 9–23.

Goodman, R. P., Schaap, I. A. T., Tardin, C. F., Erben, C. M., Berry, R. M., Schmidt, C. F., et al. (2005). Rapid chiral assembly of rigid DNA building blocks for molecular nanofabrication. *Science*, *310*, 1661–1665.

Pedram, A., Razandi, M., Aitkenhead, M., Hughes, C. C. W., & Levin, E. R. (2002). Integration of the non-genomic and genomic actions of estrogen. Membrane-initiated signaling by steroid to transcription and cell biology. *The Journal of Biological Chemistry*, *277*, 50768–50775.

Shih, W. M., Quispe, J. D., & Joyce, G. F. (2004). A 1.7-kilobase single-stranded DNA that folds into a nanoscale octahedron. *Nature*, *427*, 618–621.

Surana, S., Bhatia, D., & Krishnan, Y. (2013). A method to study in vivo stability of DNA nanostructures. *Methods*, *64*, 94–100.

Van Thienen, T. G., Lucas, B., Flesch, F. M., van Nostrum, C. F., Demeester, J., & De Smedt, S. C. (2005). On the synthesis and characterization of biodegradable dextran nanogels with tunable degradation properties. *Macromolecules*, *38*, 8503–8511.

Veetil, A. T., Chakraborty, K., Xiao, K., Minter, M. R., Sisodia, S. S., & Krishnan, Y. (2017). Cell-targetable DNA nanocapsules for spatiotemporal release of caged bioactive small molecules. *Nature Nanotechnology*, *12*, 1183–1189.

Veetil, A. T., Jani, M. S., & Krishnan, Y. (2018). Chemical control over membrane-initiated steroid signaling with a DNA nanocapsule. *Proceedings of the National Academy of Sciences of the United States of America*, *115*, 9432–9437.

CHAPTER EIGHT

Imaging nascent transcription in wholemount vertebrate embryos to characterize zygotic genome activation

Hui Chen[a], Matthew C. Good[a,b,*]

[a]Department of Cell and Developmental Biology, Perelman School of Medicine, University of Pennsylvania, Philadelphia, PA, United States

[b]Department of Bioengineering, School of Engineering and Applied Science, University of Pennsylvania, Philadelphia, PA, United States

[*]Corresponding author: e-mail address: mattgood@pennmedicine.upenn.edu

Contents

Abstract

A major event in early embryo development is the awakening of the embryonic genome, a process of large-scale transcriptional induction termed zygotic genome activation (ZGA). To understand how ZGA is controlled temporally and spatially, tools are required to image and quantify nascent transcription in wholemount embryos. In this chapter, we describe a metabolic labeling approach that leverages 5-ethynyl uridine (5-EU) incorporation into newly transcribed RNAs. Subsequently, click chemistry is used to conjugate these nascent transcripts to fluorophores for wholemount confocal imaging or biotin for RNA sequencing. Such an approach facilitates direct visualization of the

Methods in Enzymology, Volume 638
ISSN 0076-6879
https://doi.org/10.1016/bs.mie.2020.03.002

global transcriptional state of each cell during early embryogenesis and provides a spatial map of gene expression activity. We describe this procedure for imaging nascent transcription in a vertebrate embryo *Xenopus laevis*, and use it as our model the onset of large-scale ZGA. Unlike cell culture systems in which 5-EU can be added to the media, metabolic labeling in *Xenopus* embryos requires microinjection in one-cell or two-cell stage embryos. This method is a powerful tool to quantify the nascent transcriptome at a single-cell level and to dissect mechanisms that control ZGA. We propose that this methodology can be applied broadly in other embryonic systems, and demonstrate the feasibility using zebrafish cleavage stage embryos. Finally, we demonstrate how to sequence the nascent transcriptome via 5-EU incorporation and separation of zygotic vs maternal RNAs. Altogether, our generalizable methodology will facilitate new insights into gene regulation and spatial patterning of ZGA during early embryogenesis.

1. Introduction

Following fertilization, a vertebrate embryo undergoes autonomous cell divisions regulated by maternal factors loaded in the egg and in many systems the zygotic genome in dormant. After a define period, the embryo initiates large-scale transcriptional activation as part of the maternal-to-zygotic transition (MZT) (Schier, 2007; Vastenhouw, Cao, & Lipshitz, 2019). The induction of the embryonic genome is termed zygotic genome activation (ZGA) which is essential for metazoan embryo development (Jukam, Shariati, & Skotheim, 2017). The timing for ZGA is tightly controlled within a species; however, it varies dramatically among species. The period following fertilization until ZGA onset ranges from several hours in zebrafish and *Xenopus* embryos to days for human embryos (Lee, Bonneau, & Giraldez, 2014). Despite extensive research on ZGA (Schulz & Harrison, 2019), how an embryo coordinates the spatial and temporal onset of ZGA in blastomeres remains a fundamental question in developmental biology.

A challenge in characterizing ZGA is quantitative measurement of nascent zygotic gene expression in wholemount embryos. Various methods have been successfully used to detect expression of zygotic genes during early development of multiple model organisms. Generally, these methods can be divided into two major categories—sequencing-based and imaging-based approaches. One major method is based on detecting zygotic gene expression via gene profiling. Traditionally, this has been carried out by sequencing RNA isolated from entire embryos at different stages of development for many species, including human (Gao et al., 2018;

Liu et al., 2019), mouse (Lu et al., 2016), zebrafish (Aanes, Collas, & Alestrom, 2014; White et al., 2017), *Xenopus* (Gentsch, Owens, & Smith, 2019). An advantage of sequencing-based methods is that they are sensitive and can detect many, if not all, zygotic genes at the same time, which is helpful for characterizing gene expression patterns at different stages of development. However, one significant shortcoming is that the expression level of each gene is derived from a population average of many cells and thus lacks information on cell-to-cell variation. Furthermore, these methods sacrifice a spatial view of ZGA across embryonic blastomeres. Separately, metabolic labeling of nascent transcripts can be performed using uridine analogs such as 4-thiouridine (4-SU) and 5-ethynyl uridine (5-EU), as has been done in *Drosophila* (Kwasnieski, Orr-Weaver, & Bartel, 2019), zebrafish (Chan et al., 2019; Heyn et al., 2014) and *Xenopus* (Gentsch et al., 2019). Moreover, single-cell RNA-seq (scRNA-seq) has been performed on early embryos from human and mouse (Xue et al., 2013), and for embryos at relatively late stages of development from zebrafish (Farrell et al., 2018; Wagner et al., 2018) and *Xenopus* (Briggs et al., 2018). Although spatial relationships between individual cells can be inferred by computational reconstruction on known gene expression patterns, a spatial map cannot be built with high confidence for an embryo containing thousands of cells (Satija, Farrell, Gennert, Schier, & Regev, 2015). Additionally, the large sizes of blastomeres in early stages of embryos are poorly suited for scRNA-seq technologies.

The second major category of methods for detecting zygotic gene expression is based on imaging the expression of single gene or a handful of genes. Expression in fixed embryonic cells can be detected by single-molecule fluorescent in situ hybridization (smFISH) using gene-specific fluorescent probes and imaging. Such strategies have been successfully used in early embryos from *Drosophila* (Little & Gregor, 2018; Trcek, Lionnet, Shroff, & Lehmann, 2017), zebrafish (Stapel, Zechner, & Vastenhouw, 2017) and mouse (Ranisavljevic, Okamoto, Heard, & Ancelin, 2017; Xie, Timme, & Wood, 2018). However, such approaches are difficult to scale up to large, wholemount embryos. Separately, dynamics of gene transcription can be imaged live using RNA containing MS2 sequences that bind to MS2 coat protein-green fluorescent protein (GFP). Such strategies work well for single genes across narrow depths of cytoplasmic volume, such as in the syncytium of *Drosophila* embryos (Garcia, Tikhonov, Lin, & Gregor, 2013) and cortical cells in zebrafish (Campbell, Chao, Singer, & Marlow, 2015). In addition, live imaging using a catalytically dead Cas9 (dCas9) protein fused to GFP that recognizes target RNA through base pairing with a guide RNA (gRNA) has been used to detect transcription of zygotic genes expressed at a high level in early zebrafish embryos (Chan et al., 2019). In general, a limitation of these single-gene based methods is that they do not provide a global readout of transcriptional activity and live imaging can only be performed with a subset of cells and not whole embryos, particularly large vertebrate embryo models.

We developed a method to visualize and quantify large-scale zygotic transcription at the single-cell level across entire large vertebrate embryos using *Xenopus laevis* a model (Chen, Einstein, Little, & Good, 2019). This approach of metabolically labeling nascent zygotic transcripts using 5-ethynyl uridine (5-EU) leverages click chemistry to conjugate a fluorophore to EU-RNA for confocal imaging of wholemount embryos. The in situ imaging of transcripts in embryos maintains precise spatial information at cellular and tissue level and allows detecting the integrated transcriptional activity of all zygotic genes in each single cell. Such a wholemount strategy reveals spatial and temporal patterns of ZGA regulation that are often missed by other methods. We further verified this methodology for detecting nascent transcription to image large-scale ZGA in zebrafish embryos, suggesting the broad applicability of this method to other embryonic models. Importantly, our approach is also compatible with RNA-seq for characterization of the nascent transcriptome. Altogether, this methodology overcomes a number of limitations of previous techniques and offers a new tool to interrogate patterns of nascent transcription in embryos such as during embryonic genome activation.

2. Protocol

2.1 Labeling nascent transcripts in early embryogenesis using 5-ethynyl uridine (5-EU)

Various studies across multiple model organisms have shown that uridine analogues, e.g., 5-bromouridine (BrU) (Aoki, Worrad, & Schultz, 1997; Kageyama, Nagata, & Aoki, 2004) and 4-thioluridine (4-SU) (Heyn et al., 2014; Heyn & Neugebauer, 2017), become phosphorylated upon delivery to embryonic blastomeres and subsequently incorporate into newly transcribed RNAs. Thus, these metabolic labeling approaches have been used to measure nascent transcripts during early development. However, the requirement of an antibody to detect BrU and the presence of endogenous thiols in embryos limit the utility of 5-BrU and 4-SU for imaging embryonic genome activation at the single-cell level in large wholemount embryos.

Recently, another uridine analog 5-ethynyl uridine (5-EU) was shown to overcome the limitations noted above for labeling newly transcribed RNA in cultured cells (Jao & Salic, 2008). 5-EU is readily taken up from the medium by living cells and after phosphorylation becomes 5-ethynyl uridine triphosphate (5-EUTP), which is specifically incorporated into newly made RNAs but not DNA. The alkyne group in 5-EU is not naturally occurring and thus it can be very selectively and efficiently reacted with azides by a copper (I)-catalyzed alkyne-azide cycloaddition reaction, termed a click reaction (Jao & Salic, 2008). Additionally, this specific reaction ensures that background reactivity is very low. Nascent EU-RNA can be conjugated to fluorescent azides for imaging transcriptional activity at the onset of ZGA in individual cells of wholemount embryos via confocal imaging (Chan et al., 2019; Chen et al., 2019; Jao & Salic, 2008). Alternatively, the nascent transcriptome can be sequenced via biotinylation of EU-RNA and streptavidin bead purification (Chan et al., 2019; Chen et al., 2019; Kwasnieski et al., 2019).

We have developed a methodology to detect nascent zygotic transcripts in single cells of wholemount *Xenopus laevis* cleavage stage embryos (Fig. 1A) (Chen et al., 2019). The steps are broken down as follows, corresponding to the subsections of this protocol: (1A) microinjection of 5-EU into one-cell embryos and (1B) fixing blastula embryos at developmental times corresponding to the onset of large-scale, (2) performing a click chemistry reaction to conjugate a fluorophore to nascent EU-RNA, (3) additional DNA dye and antibody staining, and (4) confocal imaging of cleared wholemount embryos. The use of metabolic labeling with 5-EU allows for selective detection of nascent transcripts within a larger pool of maternal RNAs, which are highly abundant in the early embryo (Fig. 1B). Although embryos are fixed to allow for conjugating fluorophores with click chemistry, the rapid divisions of embryonic blastomeres enable the transcriptional output of a cell to be approximated by EU-RNA signal in the nucleus. Upon synthesis, this nascent EU-RNA enriches in the interphase nucleus (Fig. 1C) due to insufficient time for EU-RNA to be exported to cytoplasm; the interphase duration is approximately 20 min during cleavage stage. At mitosis and nuclear envelop breakdown, this RNA is dispersed throughout cell (Fig. 1C). This feature enables quantitation the transcriptional activity in the current cell cycle by integrating total EU-RNA signal in the nucleus. Separately, transcriptional history can be calculated by measuring the total accumulated EU-RNA in the cell. In the case of *Xenopus laevis*, wholemount embryos fixed at 5 h post-fertilization (hpf) (around embryonic

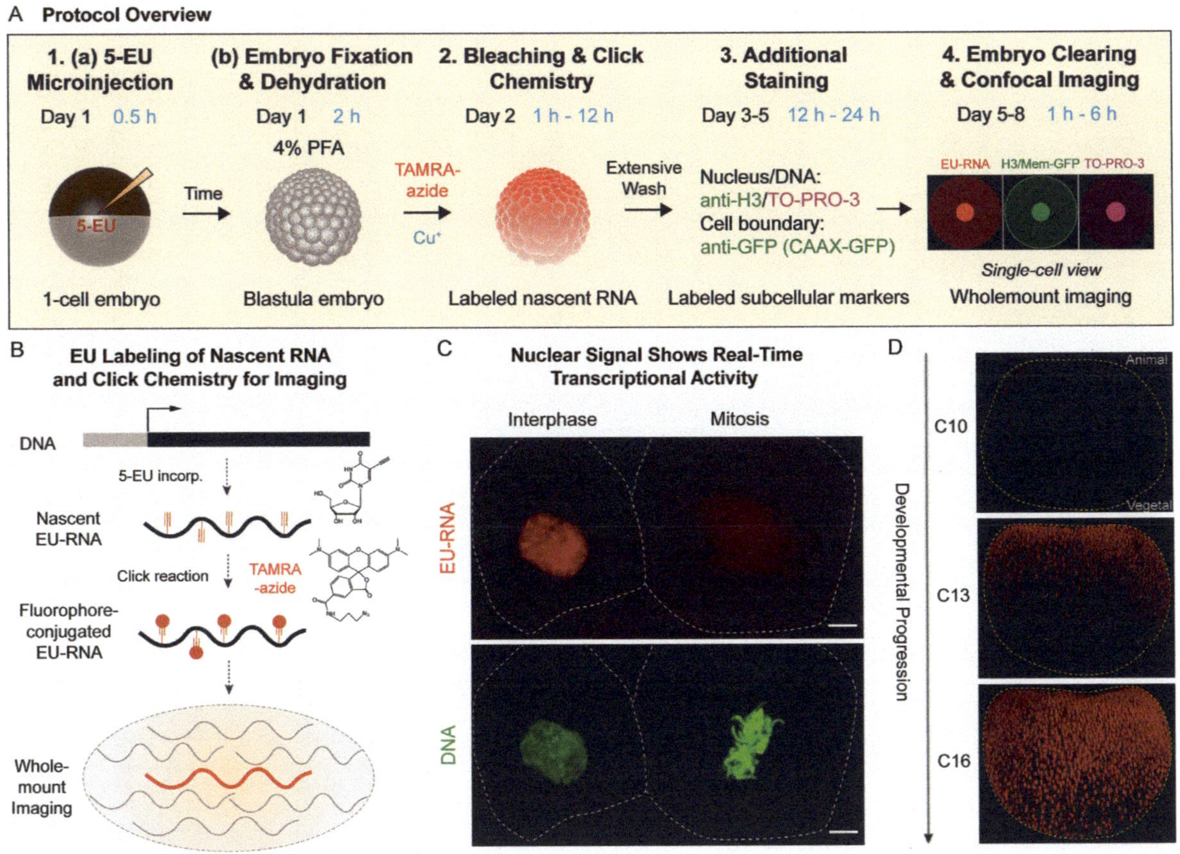

Fig. 1 See legend on opposite page.

cleavage 10 or C10), 7 hpf (around C13), 9 hpf (around C16) display a spatially graded pattern of single-cell transcriptional activation (Fig. 1D) (Chen et al., 2019).

For EU-RNA imaging in any embryonic system, controls must be performed to demonstrate the sensitivity and specificity of the nascent gene expression. Although 5-EU can be taken up by cultured cells in media, it is not readily absorbed by *Xenopus laevis* embryos and therefore microinjection is required to deliver 5-EU. The final concentration of 5-EU inside the embryo is extremely critical as low concentrations of 5-EU may reduce the sensitivity of fluorescent labeling for imaging (Fig. 2A and B) and high concentrations of 5-EU are toxic to normal developmental progression. We recommend always performing a control with co-microinjection of 5-EU with α-amanitin, an inhibitor of RNA polymerase II, to ensure that nuclear EU-RNA signal is selective, representing bona-fide nascent transcripts (Fig. 2C and D). The EU-labeling method is sensitive enough to image burst of transcription—seen as spots—at early stages of ZGA onset (Fig. 3). For analysis of large-scale genome activation, we often set a threshold minimum amount of transcript accumulated in the blastomere nucleus that corresponds to more widespread zygotic transcription (e.g., see nuclei at C12 in Fig. 3).

Fig. 1 Method overview for EU-RNA imaging. (A) Schematic summary for labeling and visualizing nascent RNA transcripts in wholemount embryos during early development of *Xenopus laevis*. Numbering corresponds to the protocol sections in main text. 1A: microinjection of one-cell embryos with 5-ethynyl uridine (5-EU); 1B: fixation of blastula embryos in 4% paraformaldehyde and dehydration in methanol for long-term storage; 2: following bleaching, nascent RNAs in an embryo conjugated to a fluorescent azide using a click reaction; 3: Staining DNA is required for image segmentation, which can be further augmented with additional immunostaining; 4: Clearing embryos for confocal imaging. Example nascent EU-RNA (red), nucleus/cell boundary (green) and DNA (magenta). (B) Schematic of EU-RNA labeling of early embryogenesis: injected 5-EU converts to 5-EUTP and incorporates into nascent but not maternal transcripts; TAMRA-azide is conjugated to EU-RNA by click chemistry. Nascent zygotic transcripts can be visualized; maternal RNAs are unlabelled. (C) Transcriptional output is measured by accumulated EU-RNA in nucleus. Example of neighboring interphase cell and a mitotic cells, in an embryo at 8–9 h post-fertilization (hpf). Scale bar, 5 μm. Dashed white lines indicate cell boundary. (D) Nascent EU-RNA accumulation as a function of embryo development for wholemount *Xenopus laevis* embryos. Embryonic cleavage 10 (C10) at ~5 hpf, prior to large-scale ZGA, embryonic cleavage C13 at ~7–7.5 hpf with partial onset of large-scale ZGA, and embryonic cleavage C16 at 9 hpf showing complete spatial onset of ZGA across embryo. Orientation: animal pole up; vegetal pole down.

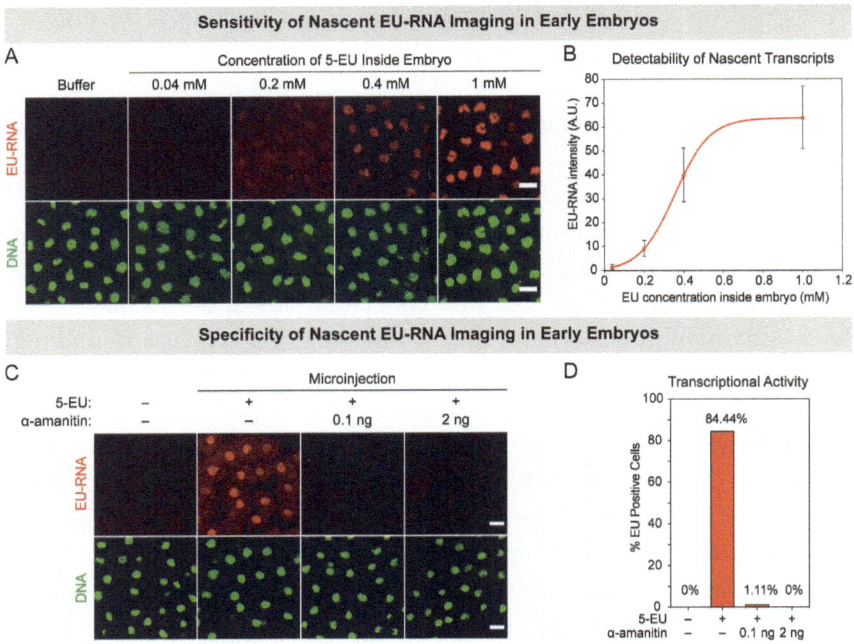

Fig. 2 Sensitivity and specificity of nascent EU-RNA imaging in early embryogenesis. (A and B) Sensitivity: dose-dependence of EU-RNA signal intensity of 5-EU concentration inside embryo. (A) *Xenopus* embryos were microinjected with buffer or indicated final concentrations of 5-EU, fixed at 10 hpf and processed to image nascent EU-RNA (red) and DNA staining (green). (B) Quantification of nuclear EU-RNA intensity (A.U., arbitrary units) after cytoplasmic background subtraction. Mean ± standard deviation, with a nonlinear fit for visualization purposes. N > 20 cells for each group. (C and D) Evaluating specificity of EU-RNA signal via α-amanitin co-injection. (C) *Xenopus* embryos microinjected with 5-EU alone or co-microinjected with 5-EU and 0.1 ng (to inhibit RNAPII) or 2 ng of α-amanitin (to inhibit RNAPII and RNAPIII). Embryos fixed at 8 hpf and 45 mpf (minutes post-fertilization). (D) Quantification of percentage of EU-RNA positive cells whose nuclear EU-RNA intensity exceeds am empirical threshold. N > 60 cells for each group. Scale bars, 20 μm.

In this protocol, we followed previous methods for fixing and bleaching *Xenopus* embryos (Lee, Kieserman, Gray, Park, & Wallingford, 2008; Wallingford, 2010). Fixation is required to preserve the integrity of RNAs in embryos. The choice of fixative and buffer is critical because they can impact the variability of the results. We found 4% paraformaldehyde (PFA) in 1 × MEM, a common fixative solution used in *in situ* hybridization (IHS), works most efficiently for downstream click reaction, DNA staining and immunostaining for *Xenopus laevis* blastula embryos. Bleaching is critical

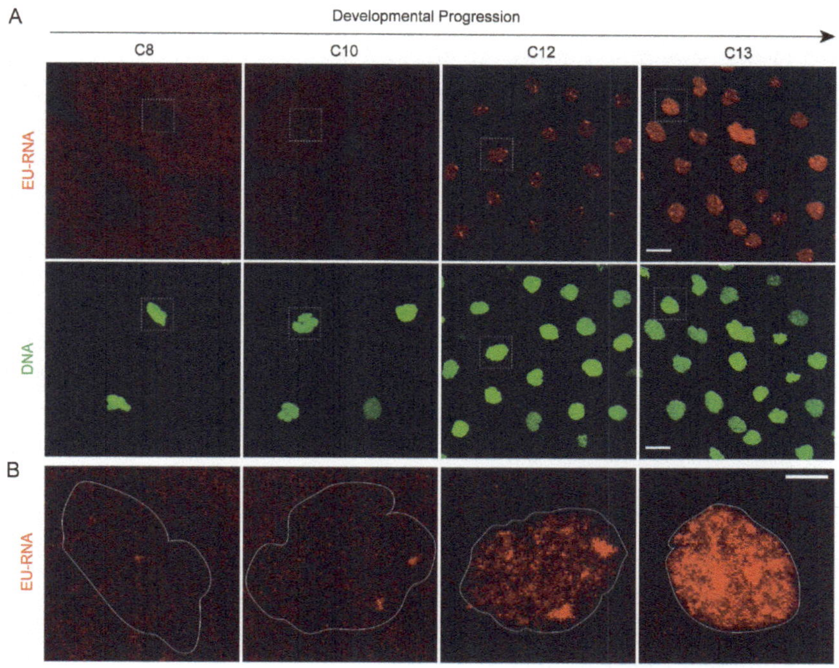

Fig. 3 Temporal onset of ZGA and levels of nascent transcription. (A) *Xenopus* embryos microinjected with 5-EU and fixed at timepoints before and surrounding ZGA. EU-RNA (red) and DNA (green). Maximum intensity projections of confocal stacks at 63 × magnification visualizing the embryo animal pole. Scale bars, 20 μm. (B) Zoomed, enlarged view of EU-RNA in nuclei in the dashed white boxes from panels in (A). Distinct levels of nascent transcription are apparent from weak diffuse signal early, to transcriptional foci, to nascent transcripts filling nucleus. Outline of nucleus in white. Scale bar, 5 μm.

for eliminating the light-absorbing pigments present in the animal pole. Lastly, clearing embryos in BABB, a mixture of benzyl alcohol (BA) and benzyl benzoate (BB), drastically improves image quality for many types of embryos.

2.1.1 Materials

1. Fertilized *Xenopus laevis* eggs. We obtained *Xenopus* embryos by in vitro fertilization (IVF). All animal experiments were performed according to the Animal Use Protocol approved by the University of Pennsylvania Animal Care and Use Committee.
2. 4-(2-Hydroxyethyl)piperazine-1-ethanesulfonic acid (HEPES) (Sigma-Aldrich, cat. no. H3375)
3. Ethylenediaminetetraacetic acid (EDTA) (Thermo Fisher Scientific, cat. no. BP120–500)

4. Ethylenebis (oxyethylenenitrilo)tetraacetic acid (EGTA) (Thermo Fisher Scientific, cat. no. AC409915000)
5. Potassium chloride (KCl) (Thermo Fisher Scientific, cat. no. P217500)
6. Magnesium chloride ($MgCl_2$) (Thermo Fisher Scientific, cat. no. M33-500)
7. Magnesium sulfate ($MgSO_4$) (Thermo Fisher Scientific, cat. no. M65-500)
8. Calcium chloride ($CaCl_2$) (Sigma-Aldrich, cat. no. C5080)
9. MOPS (Thermo Fisher Scientific, cat. no. AC172631000)
10. Sodium hydroxide (NaOH) (Thermo Fisher Scientific, cat. no. S318-1)
11. Hydrochloric acid (HCl) (Thermo Fisher Scientific, cat. no. A144SI212)
12. Bovine serum albumin (BSA) (Sigma-Aldrich, cat. no. A3059)
13. Triton X-100 (Sigma-Aldrich, cat. no. 10789704001)
14. L-cysteine (Sigma-Aldrich, cat. no. C7352)
15. Ficoll® 400 (Sigma-Aldrich, cat. no. F4375)
16. RNase-free water (Thermo Fisher Scientific, cat. no. BP561-1)
17. Methanol (Thermo Fisher Scientific, cat. no. A4524)
18. Methanol, anhydrous (Sigma-Aldrich, cat. no. 322415)
19. Dimethyl sulfoxide (DMSO), anhydrous (Thermo Fisher Scientific, cat. no. D12345)
20. Hydrogen peroxide (Sigma-Aldrich, cat. no. H1009)
21. Formamide (Thermo Fisher Scientific, cat. no. AC181090010)
22. 20 × SSC (Mediatech, cat. no. MT46-020-CM)
23. 5-Ethynyl uridine (5-EU) (Thermo Fisher Scientific, cat. no. E10345)
24. 16% Paraformaldehyde (PFA) (EMS, cat. no. 15710-S)

2.1.2 Equipment

1. Dissection stereomicroscope (Leica, model no. MZ6)
2. PLI-100 picoliter microinjector (Medical Systems Corp., Greenvale, NY)
3. Glass needle (Sutter Instrument, cat. no. BF100-58-10)
4. Micropipette puller (Sutter Instrument, model no. P-87)
5. Glass Petri dish, 6 cm (Corning, cat. no. 3160-60)
6. Plastic transfer pipette (Thermo Fisher Scientific, cat. no. 13-711-9AM)
7. Glass chamber for microinjection
8. Glass vials with caps, 2 mL (Thermo Fisher Scientific, cat. no. C4010-17)
9. Tube rotator (Thermo Fisher Scientific, cat. no. 88881001)

2.1.3 Protocol

Before starting the protocol, prepare the following solutions required for this section.

1. *0.1× Marc's Modified Ringer's (MMR)*: To prepare 0.1 × MMR, dilute 1 part of 20 × MMR (100 mM HEPES pH 7.8, 2 mM EDTA, 2 M NaCl, 40 mM KCl, 20 mM $MgCl_2$, and 40 mM $CaCl_2$) in 199 parts of Milli-Q® water. Store at room temperature up to 1 month.
2. *Dejelly solution*: To prepare dejelly solution, mix 1 g of L-cysteine and 200 μL of 10 N NaOH and bring up to 30 mL with Milli-Q® water. Prepare fresh solution before experiment.
3. *3% Ficoll (wt/vol) in 0.5× MMR*: To prepare 3% Ficoll/0.5 × MMR, mix 10 mL of 15% Ficoll (15% Ficoll® is prepared by dissolving 37.5 g of Ficoll in 250 mL Milli-Q water followed by autoclave) and 1.25 mL of 20 × MMR and bring up the final volume to 50 mL with Milli-Q water. Store at 4 °C for up to 1 month.
4. *1× TBS, pH 7.6*: To prepare 10 × TBS, dissolve 2.42 g of Tris base and 8.77 g of NaCl in 50 mL of RNase-free water. Adjust pH to 7.6 with HCl, filter sterilize and store at 4 °C. To prepare 1 × TBS (pH 7.6), add 1 part of 10 × TBS (pH 7.6) and 9 parts of RNase-free water. Store at 4 °C for up to 6 months.
5. *1× TBST, pH 7.6*: To prepare 1 × TBST, add 50 μL Triton X-100 to 50 mL 1 × TBS (pH 7.6) prepared by RNase-free water to generate a 0.1% (vol/vol) detergent solution. Store at 4 °C for up to 2 weeks.
6. *4% PFA/1× MEM*: To prepare 4% PFA/1 × MEM, mix 1 vial (10 mL) of 16% PFA stock and 4 mL of 10 × MEM (1000 mM MOPS pH 7.4, 20 mM EGTA, and 10 mM $MgSO_4$) and add RNase-free water to 40 mL. Prepare fresh fixative before experiment.
7. *Bleaching solution*: Bleaching solution is 0.5 × SSC containing 5% (vol/vol) formamide and 2% (vol/vol) H_2O_2. To prepare 10 mL of bleaching solution, mix 250 μL of 20 × SSC, 500 μL of formamide, 670 μL of H_2O_2 and 8.58 mL of RNase-free water. Prepare fresh solution right before use.
8. *5-Ethynyl uridine (5-EU) (100 mM stock)*: To prepare 100 mM 5-EU stock, dissolve 5 mg of 5-EU in 186.4 μL of 1 × TBS (pH 7.6) prepared by RNase-free water. Mix thoroughly and make aliquots of 2 μL. Store at −80 °C for up to 12 months.

2.1.3.1 Microinjection of 5-EU in embryos

1. Preparation of embryos (see *Note 1*): Follow a standard protocol for in vitro fertilization (IVF) of eggs in a glass dish and subsequent development in 0.1 × MMR (Sive, Grainger, & Harland, 2007). At ~30 min post-fertilization (mpf), dejelly the embryos. Pour off 0.1 × MMR and

add the freshly prepared dejelly solution to cover the embryos. Rinse briefly and discard the dejelly solution. Add the dejelly solution again and incubate for 3–5 min at room temperature; gently swirling several times. Stop the incubation once the jelly coats are obviously detached from embryos by pouring off the dejelly solution and replacing with 0.1 × MMR. Wash the embryos with 0.1 × MMR five times. Keep the dejellied embryos in 0.1 × MMR for further development.

2. Transfer ~50 embryos (see *Note 2*) using plastic transfer pipettes to a glass chamber containing 3% Ficoll/0.5 × MMR, in preparation for microinjection. Place the embryos in the middle of the chamber.
3. EU microinjection calculation: Thaw an aliquot of 100 mM 5-EU stock on ice. Dilute the 100 mM 5-EU stock in 1 × TBS (pH 7.6) or any other buffer suitable for microinjection, based on three considerations: (i) embryo volume, (ii) desired EU concentration, (iii) stage of embryos. For example, the typical desirable maximum volume for microinjection in embryos of *Xenopus laevis* and zebrafish is 10 and 2 nL, respectively. Further, the average embryo volume for *Xenopus laevis* and zebrafish is ~1 and ~0.3 μL, respectively. The final concentration of 5-EU inside an embryo significantly affects the sensitivity of labeling nascent RNA (Fig. 2A and B). While increasing 5-EU concentration increases the detectability of EU-RNA in cells, microinjection of high concentrations of 5-EU is toxic to embryos and thus should be avoided. In our experience, to balance the detectability and toxicity the optimal range of 5-EU concentration inside embryo is 0.4–0.6 mM. A typical experiment for *Xenopus* is to microinject 5-EU to achieve a final concentration of 0.5 mM inside the embryo (see *Note 3*).
4. Fill needle and microinject embryos: Fill glass needle ~1 μL of appropriate 5-EU dilution, and microinject 10 nL into *Xenopus laevis* one-cell embryos using a picoliter microinjector. For more details follow a standard protocol for performing microinjections (Sive, Grainger, & Harland, 2010).
5. Incubate embryos: After microinjection, carefully transfer the embryos to a new glass dish containing 3% Ficoll/0.5 × MMR to stabilize injected embryos. Incubate for 1–2 h, then carefully transfer the embryos to a new glass dish with 0.1 × MMR to continue embryonic development. Note that elongated incubation with Ficoll may interfere with normal developmental timing.

2.1.3.2 Embryo fixation

6. Collect and fix 5-EU-microinjected embryos using a plastic transfer pipette to select ~10–20 embryos at appropriate stages. For *Xenopus laevis* we collect from 5 to 9 hpf at room temperature. To prevent excess buffer transfer, let embryos settle down to the bottom of tip via gravity (see *Note 4*). Gently transfer the embryos dropwise into 2-mL glass vials filled with 4% PFA/1 × MEM solution in a chemical hood. Cap vials tightly.
7. To complete fixation, rotate the glass vials at a low speed of 10 rotations per minute (rpm) for a minimum of 2 h at room temperature or for overnight at 4 °C.
8. Dehydrate embryos: After fixation, carefully remove the fixative in a chemical hood without disturbing the embryos. Fill the glass vials with 100% methanol and close the caps tightly. Rotate the glass vials for 5 min on a rotator and then replace with fresh methanol. Repeat for a total of at least three times to completely dehydrate the embryos.
9. Store the embryos at −20 °C for up to 6 months.

2.1.3.3 Bleaching embryos

10. Rehydrate embryos by replacing methanol in the glass vials sequentially with solutions of 75% methanol, 50% methanol, 25% methanol and 0% methanol in 0.5 × SSC. For each new solution, rotate the glass vials for 5 min at low speed. Wash embryos with 0.5 × SSC at least two more times. Note that embryos must be completely rehydrated before proceeding.
11. Bleach embryos: Carefully remove 0.5 × SSC in the glass vials from the previous step. Add 2 mL of the bleaching solution, i.e., 5% (vol/vol) formamide and 2% (vol/vol) H_2O_2 in 0.5 × SSC, and close the caps of glass vials tightly. Lay down the glass vials in a box the surface of which is coated with aluminum foil. Place the box under bright light source from a bulb. Adjust the distance between the bulb and glass vials to ensure embryos receive sufficient light without overheating. Bleach embryos for 5–6 h (see *Note 5*).
12. SSC wash: Carefully remove the bleaching buffer from the glass vials by using pipettes (see *Note 6*) and make sure to avoid touching the embryos. Fill up the glass vials with 0.5 × SSC. Rotate for 5 min at low speed. Repeat 0.5 × SSC wash and rotation.

13. TBST wash: Carefully remove 0.5 × SSC and fill up glass vials with 1 ×TBST (pH 7.6). Rotate for 30 min at low speed. Carefully remove 1 × TBST (pH 7.6) and repeat washing embryos by rotation with 1 × TBST (pH 7.6) for a minimum total of 4 h; replacing 1 × TBST (pH 7.6) every 30 min. Note that the embryos must be thoroughly washed because residue from the bleaching solution can interfere with the efficiency of the click chemistry reaction.
14. TBS wash: Carefully remove 1 × TBST (pH 7.6) and fill up the glass vials with 1 × TBS (pH 7.6). Rotate the glass vials for 5 min at low speed. Carefully remove 1 × TBS (pH 7.6) and repeat wash. Finally, completely remove any residual 1 × TBS (pH 7.6) using a P10 pipette. Do so carefully to avoid causing damage to embryos. The embryos are ready for use in next steps but they can also be dehydrated in methanol and stored for later processing (see *Note* 7).

2.2 Click chemistry: Conjugation of fluorophores to nascent transcripts, EU-RNAs

In the following steps, the nascent EU-RNA transcripts are conjugated with fluorescent azide, such as tetramethylrhodamine (TAMRA)-azide, by a copper (I)-catalyzed alkyne-azide cycloaddition reaction, also known as a click reaction (Jao & Salic, 2008). Before starting this reaction, it is important that any residual bleaching solution should be completely removed by extensive washing. Although click reactions are usually fast (e.g., within 30 min) for fixed culture cells, the large size of *Xenopus* embryos requires extensive reaction time (e.g., up to 12 h) to completely label RNAs. Short incubation time (e.g., 1 h) lead to labeling only a cortical layer of cells. Alternatively, extended incubation (e.g., over 12 h) can lead to higher background due to nonspecific labeling. Thus, the optimal timing click reaction should be determined for any specific embryo and will likely scale with embryo size. Note that wash times after click reaction should also be extended in large embryos.

2.2.1 Materials

1. Tetramethylrhodamine (TAMRA)-azide (Abcam, cat. no. ab146486)
2. Copper sulfate ($CuSO_4$) (Sigma-Aldrich, cat. no. 61230)
3. Ascorbic acid (Sigma-Aldrich, cat. no. A7506)

2.2.2 Equipment

1. Variable speed nutator (Thermo Fisher Scientific, cat. no. S06622)

2.2.3 Protocol

Before starting the protocol, prepare the following solutions required in this section. Note that all following solutions and reagents should be ideally prepared using RNase-free conditions.

1. *Tris-HCl, pH 8.5 (1 M stock)*: To prepare Tris-HCl, pH 8.5 (1 M stock), dissolve 6 g of Tris base in 50 mL of RNase-free water. Adjust pH to 8.5 with HCl, filter sterilize and store at 4 °C for up to 6 months.
2. *TAMRA-azide (10 mM stock)*: To prepare 10 mM TAMRA-azide stock, dissolve 5 mg of TAMRA-azide in 975 μL anhydrous DMSO. Mix thoroughly and make aliquots of 10 μL. Store at −80 °C for up to 12 months.
3. *$CuSO_4$ (1 M stock)*: To prepare 1 mL of $CuSO_4$ (1 M stock), dissolve 160 mg of $CuSO_4$ in 1 mL RNase-free water. Mix thoroughly. Store at 4 °C for up to 12 months.
4. *Ascorbic acid (1 M stock)*: To prepare 1 mL of ascorbic acid (1 M stock), dissolve 176 mg of ascorbic acid in 1 mL RNase-free water. Mix thoroughly. Prepare fresh solution right before use.

2.2.3.1 Conjugating EU-RNA with fluorophore by click reaction

1. Prepare click reaction mix: Sequentially add the stock solutions of the following reagents to a 1.5 mL microcentrifuge tube at final concentrations of 100 mM Tris-HCl (pH 8.5), 1 mM $CuSO_4$, 25 μM TAMRA-azide, and 100 mM ascorbic acid. It is important that this reaction mix be made fresh each time; always add ascorbic acid in last step (see *Note 8*). For example, to prepare a reaction mix of 2 mL, sequentially add 1593 μL of RNase-free water, 200 μL of 1 M Tris-HCl (pH 8.5), 2 μL of 1 M $CuSO_4$, 5 μL of TAMRA-azide, and 200 μL of 1 M ascorbic acid. Gently mix well by pipetting up and down several times.
2. Click reaction: Add 200–500 μL click reaction mix to each glass vial containing embryos from the last step of previous section. Wrap each glass vial with aluminum foil to prevent light exposure. Incubate embryos for 12 h at room temperature on a nutator in the dark (see *Note 9*).
3. Stop reaction and briefly wash embryos. Carefully remove the click reaction mix and fill up the glass vials with 1 × TBST (pH 7.6). Wrap vials with aluminum foil and rotate embryos for 5 min at low speed in the dark. Repeat washing embryos with 1 × TBST (pH 7.6) once more.
4. Extensively wash embryos: Wash embryos with 1 × TBST (pH 7.6) for entire day at room temperature by rotating the glass vials on a rotator at low speed in the dark. Change 1 × TBST (pH 7.6) for every 1–2 h. A minimum of six washes is required (see *Note 10*). Extent of washing

can be gauged by decreasing coloration of embryo. For larger embryos, additionally wash with 1 × TBST (pH 7.6) overnight at 4 °C at low speed in the dark, and then wash again for an entire day, changing 1 × TBST (pH 7.6) for every 1–2 h.

2.3 Staining: Nuclear staining using DNA dyes and optional immunostaining

In this section, we describe staining of DNA and general steps for immunocytochemistry following click reaction for labeling EU-RNA in embryos. We adapt the methods from a previously described protocol (Lee et al., 2008). Staining of nuclear DNA with TO-PRO-3 facilitates nucleus segmentation in later steps. Note that complete washing from click reaction step is required before immunostaining and/or DNA staining with dyes; inadequate washing often leads poor staining.

2.3.1 Materials

1. Goat serum (Abcam, cat. no. ab7481)
2. Anti-Histone H3 antibody (Abcam, cat. no. ab1791)
3. Goat anti-rabbit Alex Fluor® 488 secondary antibody (Thermo Fisher Scientific, cat. no. A27034)
4. TO-PRO-3 (Thermo Fisher Scientific, cat. no. T3605)

2.3.2 Equipment

1. Variable speed nutator (Thermo Fisher Scientific, cat. no. S06622)

2.3.3 Protocol

Before starting the protocol, prepare the following solution required in this section.

Blocking solution: Blocking solution is 10% (vol/vol) goat serum and 0.2% (wt/vol) BSA in 1 × TBST (pH 7.6). To prepare 10 mL of blocking solution, add 1 mL of goat serum and 20 mg of BSA in 9 mL of 1 × TBST (pH 7.6). Prepare fresh solution right before use or store at −20 °C for up to 1 month.

If there is no immunostaining, directly jump to step 4 for DNA staining.

1. Block embryos: Carefully remove 1 × TBST (pH 7.6) from the glass vials and add 200–500 μL of the blocking solution (10% (vol/vol) goat serum and 0.2% (wt/vol) BSA in 1 × TBST, pH 7.6). Wrap the vials with aluminum foil and incubate for 2–3 h at room temperature on a nutator at low speed in the dark (see *Note 11*).

2. Primary antibody incubation: carefully remove the blocking solution from the glass vials using pipettes. Dilute an antibody (such as anti-histone H3 antibody at 1:1000) in the blocking solution. Add 200–500 μL of the diluted anti-histone H3 antibody to each glass vial for 10–20 embryos. Wrap the glass vials with aluminum foil and incubate overnight at 4 °C on a nutator at low speed in the dark.
3. Washing embryos: carefully remove the antibody solution from the glass vials and fill up with 1 × TBST (pH 7.6). Rotate embryos for 5 min then carefully remove 1 × TBST (pH 7.6). Repeat washing embryos with 1 × TBST (pH 7.6) once more. Wash embryos with 1 × TBST (pH 7.6) for all day at room temperature by rotating the glass vials on a rotator. Change 1 × TBST (pH 7.6) every 1–2 h. A minimum of six washes is required.
4. Secondary antibody incubation and/or DNA staining: Dilute a secondary antibody such as goat anti-rabbit Alex Fluor 488 secondary antibody and/or TO-PRO-3 DNA Stain at 1:500 in the blocking solution. Add 200–500 μL of the secondary antibody and/or TO-PRO-3 mix to each glass vial. Wrap the glass vials with aluminum foil and incubate embryos for overnight at 4 °C on at low speed in the dark.
5. Final extensive wash: carefully remove the secondary antibody and/or TO-PRO-3 mix from the glass vials and fill up with 1 × TBST (pH 7.6) to quickly rinse by rotating embryos for 5 min; repeat rinsing embryos with 1 × TBST (pH 7.6) once more. Make sure to wash embryos with 1 × TBST (pH 7.6) all day via rotation at room temperature. Change 1 × TBST (pH 7.6) every 1–2 h. A minimum of six washes is required.

For general troubleshooting of subsections on click chemistry and staining, see *Note 12*.

2.4 Confocal imaging of nascent transcription in cleared, wholemount embryos

The preparation of *Xenopus* embryos for confocal imaging has been described in detail (Wallingford, 2010). Due to their large size and yolk, *Xenopus* embryos are difficult to image without first clearing the embryos. The clearing reagent BABB, 1 part of benzyl alcohol (BA) mixed with 2 parts of benzyl benzoate (BB), is used for clearing thick tissues and works broadly for many types of embryos. To obtain best results, it is recommended to clear embryos for 24–48 h before imaging. Inadequate clearing may result in uneven signals in different regions of embryos while excessive clearing may cause the embryo to fragment or diminish fluorescent signals.

To image the same embryo from different sides, e.g., the animal pole and the vegetal pole of a *Xenopus* embryo, it is preferable that imaging chamber be able to be flipped without turning internal contents. To build such a chamber, we stick hollowed 3 M VHB™ tape on glass covers. The thickness of 3 M VHB™ tape is ~1.14 mm, which is very close to the average diameter of a *Xenopus* embryo (~1.2 mm). Embryos can be situated properly inside the chamber with desired orientation, i.e., animal pole on the top and vegetal pole at the bottom. After imaging one side, the chamber can be flipped by 180° and the embryo can be imaged again from the other side. For *Xenopus laevis* wholemount imaging, we use a laser scanning confocal microscope and 10 × air objective to fully image the entire embryo or a 25 × oil objective and zoom 0.6 × combined with image stitching.

2.4.1 Materials

1. Methanol (Thermo Fisher Scientific, cat. no. A4524)
2. Methanol, anhydrous (Sigma-Aldrich, cat. no. 322415)
3. Benzyl alcohol (Sigma-Aldrich, cat. no. 305197)
4. Benzyl benzoate (ACROS Organics, cat. no. 105860010)
5. *Clearing reagent (BABB)*: To prepare clearing reagent BABB, mix benzyl alcohol (BA) and benzyl benzoate (BB) at a ratio of 1:2 using glass transfer pipets. Store at room temperature for up to 2 weeks.

2.4.2 Equipment

1. Cover glass, No. 1, 22mm × 40 mm (VWR, cat. no. 48393048)
2. VHB™ Tape (3 M, cat. no. GPH-110GF)
3. Glass Pasteur pipets (VWR, cat. no. 14673-010)
4. Razor blade (Thermo Fisher Scientific, cat. no. 18-100-970)
5. Dissecting forceps (Thermo Fisher Scientific, cat. no. S08096)
6. Confocal microscopes (Zeiss LSM 710, Plan-Apochromat 10 ×/ numerical aperture (NA) 0.45 objective, Plan-Apochromat 25 ×/NA 0.8 and 63 ×/NA 1.4 immersion oil objectives)

2.4.3 Protocol

2.4.3.1 Clearing embryos

1. Dehydrate embryos: carefully remove 1 × TBST (pH 7.6) from the glass vials and fill up with 100% methanol. Rotate embryos for 5 min, then carefully remove methanol and repeat washing embryos with methanol for a total of at least three times.

2. Completely dehydrate embryos: Further dehydrate embryos with anhydrous methanol at least three times following procedures in the previous step. Note that embryos must be completely dehydrated before clearing. Embryos can be stored in methanol at −20 °C overnight or for longer time before further processing.
3. Clear embryos: completely remove anhydrous methanol from the glass vials. Add 500 μL of the clearing reagent (benzyl alcohol (BA): benzyl benzoate (BB) = 1:2) into each glass vial. Wait for ~10 min and let embryos settle down to the bottom of the vial. Carefully remove the clearing reagent and replace with fresh clearing agent. Note: embryos are nearly invisible after clearing so extra care must be taken to avoid losing them during pipetting. Repeat replacement of clearing reagent two–three times until no schlieren lines are visible. For complete clearing, keep embryos in the clearing reagent for 24–48 h before imaging, changing the clearing reagent for multiple times. Note that embryos in clearing reagent can be stored for short term at 4 °C; however, long-time storage may result in a reduction in fluorescence signal and fragmentation of embryos.

2.4.3.2 Preparing imaging chambers and confocal imaging

4. To make tape chambers, cut the 3M VHB™ tape into squares of 10 mm × 10 mm. Draw a square of 0.8 cm × 0.8 cm at the center of the tape using a razor blade and remove the center piece using dissecting forceps, which leaves a frame of tape. Transfer the frame of tape and let it adhere to the surface of a 20 mm × 40 mm cover-glass. Gently press the corners to make sure that the tape tightly adheres to the cover-glass. Peel off the thin plastic film on the tape using dissecting forceps.
5. Carefully transfer embryos from the glass vials in step 3 into tape chambers. Disperse embryos evenly inside the frame of tape. Gently place a second cover-glass on top to assemble the chamber to be a sandwich with embryos between two cover-glasses. Avoid air bubbles during this process.
6. Confocal microscopy: Under typical imaging conditions on a Zeiss LSM 710 we use lasers 561 nm (0.15% power), 633 nm (10% power) and 488 nm (10% power), without saturating signals. To image whole embryos, use the Plan-Apochromat 10 ×/numerical aperture (NA) 0.45 objective. To image embryos at high resolution, use the Plan-Apochromat 25 ×/NA 0.8 or 63 ×/NA 1.4 immersion oil objective. Use the following settings to acquire image stacks: scan mode = frame,

frame size = 1024 × 1024, speed = 9, averaging number = 2, bit depth = 8-bit, direction = bidirectional, method = mean, zoom = 0.6. To collect z-stacks, adjust the positions of objective and set the first (bottom) and last (top) images of the embryo. Set interval = 5 μm (for 10 ×), 2 μm (for 25 × immersion oil) or 1 μm (for 63 × immersion oil).

2.5 Image processing and analysis

Image stacks collected from confocal microscopy can be used for wholemount three-dimensional (3D) image segmentation to quantify levels of nascent transcription. We found that nucleus segmentation with background subtraction gave the most direct readout of transcriptional activity in blastomeres during cleavage stages. We found it most convenient to process and segment images using the Imaris 9.2 software, but similar approaches can be carried out using ImageJ or Matlab. The corrections we perform for large embryos include signal attenuation due to thick embryos and an optical correction for the 10 × objective due to refractive index mismatches (~ 1.5 and 1 for BABB vs air, respectively). Nuclei are easily segmented based on DNA stain or immunstaining with nuclear protein such as antibody to histone H3. Cell diameter of relatively round blastomeres was estimated using inner nuclear distances between object centroids. Cells at interphase can be separated from mitotic cells using the sphericity parameter, the threshold of which should be empirically determined.

2.6 Broader impacts and future directions

2.6.1 Generalizability to embryonic systems: Imaging ZGA in zebrafish blastula embryos

To determine whether our method for imaging zygotic genome activation can be applied to other embryonic systems, we tested two approaches in blastula embryos from the zebrafish, *Danio rerio*. First, we microinjected 5-EU in one-cell stage zebrafish embryo and fixed them at 3–4 hpf when large-scale genome activation occurs (Fig. 4A). After processing using the protocol described above, we found that nascent transcripts could be detected in single cells of wholemount embryos similar to what we observed for *Xenopus* (Fig. 4B). Second, we tested whether zebrafish embryos can be incubated with 5-EU in the medium, instead of microinjection, as works for cell culture (Jao & Salic, 2008). We removed the chorion of zebrafish embryos at one-cell stage, incubated them with 1 mM EU in the medium starting from 1 to 4 hpf (Fig. 4C), and then fixed them. We found that

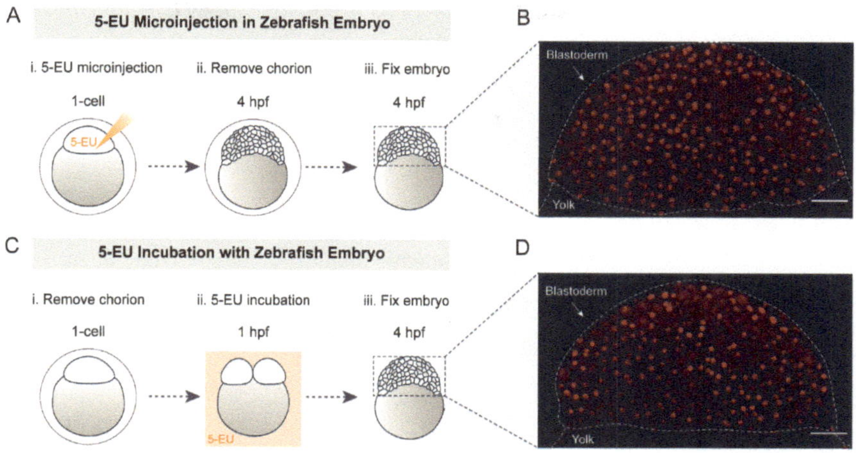

Fig. 4 Labeling and imaging nascent transcripts in zebrafish embryos. (A) Schematic of 5-EU microinjection in the blastodisc at one-cell stage. Chorion was removed prior to fixation at 4 hpf. (B) Image of 5-EU in blastoderm from 5-EU injection. (C) Schematic of zebrafish embryos incubated with 5-EU following chorion removal. (D) Image of 5-EU in blastoderm from 5-EU incubation. Confocal imaging at 25 × magnification. Maximum intensity projections of Z-stacks for embryos (only blastoderms are shown). The dashed lines indicate the shape and position of blastoderm relative to yolk. Scale bars, 50 μm.

nascent transcripts could be detected in single cells in wholemount embryos, the signal and pattern of which was similar to embryos with EU microinjection (Fig. 4D). This and other studies (Chan et al., 2019; Kwasnieski et al., 2019) show that 5-EU can used for tracking zygotic genome activation in a variety of embryonic systems. The choice of 5-EU microinjection or incubation should be determined empirically for any new embryo of interest.

2.6.2 Sequencing the nascent transcriptome via biotinylation of EU-RNA

EU-RNA imaging provides a single-cell and spatial perspective of large-scale genome activation during early embryogenesis. However, this approach lacks information on the identify of transcripts produced. Importantly, metabolic labeling with 5-EU can be directed toward (1) in vivo imaging via conjugation to an azide-flourophore and (2) in vitro characterization of the nascent transcriptome via RNA-seq via conjugation with azide-biotin (Fig. 5) (Chan et al., 2019; Chen et al., 2019; Kwasnieski et al., 2019). To generate and analyze the nascent transcriptome, total RNA is isolated from EU-microinjected embryos, and EU-RNA is biotinylated and then

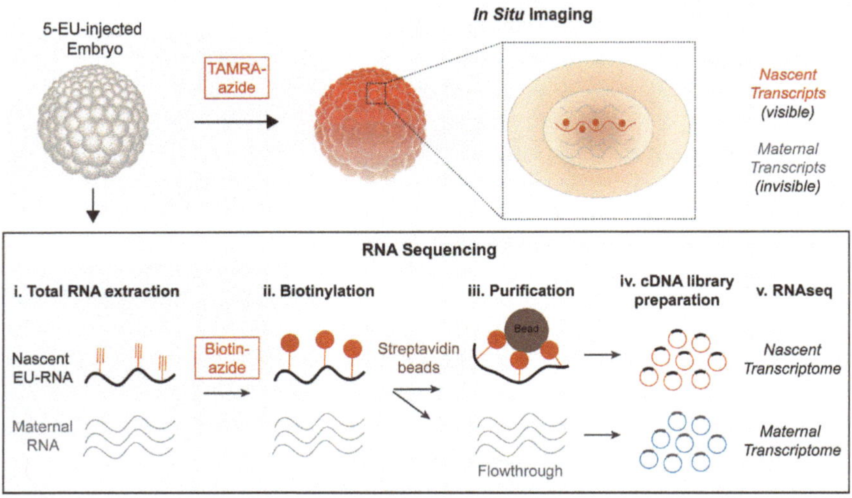

Fig. 5 Sequencing EU-RNA to characterize the nascent transcriptome. 5-EU injected embryos can either be fixed and labeled for wholemount imaging of nascent transcription at a single-cell level, or RNA can be isolated from embryos, biotinylated and separated from unlabeled RNA to generate nascent transcriptome libraries for sequencing. Unbound RNA flowthrough defines the maternal transcriptome.

purified using streptavidin beads (Fig. 5). Nascent EU-RNA pulled down by beads is reversed transcribed and the cDNA is used for making nascent transcriptome libraries for RNA-seq. Conversely, the flowthrough RNA that does not bind to beads is reversed transcribed and used to make maternal transcriptome libraries (Kwasnieski et al., 2019). Thus, using 5-EU metabolic labeling, in addition to wholemount imaging, the dynamics of both maternal and zygotic transcripts can be measured during early embryo development.

3. Summary

How the zygotic genome is activated during embryogenesis is a fundamental question in biology. This chapter describes a novel method to image nascent transcription in single cells of wholemount early embryos and to characterize their nascent transcriptomes. This approach provides temporal and spatial information on the pattern of ZGA, and it can be applied to a variety of model embryonic systems. This method will help to advance our understanding of gene regulation in complex tissues such as the blastula early embryo.

4. Notes

1. Egg quality and IVF: The quality of eggs and embryos is extremely important for all downstream processing. Ideally, embryos from IVFs with over 90% efficiency should be used in this protocol. To achieve high efficiency, freshly prepare buffers and sperm solutions used for IVF. Additionally, we recommend screening eggs from multiple frogs to identify clutch with highest quality of IVF.
2. Embryo budgeting: The number of embryos microinjected is dependent on the purpose of the experiment. As a rule of thumb for imaging, a minimum of five–six embryos should be imaged for each stage in treatment and experimental groups. To achieve this we recommend microinjecting at least 10 embryos to image 5–6 of highest quality. Some embryos can be damaged or lost during later processing.
3. 5-EU intracellular concentration: When first identifying the optical level of 5-EU for embryos from a new species, two factors should be determined empirically: sensitivity and toxicity. A final concentration of 5-EU at 0.2–0.6 mM often works for most embryos, including those from zebrafish, *Xenopus tropicalis* and *Xenopus laevis*. Higher concentrations produce higher signals for imaging, but concentrations above 1 mM often leads to toxicity and slow embryo development.
4. To avoid diluting the fixative, add embryos to the glass vial, while transferring only a minimal volume of media. Embryos will enrich in the tip of plastic transfer pipette due to gravity by waiting for 5–10 s. Excess buffer transferred will slow down the rate of fixation. If necessary, one can replace the fixative with fresh fixative to ensure rapid and complete fixation.
5. Bleaching: Use bleaching solution prepared freshly, immediately prior to the experiment. When adding the bleaching solution to the glass vials, be careful not to fill to very top as exposure of H_2O_2 to light for long periods of time will generate air bubbles that can cause vials to explode. Be careful not to overheat embryos because it may cause damage that worsens image quality. The duration of bleaching depends on the visibility of pigments in the animal pole of embryos. Bleaching can be ended when all pigments are whitened. Insufficient bleaching can lead to light absorption during fluorescence imaging.
6. Floating embryos: Following bleaching, take extra care when replacing buffer surrounding the embryos. Small air bubbles generated during

bleaching can cause the embryos to float which can make buffer removal more challenging. Work slowly and carefully. Keep in mind that embryos may become softened after extended bleaching and can be easily destroyed by shear forces during pipetting.

7. Pause step: Completely dehydrated embryos in 100% methanol can be stored long-term at −20 °C. This enables one to process separate sets of embryos at different times and then to image them all at one.
8. Click reaction details: Always freshly prepare the click reaction mix right before use. Also, avoid using previously used aliquots of reagents. Add ascorbic acid last to the reaction mix. We store 5-EU aliquots in aqueous solution of TBS or water at 100 mM stock concentration at −80 °C. Use of DMSO as solvent can slow embryos development (e.g., microinjection of DMSO at 0.5% may extend cleavage stage cell cycle or lead to reduced embryogenesis efficiency).
9. Volume of click reaction: The volume of click reaction mix to be added in each vial depends on the number of embryos inside. Minimally, one wants a volume sufficient for submerging all embryos inside the vial. The incubation time should be empirically determined for embryos from different species. Generally, the size of embryos affects the timing for click reaction, i.e., large embryos requires longer incubation than small embryos. Insufficient incubation may result in no or little labeling of EU-RNA inside the embryo, while over-incubation may cause higher nonspecific background. In our experience with *Xenopus laevis*, ~1.2 mm diameter embryos, we used 200–500 μL of reaction volume in a glass vial for 10–20 embryos, and incubated the reaction for ~12 h. For larger embryos we would perhaps scale up reaction volumes and times; for smaller embryos we would shorten reaction times.
10. Importance of washing: Complete washing of embryos following the click reaction is required for successful downstream processing, including immunostaining and DNA staining. An important indicator of complete washing is that no color remains in the embryos from fluorescent azides, e.g., red from TAMRA-azide. However, we still wash embryos for an extended period such as another day once red TAMRA-azide signal is gone. Insufficient washing may result in high background. Therefore, embryos must be extensively and adequately washed before further processing. Expect to extend washing times listed in this protocol for embryos >2 mm in diameter.

11. Blocking of embryos: The volume of the blocking buffer to be added in each vial should be sufficient to submerge all embryos inside the vial. For large-sized embryos, e.g., >1 mm in diameter, the incubation can also be performed overnight at 4 °C.

12. *Troubleshooting.* Always prepare buffers for processing fixed embryos with RNase-free water. If DNA staining is weak or nonspecific, try making a fresh preparation of MEMFA for fixation step. If EU or DNA staining appear somewhat nonspecific, particularly in cortical layers of embryo, it may be helpful to increase washing time (particularly if embryos are larger than those of *Xenopus laevis*). If embryos are being damaged during processing make sure to always fill glass tubes to the very top so that large air pockets are not present during rotation.

Acknowledgments

We thank the Cell and Developmental Biology Microscopy Core at the University of Pennsylvania for technical support in confocal imaging, National *Xenopus* Resource (NXR) for training and guidance on frogs and the Dr. Mary Mullins lab for providing zebrafish embryos. We are grateful to Dr. John Wallingford and Dr. Lance Davidson for their pioneering techniques in imaging early *Xenopus* embryos. This work was supported in part by Burroughs Wellcome Fund, Charles E. Kaufman Foundation, the March of Dimes and the National Institute of General Medical Sciences (R35GM12874) (M.C.G.).

References

Aanes, H., Collas, P., & Alestrom, P. (2014). Transcriptome dynamics and diversity in the early zebrafish embryo. *Briefings in Functional Genomics*, *13*(2), 95–105. https://doi.org/10.1093/bfgp/elt049.

Aoki, F., Worrad, D. M., & Schultz, R. M. (1997). Regulation of transcriptional activity during the first and second cell cycles in the preimplantation mouse embryo. *Developmental Biology*, *181*(2), 296–307. https://doi.org/10.1006/dbio.1996.8466.

Briggs, J. A., Weinreb, C., Wagner, D. E., Megason, S., Peshkin, L., Kirschner, M. W., et al. (2018). The dynamics of gene expression in vertebrate embryogenesis at single-cell resolution. *Science*, *360*(6392). eaar5780https://doi.org/10.1126/science.aar5780.

Campbell, P. D., Chao, J. A., Singer, R. H., & Marlow, F. L. (2015). Dynamic visualization of transcription and RNA subcellular localization in zebrafish. *Development*, *142*(7), 1368–1374. https://doi.org/10.1242/dev.118968.

Chan, S. H., Tang, Y., Miao, L., Darwich-Codore, H., Vejnar, C. E., Beaudoin, J. D., et al. (2019). Brd4 and P300 confer transcriptional competency during zygotic genome activation. *Developmental Cell*, *49*(6), 867–881. e868. https://doi.org/10.1016/j.devcel.2019.05.037.

Chen, H., Einstein, L. C., Little, S. C., & Good, M. C. (2019). Spatiotemporal patterning of zygotic genome activation in a model vertebrate embryo. *Developmental Cell*, *49*(6), 852–866. e857. https://doi.org/10.1016/j.devcel.2019.05.036.

Farrell, J. A., Wang, Y., Riesenfeld, S. J., Shekhar, K., Regev, A., & Schier, A. F. (2018). Single-cell reconstruction of developmental trajectories during zebrafish embryogenesis. *Science*, *360*(6392), eaar3131. https://doi.org/10.1126/science.aar3131.

Gao, L., Wu, K., Liu, Z., Yao, X., Yuan, S., Tao, W., et al. (2018). Chromatin accessibility landscape in human early embryos and its association with evolution. *Cell*, *173*(1), 248–259. e215. https://doi.org/10.1016/j.cell.2018.02.028.

Garcia, H. G., Tikhonov, M., Lin, A., & Gregor, T. (2013). Quantitative imaging of transcription in living Drosophila embryos links polymerase activity to patterning. *Current Biology*, *23*(21), 2140–2145. https://doi.org/10.1016/j.cub.2013.08.054.

Gentsch, G. E., Owens, N. D. L., & Smith, J. C. (2019). The spatiotemporal control of zygotic genome activation. *iScience*, *16*, 485–498. https://doi.org/10.1016/j.isci.2019.06.013.

Heyn, P., Kircher, M., Dahl, A., Kelso, J., Tomancak, P., Kalinka, A. T., et al. (2014). The earliest transcribed zygotic genes are short, newly evolved, and different across species. *Cell Reports*, *6*(2), 285–292. https://doi.org/10.1016/j.celrep.2013.12.030.

Heyn, P., & Neugebauer, K. M. (2017). Purification of zygotically transcribed RNA through metabolic labeling of early zebrafish embryos. *Methods in Molecular Biology*, *1605*, 121–131. https://doi.org/10.1007/978-1-4939-6988-3_8.

Jao, C. Y., & Salic, A. (2008). Exploring RNA transcription and turnover in vivo by using click chemistry. *Proceedings of the National Academy of Sciences of the United States of America*, *105*(41), 15779–15784. https://doi.org/10.1073/pnas.0808480105.

Jukam, D., Shariati, S. A. M., & Skotheim, J. M. (2017). Zygotic genome activation in vertebrates. *Developmental Cell*, *42*(4), 316–332. https://doi.org/10.1016/j.devcel.2017.07.026.

Kageyama, S., Nagata, M., & Aoki, F. (2004). Isolation of nascent messenger RNA from mouse preimplantation embryos. *Biology of Reproduction*, *71*(6), 1948–1955. https://doi.org/10.1095/biolreprod.104.031906.

Kwasnieski, J. C., Orr-Weaver, T. L., & Bartel, D. P. (2019). Early genome activation in Drosophila is extensive with an initial tendency for aborted transcripts and retained introns. *Genome Research*, *29*(7), 1188–1197. https://doi.org/10.1101/gr.242164.118.

Lee, M. T., Bonneau, A. R., & Giraldez, A. J. (2014). Zygotic genome activation during the maternal-to-zygotic transition. *Annual Review of Cell and Developmental Biology*, *30*, 581–613. https://doi.org/10.1146/annurev-cellbio-100913-013027.

Lee, C., Kieserman, E., Gray, R. S., Park, T. J., & Wallingford, J. (2008). Whole-mount fluorescence immunocytochemistry on Xenopus embryos. *CSH Protocols*, *2008*, 1–4. https://doi.org/10.1101/pdb.prot4957. pdb.prot4957.

Little, S. C., & Gregor, T. (2018). Single mRNA molecule detection in Drosophila. *Methods in Molecular Biology*, *1649*, 127–142. https://doi.org/10.1007/978-1-4939-7213-5_8.

Liu, L., Leng, L., Liu, C., Lu, C., Yuan, Y., Wu, L., et al. (2019). An integrated chromatin accessibility and transcriptome landscape of human pre-implantation embryos. *Nature Communications*, *10*(1), 364. https://doi.org/10.1038/s41467-018-08244-0.

Lu, F., Liu, Y., Inoue, A., Suzuki, T., Zhao, K., & Zhang, Y. (2016). Establishing chromatin regulatory landscape during mouse preimplantation development. *Cell*, *165*(6), 1375–1388. https://doi.org/10.1016/j.cell.2016.05.050.

Ranisavljevic, N., Okamoto, I., Heard, E., & Ancelin, K. (2017). RNA FISH to study zygotic genome activation in early mouse embryos. *Methods in Molecular Biology*, *1605*, 133–145. https://doi.org/10.1007/978-1-4939-6988-3_9.

Satija, R., Farrell, J. A., Gennert, D., Schier, A. F., & Regev, A. (2015). Spatial reconstruction of single-cell gene expression data. *Nature Biotechnology*, *33*(5), 495–502. https://doi.org/10.1038/nbt.3192.

Schier, A. F. (2007). The maternal-zygotic transition: Death and birth of RNAs. *Science*, *316*(5823), 406–407. https://doi.org/10.1126/science.1140693.

Schulz, K. N., & Harrison, M. M. (2019). Mechanisms regulating zygotic genome activation. *Nature Reviews Genetics*, *20*(4), 221–234. https://doi.org/10.1038/s41576-018-0087-x.

Sive, H. L., Grainger, R. M., & Harland, R. M. (2007). *Xenopus laevis* in vitro fertilization and natural mating methods. *CSH Protocols*, *2007*. https://doi.org/10.1101/pdb.prot4737. pdb.prot4737.

Sive, H. L., Grainger, R. M., & Harland, R. M. (2010). Microinjection of Xenopus embryos. *Cold Spring Harbor Protocols*, *2010*(12), 1348–1349. https://doi.org/10.1101/pdb.ip81.pdb.ip81.

Stapel, L. C., Zechner, C., & Vastenhouw, N. L. (2017). Uniform gene expression in embryos is achieved by temporal averaging of transcription noise. *Genes & Development*, *31*(16), 1635–1640. https://doi.org/10.1101/gad.302935.117.

Trcek, T., Lionnet, T., Shroff, H., & Lehmann, R. (2017). mRNA quantification using single-molecule FISH in Drosophila embryos. *Nature Protocols*, *12*(7), 1326–1348. https://doi.org/10.1038/nprot.2017.030.

Vastenhouw, N. L., Cao, W. X., & Lipshitz, H. D. (2019). The maternal-to-zygotic transition revisited. *Development*, *146*(11), dev161471. https://doi.org/10.1242/dev.161471.

Wagner, D. E., Weinreb, C., Collins, Z. M., Briggs, J. A., Megason, S. G., & Klein, A. M. (2018). Single-cell mapping of gene expression landscapes and lineage in the zebrafish embryo. *Science*, *360*(6392), 981–987. https://doi.org/10.1126/science.aar4362.

Wallingford, J. B. (2010). Preparation of fixed Xenopus embryos for confocal imaging. *Cold Spring Harbor Protocols*, *2010*(5), 1–7. https://doi.org/10.1101/pdb.prot5426. pdb.prot5426.

White, R. J., Collins, J. E., Sealy, I. M., Wali, N., Dooley, C. M., Digby, Z., et al. (2017). A high-resolution mRNA expression time course of embryonic development in zebrafish. *eLife*, *6*. e30860. https://doi.org/10.7554/eLife.30860.

Xie, F., Timme, K. A., & Wood, J. R. (2018). Using single molecule mRNA fluorescent in situ hybridization (RNA-FISH) to quantify mRNAs in individual murine oocytes and embryos. *Scientific Reports*, *8*(1), 7930. https://doi.org/10.1038/s41598-018-26345-0.

Xue, Z., Huang, K., Cai, C., Cai, L., Jiang, C. Y., Feng, Y., et al. (2013). Genetic programs in human and mouse early embryos revealed by single-cell RNA sequencing. *Nature*, *500*(7464), 593–597. https://doi.org/10.1038/nature12364.

CHAPTER NINE

Engineering reversible cell-cell interactions with chemical biology

Clifford M. Csizmar, Carston R. Wagner*
Department of Medicinal Chemistry, University of Minnesota, Minneapolis, MN, United States
*Corresponding author: e-mail address: wagne003@umn.edu

Contents

Abstract

Techniques to direct cell-cell interactions have advanced our understanding of fundamental biology and opened new avenues in tissue engineering, regenerative medicine, and immunotherapy. This is often achieved by introducing new targeting ligands to the cell membrane, which can be accomplished through both genetic and nongenetic approaches. While both offer advantages, nongenetic modifications tend to be faster to produce, innocuous to the modified cell, and potentially reversible. This chapter will

Methods in Enzymology, Volume 638
ISSN 0076-6879
https://doi.org/10.1016/bs.mie.2020.02.019

outline nongenetic methods that have been used to control intercellular interactions—namely hydrophobic insertion, chemical modification, liposome fusion, metabolic engineering, and enzymatic remodeling—and provide protocols that can serve as a starting point for future applications.

1. Introduction

The existence of multicellular organisms requires that constituent cells communicate and interact with each other in a controlled and carefully regulated fashion. In humans, the breakdown of this cellular coordination results in a variety of diseases, ranging from autoimmune disorders to metastatic malignancies. Accordingly, intense effort has been placed on developing techniques to artificially direct cell-cell interactions to enable therapeutic interventions in tissue engineering, regenerative medicine, and cellular immunotherapy. To this end, multiple approaches to introduce nonnative receptors to cell surfaces have been developed (Csizmar, Petersburg, & Wagner, 2018).

Perhaps the most notable example is the chimeric antigen receptor (CAR) platform, where a patient's own T cells are genetically engineered to express a novel receptor that recognizes a tumor-associated antigen (Sadelain, Rivière, & Riddell, 2017). While CAR T cells have been clinically efficacious for the treatment of hematologic malignancies, there are, understandably, some limitations to the approach that have hindered the use of CAR technology in certain settings. For example, the genetic manipulation that underlies CAR introduction has been difficult to apply to sensitive stem cell populations (Nowakowski, Andrzejewska, Janowski, Walczak, & Lukomska, 2013). Moreover, the permanence of the modification has limited the ability to use CAR T cells to treat solid tumors, especially those malignancies for which tumor specific antigens have not been identified (Lamers et al., 2013). Finally, it takes considerable time (2–3 weeks) to prepare therapeutic volumes of CAR T cells, and this timeline is not amenable to more acute clinical scenarios. While efforts are underway to improve the CAR T cell platform in each of these domains, these issues sparked early motivation to pursue alternative, nongenetic approaches to introduce synthetic ligands to cell surfaces.

Several fields are poised to benefit from advances in nongenetic cell membrane engineering for the purposes of directing cell-cell interactions (Csizmar, Petersburg, & Wagner, 2018). Tissue engineering requires that cell types and populations be precisely patterned in order to recreate a functional anatomic motif. The ability to quickly and transiently modify cells grown in vitro with synthetic receptors has enabled the preparation of complex tissues and organoids. Likewise, the use of stem cells for regenerative medicine often requires that the cells successfully traffic to the necessary supportive niche or location of injury (Karp & Leng Teo, 2009). For example, the use mesenchymal stem cells (MSCs) has been explored in the context of myocardial infarction, where is has been hypothesized that delivery of MSCs to ischemic cardiac tissue could promote tissue regeneration instead of fibrosis (Cashman, Gouon-Evans, & Costa, 2013). However, the ability of MSCs to either home to or be retained within the ischemic region is limited, and the use of membrane anchored-ligands to promote this localization has been investigated (Kean et al., 2012). Finally, tumor-targeting receptors have been installed on T cell membranes in order to rapidly prepare them to recognize malignant cells (Gabrielse et al., 2014). Using nongenetic techniques allows for large numbers of effector cells to be rapidly modified in a manner of hours as opposed to weeks; moreover, because the modification is transient, there is the potential for more refined control over the duration of the therapeutic effect and thus the opportunity to quickly abrogate the therapy should adverse events arise.

These examples highlight what are considered to be the ideal properties of such nongenetic membrane engineering platforms (Csizmar, Petersburg, & Wagner, 2018). That is, such systems should be stable for a meaningful period of time, universally applicable to diverse cell types, innocuous with respect to the cell being modified, and reversible should the need arise. These features ought to enhance both the utility of such systems at the bench as well as the clinical translatability of the approach. To that end, this chapter will focus on nongenetic membrane modification systems that incorporate some of these features for the purposes of directing reversible interactions between distinct cells.

2. Nongenetic approaches to engineering cell-cell interactions

Numerous nongenetic approaches to engineer cell membranes with artificial receptors have been developed. These techniques typically fall into one of five general categories: hydrophobic insertion, chemical modification, liposome fusion, metabolic engineering, or enzymatic remodeling (Fig. 1). Examples of how each have been applied to the direction of cell-cell interactions will be given here, with a particular emphasis on those

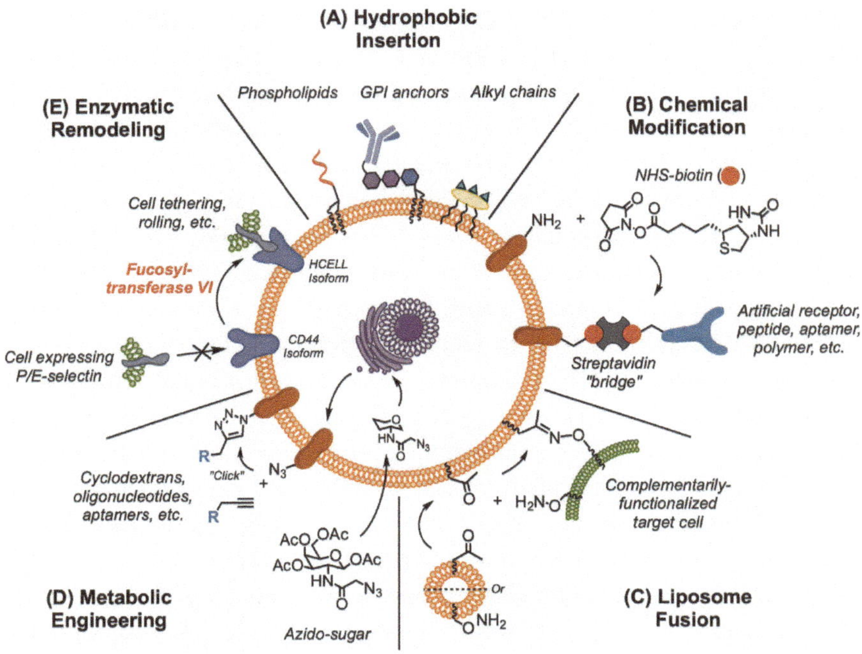

Fig. 1 Nongenetic membrane engineering approaches to direct intercellular interactions. Several nongenetic methods for cell membrane modification have been used to direct cell-cell interactions. (A) Hydrophobic groups will spontaneously insert into the lipid bilayer, tethering any conjugated cargo or ligands to the cell surface. (B) Functional groups naturally present on the cell surface can be reacted with chemical moieties to covalently tether a variety of species; in many cases, biotin is installed and streptavidin is used as a "bridge" to link other ligands to the cell surface. (C) Liposomes bearing various functional groups can be spontaneously fused with the cell membrane, thereby decorating the surface with the respective modifications. (D) Cells can be grown in media supplemented with functionalized sugar analogs, enabling the incorporation of bioorthogonal groups into surface glycoproteins; these groups can then be conjugated to exogenous ligands. (E) Enzymes can be used to modify naturally-existing proteins and glycans on the cell surface, creating divergent isoforms or tethering synthetic targeting elements. *Figure and caption minimally adapted from Csizmar, C. M., Petersburg, J. R., & Wagner, C. R. (2018). Programming cell-cell interactions through nongenetic membrane engineering.* Cell Chemical Biology, *25(8), 931–940, copyright 2018, with permission from Elsevier.*

modifications that are reversible and thus provide some degree of temporospatial control over the resultant interactions.

2.1 Hydrophobic insertion

Hydrophobic insertion (Fig. 1A) takes advantage of the thermodynamically favorable and thus spontaneous insertion of hydrophobic species into the cell's lipid bilayer membrane (Fernandez-Vidal, White, & Ladokhin, 2011; Gorfe, Baron, & McCammon, 2008; Jeong et al., 2013). Taking cues from nature, investigators have employed a variety of hydrophobic motifs to anchor artificial receptors to the cell membrane, including phospholipids (Csizmar, Petersburg, Hendricks, et al., 2018), alkyl chains (Jeong et al., 2013), and glycosylphosphatidylinositol (GPI) anchors (Hamdy et al., 2005). These species are often attached to a targeting element through various conjugation chemistries, including peptide couplings (Ko, Kean, & Dennis, 2009), sulfhydryl additions (Won, Patel, & Bull, 2014), or click ligations (Lu, Gao, & Guo, 2015). Rarely, hydrophobically-tagged protein ligands have been produced recombinantly (de Kruif, Tijmensen, Goldsein, & Logtenberg, 2000). Similar diversity exists among the chosen targeting element, with antibodies (and their single-chain derivatives) (de Kruif et al., 2000; Kim & Peacock, 1993), peptides (Kean et al., 2012), aptamers (Altman, Chang, Xiong, & Tan, 2013), and oligonucleotides (Selden et al., 2012) all having been used successfully. Larger scaffolds such as nanoparticles and polymers can also be anchored to cells in this fashion (Csizmar, Petersburg, Hendricks, et al., 2018; Takeo et al., 2015).

Genetic engineering of stem cells can be difficult without negatively impacting pluripotency (Nowakowski et al., 2013). Thus, hydrophobic insertion has been a popular choice to impart new functionality to MSC membranes in a rapid and innocuous manner, especially in the context of directing MSC homing. In one example, peptides engineered to recognize ischemic myocardium were synthesized with a terminal palmitic acid group (Kean et al., 2012). These peptides were then "painted" onto the surface of MSCs and, compared to unmodified control cells, increased the number of MSCs that homed to and were retained within ischemic myocardium in a mouse model of myocardial infarction (Kean et al., 2012). MSC membranes have also been functionalized with antibodies and fusion proteins targeting vascular adhesion proteins. This enabled the MSCs to interact with the vascular endothelial cells more effectively, enhancing their ability to tether and roll along the endothelium in a manner similar to extravasating leukocytes (Lo et al., 2013).

Another popular application of the hydrophobic insertion approach is immunotherapy, where large quantities of effector cells—or a model thereof—can be rapidly armed with tumor-targeting receptors. In one embodiment, phospholipid-deoxyribose nucleic acid (DNA) conjugates were used to install aptamers targeting protein tyrosine kinase 7 (PTK7) on the surface of various cells, including T and NK cells (Xiong et al., 2013). These targeted cells then recognized and bound to PTK7-expressing target cells. Importantly, the cell aggregates could subsequently be dissociated by degrading the oligonucleotide components with deoxyribose nuclease I (DNaseI), providing an enzymatic means for reversing the cell-cell interactions. An approach developed by our group utilizes a lipid-anchored, multivalent, self-assembling nanoring to embed single-chain variable fragments (scFvs) targeting epithelial cell adhesion molecule (EpCAM) onto T cell membranes (Gabrielse et al., 2014). Once activated, the T cells selectively recognized and killed EpCAM-expressing MCF-7 cells in vitro. The targeting capabilities of the T cells could be abrogated by the addition of the small molecule antibiotic, trimethoprim, which drives the dissociation of the nanoring scaffold and effectively removes the EpCAM-binding scFvs from the T cell surface.

Aside from the aforementioned enzymatic and pharmacologic mechanisms, most lipid-based membrane modifications are reversible by virtue of the temporary nature of the hydrophobic group's residence within the cell membrane. A dynamic equilibrium exists between hydrophobic groups residing within the membrane and those that have partitioned into the surrounding environment (Palte & Raines, 2012). The extent of this partitioning can be modulated to a certain degree by changing the hydrophobicity of the anchor, where an increase in the net hydrophobicity serves to stabilize the species within the membrane (Weber, Liang, Selden, Desai, & Gartner, 2014). Otherwise, such species typically have a cell surface residence time ranging from minutes to several hours (de Kruif et al., 2000; Kean et al., 2012). One strategy to increase the stability of such modifications is to increase the valency of the hydrophobic anchors. Indeed, our group has used this approach to extend the cell surface residence of phospholipid-anchored nanorings out to several days (Csizmar, Petersburg, Hendricks, et al., 2018; Gabrielse et al., 2014).

2.2 Chemical modification

Along with hydrophobic insertion, chemical modification techniques are among the most popular due to their speed, innocuity, and relative ease (Fig. 1B). These approaches take advantage of reactive functional groups—such as free amines present on lysine residues—that are naturally present on the cell surface to conjugate or otherwise attach targeting ligands directly to the cell. Such techniques typically require little-to-no preconditioning or processing of the cell, as the necessary functionality is already present on the cell membrane. Moreover, essentially any cell membrane constituent can be modified in this manner, including proteins, glycans, phospholipids, and carbohydrates, albeit all in a nonspecific manner.

As alluded to, lysine side chains are the most frequent site of modification due to their abundance on the cell surface and the ease with which they can be reacted with *N*-hydroxysuccinimide (NHS)-activated esters. Using such chemistry, cells can be nonspecifically biotinylated at these residues to provide a handle for further functionalization. Upon subsequent incubation with tetravalent streptavidin, the streptavidin binds to the biotin moieties which acts as a "bridge" to connect other biotinylated ligands. In the simplest application, groups of biotinylated cells can be connected to each other to form clusters and layers, which can be further connected to additional cell layers to build preliminary tissue architectures (Gong et al., 2013). More common, however, is the use of the streptavidin bridge approach to functionalize MSCs with targeting ligands, such as sialyl Lewis X (sLe^X) motifs (Sarkar et al., 2011) or selectin-binding aptamers (Zhao et al., 2011) that can help the MSCs interact with inflamed endothelial cells.

Other surface modification chemistries have been used infrequently. Chromic chloride coupling was used to arm erythrocytes with antibodies that recognized CD90.2, and these functionalized erythrocytes formed rosettes around CD90.2-expressing T lymphocytes (Chiarantini, Droleskey, Magnani, & DeLoach, 1992). In another example, 3-(acrylamide)phenylboronic acid (APBA) polymers were bound to cell membrane glycoproteins via covalent boronate ester bonds (Amaral & Pasparakis, 2015).

These membrane modifications tend to be highly stable due to the covalent nature of the attachment. Oligonucleotide-based ligands could theoretically be degraded with DNAase, as previously mentioned, while some polymer-based embodiments could be reversed via a decrease in temperature, which destabilized the scaffold (Amaral & Pasparakis, 2015). As will be discussed further in the next section, photo- and electrochemical reactions have been used to cleave sensitive bonds. However, all of these techniques are highly specific to the individual setting, and care should be taken to select a conjugation chemistry that is most suitable to the desired application.

2.3 Liposome fusion

In many respects, the liposome fusion approach pioneered by Yousaf and colleagues is a conceptual amalgam of the hydrophobic insertion and chemical modification techniques (Fig. 1C). A fundamental process in cell biology, their group has taken advantage of the natural ability of liposomes to spontaneously fuse with a cell's lipid bilayer to introduce a variety of species to the membrane. The liposome formation process is generally straightforward. Briefly, the desired lipid components are dissolved in a volatile solvent (e.g., chloroform), concentrated and dried under vacuum, resuspended in the desired aqueous buffer, and finally subject to an agitating stimulus (e.g., sonication) that prompts vesicle formation (Dutta, Pulsipher, Luo, Mak, & Yousaf, 2011). These liposomal vesicles can then be incubated with cells to permit their fusion with and functionalization of the membrane.

Liposome fusion has largely been developed within the context of tissue engineering. In an early example, liposomes formed from phospholipids bearing ketone and oxyamine groups were fused with mouse fibroblast cells; the modified cells were then assembled into sheets and spheroids by taking advantage of the complimentary ligation reaction between the two functional groups (Dutta, Pulsipher, Luo, & Yousaf, 2011). This same methodology was later used to pattern fibroblasts, cardiomyocytes, and human umbilical vein endothelial cells (HUVECs) into a functional three-dimensional cardiac tissue that was able to beat spontaneously and synchronously (Rogozhnikov, O'Brien, Elahipanah, & Yousaf, 2016). Enhanced control over cell assembly was also achieved by using a microfluidic chamber to introduce cells to the growing spheroid in a more orderly fashion (O'Brien, Luo, Rogozhnikov, Chen, & Yousaf, 2015).

Adaptations of this work have specifically focused on developing reversible cellular assemblies. By incorporating a photocleavable group into the oxyamine-bearing phospholipids, the ligation products could be separated via exposure to ultraviolet (UV) light (Luo, Pulsipher, Dutta, Lamb, & Yousaf, 2014). In the in vitro setting, this provided temporal and spatial control over the otherwise covalent cell assembly. In another case, human MSCs and fibroblasts were instead modified with hydroquinone and aminoxy-containing liposomes, respectively. Electrochemical oxidation of the hydroquinone to the active quinone enabled the reaction with the corresponding aminoxy groups, leading to MSC-fibroblast clustering. These interactions were able to be reversed upon reduction of the quinone back to the hydroquinine, thereby disassembling the cellular aggregates (Pulsipher, Dutta, Luo, & Yousaf, 2014).

2.4 Metabolic engineering

Metabolic engineering in the context of cell surface modifications involves supplementing the cellular growth media with an analog of a metabolic substrate such that the analog becomes naturally incorporated into the cell membrane or its constituents (Fig. 1D). Typically the synthetic substrates contain a bioorthogonal group that, once incorporated into a surface protein or glycan, can be reacted with the desired ligand. In contrast to the nonspecific chemical modification approaches discussed above, metabolic engineering affords the opportunity to specifically react only with the metabolically incorporated group, thereby sparing naturally-occurring residues (such as lysine side chains) from potentially deleterious adduct formation.

In a classic example, Jurkat cells were metabolically labeled with *N*-azideacetylmannosamine, affording the incorporation of azido sialic acid (SiaNAz) residues into cell surface glycoproteins (Gartner & Bertozzi, 2009). The SiaNAz residues were then reacted with either phosphine-conjugated single-stranded DNA (ssDNA) oligonucleotides via a Staudinger ligation or with difluorinated cyclooctyne (DIFO)-conjugated ssDNA oligonucleotides via copper-free click chemistry. Cells bearing complimentary oligonucleotide sequences were then able to assemble into aggregates, and the order of cells in the aggregate could be controlled stoichiometrically. Moreover, the cell-cell interactions were able to be reversed by either thermally denaturing the hybridized oligonucleotides at 37 °C or by degrading them with DNase.

Another reversible approach took advantage of photoisomerizable β-cyclodextran groups (Shi et al., 2016). Cell surface glycoproteins were again metabolically labeled with azido groups, and these were conjugated to pegylated β-cyclodextrans via a copper-catalyzed azide-alkyne cycloaddition reaction. The β-cyclodextrans are only capable of binding to *trans*-azobenzene, and the isoform of the azobenzene species (i.e., *cis* or *trans*) is interconvertable via UV photoisomerization. Thus, homobifunctional azobenzene species were used to bridge between cyclodextran-labeled MCF-7 cells, and these cell-cell interactions were readily reversed with UV irradiation. Similarly, azobenzene-conjugated aptamers targeting mucin 1 (MUC 1) were installed on metabolically-engineered HeLa cells and peripheral blood mononuclear cells (PBMCs); both of these cell types were then able to recognize MUC 1+ MCF-7 cells (Shi et al., 2016).

2.5 Enzymatic remodeling

As the name suggests, enzymatic remodeling involves using an enzyme to catalyze a change in a cell membrane constituent in a manner that alters the function of the ligand (Fig. 1F). Because this approach requires a specific combination of enzyme, membrane substrate, and desired functional outcome, examples in the context of induced cell-cell interactions are quite rare.

Merzaban et al., used an enzymatic remodeling technique called glycosyltransferase-programmed stereosubstitution (GPS) to enhance the adhesion and extravasation capabilities of neural stem cells (NSCs) (Merzaban et al., 2015). NSCs naturally express CD44 but lack the isoforms necessary for *E*-selectin binding. The investigators treated NSCs with the α(1,3)-linkage specific fucosyltransferase VI (FTVI), which specifically places a fucose moiety onto a terminal type 2-lactosamine unit. When that lactosamine was capped with an α(2,3)-linked sialic acid, sLe^X epitopes capable of being recognized by P- and E-selectin residues were created. These GPS-modified NSCs could then interact with HUVECs expressing E-selectin in a microfluidic chamber. Furthermore, the GPS-modified NSCs underwent enhanced trans-endothelial extravasation in a murine model of multiple sclerosis, preferentially accumulating within the brain, spleen, and liver parenchyma.

3. Engineering reversible cell-cell interactions with chemically self-assembled nanorings (CSANS)

Our approach to directing reversible interactions between cells draws inspiration from the hydrophobic insertion, chemical modification, and metabolic engineering approaches described above. The reversible nature of our system relies upon a multivalent protein scaffold known as the chemically self-assembled nanoring (CSAN; Fig. 2) (Carlson et al., 2006). CSANs are formed when bivalent dihydrofolate reductase ($DHFR^2$) fusion proteins are spontaneously oligomerized by a chemical dimerizer, bis-methotrexate (bisMTX). As methotrexate is the small molecule inhibitor of DHFR, the bisMTX acts as a bridge between $DHFR^2$ subunits and drives the oligomerization of a predominantly octameric nanoring scaffold (Li et al., 2010). CSANs can be further functionalized in two key ways: (1) by recombinantly fusing various binding entities to the $DHFR^2$ subunits; and (2) by chemically modifying the bisMTX dimerizer to incorporate reactive groups or small molecule cargo as desired. It is convenient for the targeting ligands to be protein-based—such as an scFv (Petersburg et al., 2018; Shen, Vallera, & Wagner, 2015)—as this allows for recombinant production of the requisite fusion proteins; however, this is not a requirement, and variations to incorporate nonprotein ligands can be easily conceived (Gangar, Fegan, Kumarapperuma, & Wagner, 2012; Shah et al., 2016).

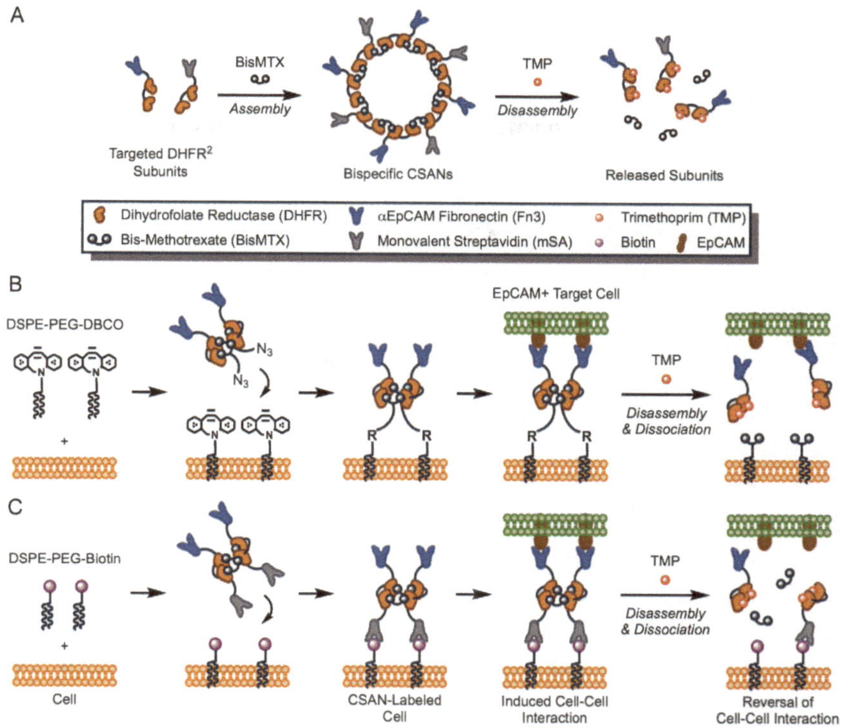

Fig. 2 Cell surface engineering with chemically self-assembled nanorings (CSANs). (A) CSANs are composed of targeted-DHFR^2 fusion proteins that are spontaneously oligomerized by the chemical dimerizer, bisMTX; they can be pharmacologically disassembled by the antibiotic trimethoprim. (B) DSPE-PEG$_{2000}$-DBCO moieties spontaneously insert into cell membranes and are stabilized in the lipid bilayer by the hydrophobic effect. EpCAM-targeted Fn3 CSANs oligomerized with an azide-bisMTX dimerizer are then installed on the cell surface through a copper-free, strain-promoted alkyne/azide cycloaddition. The CSAN-functionalized cells can then form targeted interactions with EpCAM+ cells, and these interactions can be reversed with trimethoprim. (C) Similarly, cells modified with DSPE-PEG$_{2000}$-biotin moieties can be functionalized with bispecific mSA/Fn3 CSANs, enabling recognition of EpCAM+ target cells. Trimethoprim-induced disassembly of the CSAN reverses the intercellular interactions. *Figure and caption minimally adapted from Csizmar, C. M., Petersburg, J. R., Hendricks, A., Stern, L. A., Hackel, B. J., & Wagner, C. R. (2018). Engineering reversible cell-cell interactions with lipid anchored prosthetic receptors.* Bioconjugate Chemistry, *29(4), 1291–1301, copyright 2018, with permission from ACS Publications.*

For the purposes of these methods, we elected to fuse either a monovalent streptavidin domain (mSA) (Lim, Huang, Pralle, & Park, 2013) or a fibronectin (Fn3) domain with engineered specificity for epithelial cell adhesion molecule (EpCAM) (Csizmar et al., 2019; Stern, Csizmar, Woldring,

Wagner, & Hackel, 2017). Meanwhile, the bisMTX moiety was synthesized to contain a bioorthogonal azide for use as a ligation handle (Csizmar, Petersburg, Hendricks, et al., 2018). Combining these components allows one to assemble multivalent and heterobifunctional CSANs capable of interfacing with multiple distinct entities (i.e., biotin, EpCAM, and alkynes).

In our workflow (Fig. 2B and C) cells are first modified with one of two phospholipid conjugates via hydrophobic insertion: (1) 1,2-distearoyl-*sn*-glycero-3-phosphoethanolamine-*N*-{biotinyl(polyethylene glycol)-2000} (DSPE-PEG$_{2000}$-biotin); or (2) 1,2-distearoyl-*sn*-glycero-3-phosphoethanolamine-*N*-{dibenzocyclooctyl(polyethylene glycol)-2000} (DSPE-PEG$_{2000}$-DBCO). Targeted CSANs are then attached to the lipid-modified cells via a noncovalent biotin/mSA interaction or a copper-free, strain-promoted alkyne/azide cycloaddition (SPAAC) involving the DBCO/azide groups, thereby functionalizing the cell with the EpCAM-binding domains. In this manner, the CSAN-functionalized cells are able to recognize and cluster with EpCAM-expressing target cells (Csizmar, Petersburg, Hendricks, et al., 2018).

The reversibility of the system comes from the fact that the bisMTX can be competed from the DHFR active site by a more potent DHFR inhibitor, trimethoprim (TMP). Thus, by introducing TMP to the system, the CSAN is disassembled and crucial components of the scaffold are dissociated from the cell surface (Fig. 2). When designed appropriately, this feature can be used to dissociate the CSAN's targeting ligands from the cell membrane and disrupt any cell-cell interactions induced by the scaffold (Csizmar, Petersburg, Hendricks, et al., 2018; Gabrielse et al., 2014).

Here, we provide our methods for: (1) modifying cell membranes with phospholipid conjugates via hydrophobic insertion; (2) conjugating protein scaffolds to these modified cells via either a protein/ligand interaction or click chemistry; (3) forming and reversing cell-cell interactions using CSANs; and (4) analyzing the cell-cell interactions by both flow cytometry and fluorescence microscopy. While these have been developed with the CSAN platform in mind, these protocols should be directly applicable to other cell modification and protein production strategies with only minor adaptation.

4. Protocols

4.1 Equipment

1. Glass microscopy slides and coverslips suitable for the intended microscope

2. Fluorescent microscope (e.g., Nikon Ti-E Wide Field Deconvolution Inverted Microscope)
3. Flow cytometer with at least two colors (e.g., BD Biosciences LSR II)
4. Flow cytometry tubes without cell strainer caps
5. Centrifuge and microcentrifuge with compatible sterile centrifuge tubes
6. Rotating incubation platform

4.2 Chemicals

1. DSPE-PEG$_{2000}$-Biotin (Avanti Polar Lipids, Cat: 880129P)
2. DSPE-PEG$_{2000}$-DBCO (Avanti Polar Lipids, Cat: 880229P)
3. Phosphate buffered saline (PBS, pH 7.4)
4. PBS with 0.1% w/v bovine serum albumin (PBSA)
5. Cell culture media and dissociation reagents appropriate for chosen cell line(s)
6. Streptavidin-AlexaFluor 488 conjugate or similar (Thermo Fisher Scientific Cat: S32354, diluted to 10 μg/mL in PBS)
7. Azide-AlexaFluor 488 conjugate or similar (Thermo Fisher Scientific Cat: A10266, diluted to 5 μM in PBS)
8. Trimethoprim, only required if reversing CSAN-induced interactions (Fisher Scientific, Cat: AAJ66646MD)
9. CellTrace Violet (Thermo Fisher Scientific, Cat: C34571)
10. CellTrace Far Red (Thermo Fisher Scientific, Cat: 34572)
11. Carboxyfluorescein succinimidyl ester (CFSE; Biolegend, Cat: 423801)
12. Paraformaldehyde, 4% in PBS
13. ProLong Gold Antifade Mountant or similar mounting agent (Thermo Fisher Scientific, Cat: P10144 or P36935 if nuclear staining with DAPI is desired)
14. Targeting ligand(s) of choice, appropriately modified with biotin, streptavidin, or azide groups depending on the planned conjugation approach

4.3 Modifying cell membranes with phospholipid conjugates via hydrophobic insertion

There are two basic approaches to modifying cells with phospholipid conjugates—an "in vitro" method where the cell culture media is supplemented with the phospholipid, and an "ex vitro" method where cells are harvested from culture and resuspended in a phospholipid-containing buffer. For most applications, the in vitro method will be more useful, as it is less technically demanding and is generally more amenable to maintaining sterility of the cell culture; still, we will describe both approaches. If desired, successful phospholipid insertion can be quickly verified via flow cytometry prior to proceeding to the conjugation steps.

Ex vitro modification (buffer method)

1. Prepare or obtain the requisite buffer solutions:
   a. PBS, pH 7.4.
   b. PBSA, pH 7.4.
   c. Phospholipid solution, wherein the desired phospholipid conjugate(s) is/are dissolved in PBS to the desired stock or working concentrations (typically 1–100 μM).
2. Grow mammalian cells to the desired confluence under conditions appropriate for your cell line or type. The volumes here assume mammalian cells were grown to ~80% confluence in a standard T75 culture flask.
3. Harvest mammalian cells and centrifuge (500 × *g*, 5 min) to pellet cells; discard media supernatant. If the cell line is adherent, either enzymatic or mechanical dissociation is acceptable from the perspective of the protocol.
4. Resuspend the cell pellet in 1 mL ice cold PBSA; transfer to a 1.5 mL microcentrifuge tube.
5. Pellet cells at 8000 × *g*, 30 s, 4 °C; discard supernatant.
6. Resuspend in 1 mL ice cold PBSA.
7. Determine cell count/viability by trypan blue exclusion or an equivalent method.
8. Resuspend cells in the phospholipid solution at a ratio of approximately 0.5×10^6 cells per 200 μL of phospholipid solution; generally, a 10–100 μM concentration of phospholipid will saturate most cell surfaces under these conditions (Csizmar, Petersburg, Hendricks, et al., 2018). Depending on the physical size of your cells, however, these concentrations and ratios may require individual optimization.
9. Incubate at room temperature for 1 h on a rotary platform to facilitate mixing.
10. Pellet cells at 8000 × *g*, 30 s, 4 °C; discard supernatant.
11. Wash cells twice with 1 mL ice cold PBS. The cells are now ready for downstream use.

Tip: Albumin contains multiple lipid-binding domains. Therefore, we avoid the use of albumin in our buffers during the hydrophobic insertion process (step 8, above) and thereafter to reduce the risk of sequestering the lipids on the carrier protein instead of inserting in the bilayer.

In vitro modification (culture method)

1. Grow mammalian cells to the desired confluence under conditions appropriate for your cell line or type.
2. Passage cells into sterile growth media supplemented with the phospholipid conjugate at the desired concentration.
3. Continue to grow/incubate cells for a minimum of 16 h.
4. Harvest mammalian cells and centrifuge (500 × *g*, 5 min) to pellet cells; discard media supernatant. If the cell line is adherent, either enzymatic or mechanical dissociation is acceptable from the perspective of the protocol.
5. Wash cells thrice in cold PBS to remove residual, uninserted phospholipid conjugates. The cells are now ready for downstream use.

Tip: Albumin contains multiple lipid-binding domains. Therefore, we avoid the use of albumin in the wash buffers (step 5, above) and thereafter to reduce the risk of further partitioning the lipids out of the cell membrane.

Verification of lipid insertion via flow cytometry (optional)

1. Resuspend 0.5×10^6 of the phospholipid-modified cells in 50 μL of primary fluorophore stain solution
   a. *For biotin-modified samples, consider a streptavidin-AlexaFluor 488 conjugate*
   b. *For DBCO-modified samples, consider an azide-AlexaFluor 488 conjugate*
2. Incubate at the necessary temperature in the dark for ≥30 min
   a. *Biotin/streptavidin samples will be labeled within 30 min at 4°C*
   b. *DBCO/azide samples will typically require prolonged incubation times at elevated temperatures* (e.g., *≥3 h at 37°C) to promote the cycloaddition reaction*
3. Pellet cells at 8000 × *g*, 30 s, 4 °C; discard the supernatant
4. Resuspend cells in 1 mL ice cold PBS
5. Repeat the wash steps 3–4 for a total of three washes in this manner
6. Strain washed cells into a flow cytometry tube; keep cells on ice until ready to analyze
7. Analyze on cells on a flow cytometer, assessing the mean fluorescence intensity in the 488 nm channel. Include appropriate positive and negative controls

4.4 Conjugating protein scaffolds to phospholipid-modified cells

Following the hydrophobic insertion of the phospholipid conjugates, the subsequent conjugation steps are relatively straightforward. Generally, the phospholipid-modified cells need only to be resuspended in a buffer containing the desired conjugate and incubated under empirically determined conditions. Depending on the nature of the system, verification of conjugation can be again assessed by flow cytometry akin to above, probing now for a tag or epitope on the conjugated cargo (if present).

Ligand conjugation (biotin/streptavidin interaction)

1. Resuspend phospholipid-modified cells (0.5×10^6) in 500 μL of 100 nM streptavidin-protein complex in PBS.
2. Incubate at 4 °C for ≥30 min.
3. Pellet cells at 8000 × *g*, 30 s, 4 °C; discard the supernatant.
4. Resuspend cells in 1 mL ice cold PBS.
5. Repeat the wash steps 3–4 for a total of three washes in this manner. The cells are now ready for downstream use.

Ligand conjugation (DBCO/azide cycloaddition reaction)

1. Resuspend phospholipid-modified cells (0.5×10^6) in 500 μL of 100 nM azide-labeled ligand in PBS.
2. Incubate at 37 °C for ≥3 h.
3. Pellet cells at 8000 × *g*, 30 s, 4 °C; discard the supernatant.
4. Resuspend cells in 1 mL ice cold PBS.
5. Repeat the wash steps 3–4 for a total of three washes in this manner. The cells are now ready for downstream use.

Tip: If cell viability decreases significantly during the prolonged incubation in PBS, cell culture media can be substituted for the buffer to facilitate cell survival.

4.5 Forming and reversing cell-cell interactions using CSANs

This section provides protocols for forming cell-cell interactions following cellular modification with targeting ligands as outlined above. The methods here—in particular the reversal strategy with trimethoprim—are optimized for the CSAN platform. However, they should provide a reasonable starting point for developing individualized workflows for other ligand systems. The next section will provide guidance on how to prepare cells in a manner that will allow visualization and analysis of the cell-cell interactions.

Formation of targeted intercellular interactions in suspension

1. Prepare ligand-modified cells as described above.
2. Resuspend ligand-modified cells with the antigen-expressing (or complimentarily labeled) target cells in the desired ratio (e.g., 1:1) in a minimal volume of PBS. For most preliminary experiments, a buffer volume of 1 mL will suffice.

3. Incubate cell mixture at 4 °C for ≥60 min on a rotary platform to facilitate mixing.
4. Pellet cells at 8000 × *g*, 30 s, 4 °C; discard the supernatant.
5. Gently resuspend cells in 1 mL ice cold PBS.
6. (Optional) Repeat the wash steps 4–5 as desired if there is concern for free or dissociated ligands that would interfere with subsequent applications.
7. Analyze cell-cell interactions by flow cytometry or microscopy as described in the next section.

Tip: If the intercellular interactions are not strong or stable enough to tolerate rapid centrifugation, the pelleting conditions can be slowed to 500 × *g*, 5 min. Do take care to gently resuspend cells during washing to avoid shearing apart cell clusters.

Formation of targeted intercellular interactions in adherent monolayers

1. Prepare ligand-targeted cells as described above.
2. Grow antigen-expressing (or complimentarily labeled) target cells on glass coverslips in a six-well plate. Alternatively, the target cells can be grown or adhered directly in the well plate (or other suitable vessel) if your microscopy system permits such analysis.
3. Remove media from the six-well plate via gentle aspiration.
4. Wash each well once with 1 mL cold PBS. Remove buffer via gentle aspiration.
5. Add 1 mL fresh, cold cell culture media to each well.
6. Add an equal number of ligand-targeted cells to each well.
7. Incubate the six-well plate static at 4 °C for 1 h.
8. Remove media and supernatant from each well by gentle aspiration.
9. Wash each well thrice with 1 mL cold PBS. Remove buffer via gentle aspiration.
10. Analyze cell-cell interactions by microscopy as described in the next section. This will likely require cell fixation and mounting of the coverslips on glass microscopy slides, both of which are described subsequently.

Reversal of CSAN-directed intercellular interactions

1. Label cells with antigen-targeted CSANs using one of the hydrophobic insertion and conjugation approaches described previously.
2. Form cell-cell interactions between CSAN-labeled cells and antigen-expressing cells as described above.
3. For cellular interactions in suspension, resuspend the cell pellet 1 mL cold PBS supplemented with 2 μM trimethoprim. Incubate at 4 °C on a rotary platform for 1 h. Gently pellet and wash dissociated cells as above.

4. For cellular interactions in monolayers, add 1 mL of cold culture media supplemented with 2 μM trimethoprim to the desired wells. Incubate the cell layers at 4 °C for 1 h. Remove media and unbound cells via gentle aspiration and wash cells with 1 mL cold PBS as above.
5. Following dissociation, proceed to the desired downstream application or analysis.

5. Analysis of the cell-cell interactions

Essentially any method for visualizing cells can be used to analyze cell-cell interactions. Due to their ease, widespread availability, and relatively low cost, flow cytometry and fluorescence microscopy are commonly used in this setting. Flow cytometry is only amenable to suspensive cells, but offers a convenient and rapid approach. Regardless, both techniques benefit from fluorescently-labeling the different cell populations prior to beginning the cell modification workflow. A convenient way to accomplish this is to use various commercial "cell trace" dyes to label the growing cell cultures 1–2 days prior to use. This is especially true if the synthetic ligand(s) installed on the cell lack fluorescent tags of their own or epitopes that could be labeled with fluorophore-conjugated antibodies. In the absence of fluorescent imaging, standard light microscopy can be used, but the cell populations may be difficult to distinguish if they are similar in size and morphology or are particularly abundant (e.g., overgrown cultures or formation of multicellular layers).

5.1 Fluorescence microscopy

1. Label the target and effector cell populations with distinct fluorescent dyes according to the manufacturer's protocol.
   a. *For example, CellTrace Far Red, CFSE, and DAPI have* minimally-*overlapping emission spectra suitable for analysis by fluorescence microscopy.*
2. (Optional) Return fluorescently-labeled cell populations to culture for ≥16 h. Some manufactures provide protocols for labeling adherent cells in culture; in this case, the adherent target cells can be labeled in vitro without the need to dissociate and then re-adhere them. If the effector cells are also adherent, it may be more convenient to begin the cell modification workflow immediately rather than return them to culture, which would require and additional dissociation step to apply them to the target cell layer.

3. Modify effector cell population with phospholipid conjugates via hydrophobic insertion, as described above.
4. Functionalize the phospholipid-modified effector cell population with a targeting ligand, as described above.
5. Form directed interactions between the effector and target cells populations, as described above.
6. (Optional) Reverse the cell-cell interactions if desired.
7. (Optional) Fix the cell samples by adding 1 mL 4% paraformaldehyde to each well and incubate at room temperature for 15 min; remove the paraformaldehyde solution via gentle aspiration. Gently wash each well thrice with 1 mL ice cold PBS.
8. Rinse each coverslip twice using 1 mL ultrapure water; blot excess liquid away from the back of the coverslip.
9. Add one drop of mounting agent (e.g., ProLong Gold Antifade Reagent with or without DAPI nuclear stain) to a glass slide and place the coverslip sample-side down onto the drop; press gently to remove air bubbles.
10. Allow the slide to cure at room temperature for ≥24 h in the dark.
11. (Optional) Seal the edges of the coverslip with clear epoxy. Store slides in the dark until ready to image.
12. Image the slides on a fluorescent microscope (such as a Nikon Ti-E Wide Field Deconvolution Inverted Microscope).
13. Assess the population for the presence of cell clusters, rosettes, or layers (Fig. 3A–C). If reversal of the interactions was attempted, there should be a corresponding decrease in these groups in the reversed sample.

Tip: There are many commercial products that replace the traditional slide and coverslip system described above. Depending on the application, these may help streamline the sample preparation and analysis process, and the user is encouraged to explore these options as desired.

5.2 Flow cytometry

1. Label the target and effector cell populations with distinct fluorescent dyes according to the manufacturer's protocol.
   a. *For example, CellTrace Violet and CFSE provide minimally-overlapping emission spectra suitable for flow cytometric analysis.*
2. (Optional) Return fluorescently-labeled cell populations to culture for ≥16 h. This can be useful for expanding the cell populations or for streamlining the remainder of the workflow.

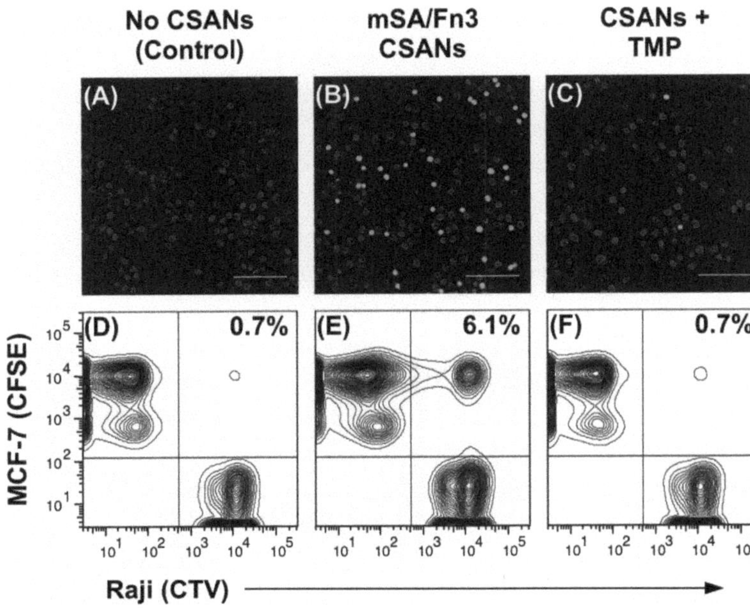

Fig. 3 Analysis of CSAN-directed cell-cell interactions. For the fluorescence microscopy experiment (top row), Raji cells were sequentially labeled with CFSE, DSPE-PEG$_{2000}$-biotin and with or without mSA/Fn3 CSANs; they were then incubated with a monolayer of EpCAM+ MCF-7 cells. (A) In the absence of CSANs, the phospholipid-modified Raji cells are unable to interact with the MCF-7 cells. (B) When functionalized with the mSA/Fn3 CSANs, the EpCAM-targeted Raji cells adhere to the MCF-7 cell monolayer. (C) The EpCAM-targeted Raji cells can be dissociated from the MCF-7 cell monolayer by disassembling the CSAN with trimethoprim (TMP). Scale bars in (a–c) represent 100 μm. For the flow cytometry experiment (bottom row), the target MCF-7 cells were labeled with CFSE while the Raji cells were labeled with CTV. Raji cells were again modified with DSPE-PEG$_{2000}$-biotin and labeled with or without mSA/Fn3 CSANs. (D) In the absence of CSANs, the phospholipid-modified Raji cells are unable to interact with the MCF-7 cells. (E) When functionalized with the mSA/Fn3 CSANs, the EpCAM-targeted Raji cells formed stable clusters with the MCF-7 cells. (F) The Raji/MCF-7 cell clusters were readily dissociated with trimethoprim. Data are representative of replicate ($n = 3$) experiments. *Figure and caption reprinted from Csizmar, C. M., Petersburg, J. R., Hendricks, A., Stern, L. A., Hackel, B. J., & Wagner, C. R. (2018). Engineering reversible cell-cell interactions with lipid anchored prosthetic receptors.* Bioconjugate Chemistry, *29(4), 1291–1301, copyright 2018, with permission from ACS Publications.*

3. Modify effector cell population with phospholipid conjugates via hydrophobic insertion, as described above.
4. Functionalize the phospholipid-modified effector cell population with a targeting ligand, as described above.

5. Form directed interactions between the effector and target cells populations, as described above.
6. (Optional) Reverse the cell-cell interactions if desired.
7. Resuspend washed cell population in a suitable volume of PBS (or other cytometer running buffer). It is recommended to keep the cell concentration reasonably low ($<5 \times 10^6$ cells/mL) to avoid clogging the cytometer with cellular aggregates.
8. Transfer the cell suspension to a flow cytometry tube; do *not* pass the cell suspension through a strainer cap or mesh, as this will mechanically shear the interacting cells.
9. Analyze the cell suspension on a flow cytometer equipped with the laser lines required for your chosen fluorophores (such as a BD Biosciences LSR II).
10. Assess the population for the presence of a double-positive cell population (e.g., a CTV+/CFSE+ population) that represents the distinctly labeled cells clustered together (Fig. 3D–F). If reversal of the interactions was attempted, there should be a corresponding decrease in this population in the reversed sample.

Tip: One should be cautious when applying forward and side scatter gates to the sample. These gates are often used to remove cell doublets and aggregates, which would be counter-productive in this setting.

6. Summary

A variety of nongenetic methods exist for functionalizing cell membranes with targeting and binding moieties. This chapter describes these approaches, with an emphasis on those that have been used to direct reversible cell-cell interactions. The methods here provide a starting point for investigators who are interested in using hydrophobic insertion or chemical modification techniques to quickly modify cell membranes for a variety of downstream applications, including tissue engineering, cell homing, and cellular immunotherapy.

Acknowledgments

The authors gratefully acknowledge support from the National Institutes of Health R21 CA185627 (C.R.W.), T32 GM008244 (C.M.C.), and the University of Minnesota Foundation.

References

Altman, M. O., Chang, Y. M., Xiong, X., & Tan, W. (2013). Modifying cellular properties using artificial aptamer-lipid receptors. *Scientific Reports, 3*, 3343.

Amaral, A. J., & Pasparakis, G. (2015). Macromolecular cell surface engineering for accelerated and reversible cellular aggregation. *Chemical Communications (Cambridge, England), 51*(99), 17556–17559.

Carlson, J. C. T., Jena, S. S., Flenniken, M., Chou, T.-F., Siegel, R. A., & Wagner, C. R. (2006). Chemically controlled self-assembly of protein nanorings. *Journal of the American Chemical Society, 128*(23), 7630–7638.

Cashman, T. J., Gouon-Evans, V., & Costa, K. D. (2013). Mesenchymal stem cells for cardiac therapy: Practical challenges and potential mechanisms. *Stem Cell Reviews, 9*(3), 254–265.

Chiarantini, L., Droleskey, R., Magnani, M., & DeLoach, J. R. (1992). In vitro targeting of erythrocytes to cytotoxic T-cells by coupling of Thy-1.2 monoclonal antibody. *Biotechnology and Applied Biochemistry, 15*(2), 171–184.

Csizmar, C. M., Petersburg, J. R., Hendricks, A., Stern, L. A., Hackel, B. J., & Wagner, C. R. (2018). Engineering reversible cell-cell interactions with lipid anchored prosthetic receptors. *Bioconjugate Chemistry, 29*(4), 1291–1301.

Csizmar, C. M., Petersburg, J. R., Perry, T. J., Rozumalski, L., Hackel, B. J., & Wagner, C. R. (2019). Multivalent ligand binding to cell membrane antigens: Defining the interplay of affinity, valency, and expression density. *Journal of the American Chemical Society, 141*(1), 251–261.

Csizmar, C. M., Petersburg, J. R., & Wagner, C. R. (2018). Programming cell-cell interactions through non-genetic membrane engineering. *Cell Chemical Biology, 25*(8), 931–940.

de Kruif, J., Tijmensen, M., Goldsein, J., & Logtenberg, T. (2000). Recombinant lipid-tagged antibody fragments as functional cell-surface receptors. *Nature Medicine, 6*(2), 223–227.

Dutta, D., Pulsipher, A., Luo, W., Mak, H., & Yousaf, M. N. (2011). Engineering cell surfaces via liposome fusion. *Bioconjugate Chemistry, 22*(12), 2423–2433.

Dutta, D., Pulsipher, A., Luo, W., & Yousaf, M. N. (2011). Synthetic chemoselective rewiring of cell surfaces: Generation of three-dimensional tissue structures. *Journal of the American Chemical Society, 133*(22), 8704–8713.

Fernandez-Vidal, M., White, S. H., & Ladokhin, A. S. (2011). Membrane partitioning: "Classical" and "nonclassical" hydrophobic effects. *The Journal of Membrane Biology, 239*(1–2), 5–14.

Gabrielse, K., Gangar, A., Kumar, N., Lee, J. C., Fegan, A., Shen, J. J., et al. (2014). Reversible re-programing of cell-cell interactions. *Angewandte Chemie (International Ed. in English), 53*(20), 5112–5116.

Gangar, A., Fegan, A., Kumarapperuma, S. C., & Wagner, C. R. (2012). Programmable self-assembly of antibody-oligonucleotide conjugates as small molecule and protein carriers. *Journal of the American Chemical Society, 134*(6), 2895–2897.

Gartner, Z. J., & Bertozzi, C. R. (2009). Programmed assembly of 3-dimensional microtissues with defined cellular connectivity. *Proceedings of the National Academy of Sciences of the United States of America, 106*(12), 4606–4610.

Gong, P., Zheng, W., Huang, Z., Zhang, W., Xiao, D., & Jiang, X. (2013). A strategy for the construction of controlled, three-dimensional, multilayered, tissue-like structures. *Advanced Functional Materials, 23*(1), 42–46.

Gorfe, A. A., Baron, R., & McCammon, J. A. (2008). Water-membrane partition thermodynamics of an amphiphilic lipopeptide: An enthalpy-driven hydrophobic effect. *Biophysical Journal, 95*(7), 3269–3277.

Hamdy, N., Goustin, A. S., Desaulniers, J. P., Li, M., Chow, C. S., & Al-Katib, A. (2005). Sheep red blood cells armed with anti-CD20 single-chain variable fragments (scFvs) fused to a glycosylphosphatidylinositol (GPI) anchor: A strategy to target CD20-positive tumor cells. *Journal of Immunological Methods, 297*(1–2), 109–124.

Jeong, J. H., Schmidt, J. J., Kohman, R. E., Zill, A. T., DeVolder, R. J., Smith, C. E., et al. (2013). Leukocyte-mimicking stem cell delivery via in situ coating of cells with a bioactive hyperbranched polyglycerol. *Journal of the American Chemical Society, 135*(24), 8770–8773.

Karp, J. M., & Leng Teo, G. S. (2009). Mesenchymal stem cell homing: The devil is in the details. *Cell Stem Cell, 4*(3), 206–216.

Kean, T. J., Duesler, L., Young, R. G., Dadabayev, A., Olenyik, A., Penn, M., et al. (2012). Development of a peptide-targeted, myocardial ischemia-homing, mesenchymal stem cell. *Journal of Drug Targeting, 20*(1), 23–32.

Kim, S. A., & Peacock, J. S. (1993). The use of palmitate-conjugated protein A for coating cells with artificial receptors which facilitate intercellular interactions. *Journal of Immunological Methods, 158*(1), 57–65.

Ko, I. K., Kean, T. J., & Dennis, J. E. (2009). Targeting mesenchymal stem cells to activated endothelial cells. *Biomaterials, 30*(22), 3702–3710.

Lamers, C. H., Sleijfer, S., van Steenbergen, S., van Elzakker, P., van Krimpen, B., Groot, C., et al. (2013). Treatment of metastatic renal cell carcinoma with CAIX CAR-engineered T cells: Clinical evaluation and management of on-target toxicity. *Molecular Therapy, 21*(4), 904–912.

Li, Q., So, C. R., Fegan, A., Cody, V., Sarikaya, M., Vallera, D. A., et al. (2010). Chemically self-assembled antibody nanorings (CSANs): Design and characterization of an anti-CD3 IgM biomimetic. *Journal of the American Chemical Society, 132*(48), 17247–17257.

Lim, K. H., Huang, H., Pralle, A., & Park, S. (2013). Stable, high-affinity streptavidin monomer for protein labeling and monovalent biotin detection. *Biotechnology and Bioengineering, 110*(1), 57–67.

Lo, C. Y., Antonopoulos, A., Dell, A., Haslam, S. M., Lee, T., & Neelamegham, S. (2013). The use of surface immobilization of P-selectin glycoprotein ligand-1 on mesenchymal stem cells to facilitate selectin mediated cell tethering and rolling. *Biomaterials, 34*(33), 8213–8222.

Lu, L., Gao, J., & Guo, Z. (2015). Labeling cell surface GPIs and GPI-anchored proteins through metabolic engineering with artificial inositol derivatives. *Angewandte Chemie (International Ed. in English), 54*(33), 9679–9682.

Luo, W., Pulsipher, A., Dutta, D., Lamb, B. M., & Yousaf, M. N. (2014). Remote control of tissue interactions via engineered photo-switchable cell surfaces. *Scientific Reports, 4*, 6313.

Merzaban, J. S., Imitola, J., Starossom, S. C., Zhu, B., Wang, Y., Lee, J., et al. (2015). Cell surface glycan engineering of neural stem cells augments neurotropism and improves recovery in a murine model of multiple sclerosis. *Glycobiology, 25*(12), 1392–1409.

Nowakowski, A., Andrzejewska, A., Janowski, M., Walczak, P., & Lukomska, B. (2013). Genetic engineering of stem cells for enhanced therapy. *Acta Neurobiologiae Experimentalis (Wars), 73*(1), 1–18.

O'Brien, P. J., Luo, W., Rogozhnikov, D., Chen, J., & Yousaf, M. N. (2015). Spheroid and tissue assembly via click chemistry in microfluidic flow. *Bioconjugate Chemistry, 26*(9), 1939–1949.

Palte, M. J., & Raines, R. T. (2012). Interaction of nucleic acids with the glycocalyx. *Journal of the American Chemical Society, 134*(14), 6218–6223.

Petersburg, J., Shen, J., Csizmar, C. M., Murphy, K. A., Spanier, J., Gabrielse, K., et al. (2018). Eradication of established tumors by chemically self-assembled nanoring (CSAN) targeted T-cells. *ACS Nano, 12*, 6563–6576.

Pulsipher, A., Dutta, D., Luo, W., & Yousaf, M. N. (2014). Cell-surface engineering by a conjugation-and-release approach based on the formation and cleavage of oxime linkages upon mild electrochemical oxidation and reduction. *Angewandte Chemie (International Ed. in English), 53*(36), 9487–9492.

Rogozhnikov, D., O'Brien, P. J., Elahipanah, S., & Yousaf, M. N. (2016). Scaffold free bioorthogonal assembly of 3-dimensional cardiac tissue via cell surface engineering. *Scientific Reports, 6*, 39806.

Sadelain, M., Rivière, I., & Riddell, S. (2017). Therapeutic T cell engineering. [Review Article], *Nature, 545*, 423.

Sarkar, D., Spencer, J. A., Phillips, J. A., Zhao, W., Schafer, S., Spelke, D. P., et al. (2011). Engineered cell homing. *Blood, 118*(25), e184–e191.

Selden, N. S., Todhunter, M. E., Jee, N. Y., Liu, J. S., Broaders, K. E., & Gartner, Z. J. (2012). Chemically programmed cell adhesion with membrane-anchored oligonucleotides. *Journal of the American Chemical Society, 134*(2), 765–768.

Shah, R., Petersburg, J., Gangar, A. C., Fegan, A., Wagner, C. R., & Kumarapperuma, S. C. (2016). In vivo evaluation of site-specifically PEGylated chemically self-assembled protein nanostructures. *Molecular Pharmaceutics, 13*(7), 2193–2203.

Shen, J., Vallera, D. A., & Wagner, C. R. (2015). Prosthetic antigen receptors. *Journal of the American Chemical Society, 137*(32), 10108–10111.

Shi, P., Ju, E., Yan, Z., Gao, N., Wang, J., Hou, J., et al. (2016). Spatiotemporal control of cell-cell reversible interactions using molecular engineering. *Nature Communications*, 7, 13088.

Stern, L. A., Csizmar, C. M., Woldring, D. R., Wagner, C. R., & Hackel, B. J. (2017). Titratable avidity reduction enhances affinity discrimination in mammalian cellular selections of yeast-displayed ligands. *ACS Combinatorial Science, 19*(5), 315–323.

Takeo, M., Li, C., Matsuda, M., Nagai, H., Hatanaka, W., Yamamoto, T., et al. (2015). Optimum design of amphiphilic polymers bearing hydrophobic groups for both cell surface ligand presentation and intercellular cross-linking. *Journal of Biomaterials Science Polymer Edition, 26*(6), 353–368.

Weber, R. J., Liang, S. I., Selden, N. S., Desai, T. A., & Gartner, Z. J. (2014). Efficient targeting of fatty-acid modified oligonucleotides to live cell membranes through stepwise assembly. *Biomacromolecules, 15*(12), 4621–4626.

Won, Y. W., Patel, A. N., & Bull, D. A. (2014). Cell surface engineering to enhance mesenchymal stem cell migration toward an SDF-1 gradient. *Biomaterials, 35*(21), 5627–5635.

Xiong, X., Liu, H., Zhao, Z., Altman, M. B., Lopez-Colon, D., Yang, C. J., et al. (2013). DNA aptamer-mediated cell targeting. *Angewandte Chemie International Edition, 52*(5), 1472–1476.

Zhao, W., Loh, W., Droujinine, I. A., Teo, W., Kumar, N., Schafer, S., et al. (2011). Mimicking the inflammatory cell adhesion cascade by nucleic acid aptamer programmed cell-cell interactions. *The FASEB Journal, 25*(9), 3045–3056.

CHAPTER TEN

Fast phosphine-activated control of protein function using unnatural lysine analogues

Joshua S. Wesalo, Alexander Deiters*
Department of Chemistry, University of Pittsburgh, Pittsburgh, PA, United States
*Corresponding author: e-mail address: deiters@pitt.edu

Contents

Abstract

Effective, general methods for conditionally activating proteins in their native biological environments are highly useful for biological studies. Since phosphines and azides are not found in pro- and eukaryotic cells, the Staudinger reduction can function as an excellent small molecule-controlled switch for protein activation. This methodology involves site-specifically incorporating azidobenyl-lysine analogues into proteins in live cells. When placed at a crucial position, these unnatural side chains block protein function until a phosphine trigger is added. We discuss methods for expressing caged proteins in bacterial and mammalian cells in high yields, and activating the proteins with an optimized phosphine trigger. We also discuss important considerations for safe and effective synthesis of these molecules. This methodology was used to translocate proteins to the nucleus and to turn-on a protein post-translational modification (SUMOylation) in living cells.

Methods in Enzymology, Volume 638
ISSN 0076-6879
https://doi.org/10.1016/bs.mie.2020.02.020

Abbreviations

2DPBM	2-(diphenylphosphino)benzamide
aaRS	aminoacyl-tRNA synthetase
CTD	C-terminal domain
DREADDs	designer receptors exclusively activated by designer drugs
HBS	HEPES-buffered saline
IMAC	immobilized metal ion affinity chromatography
LB+Kan+Tet	LB containing kanamycin (50 $\mu g\ mL^{-1}$) and tetracycline (25 $\mu g\ mL^{-1}$)
LPEI	linear polyethylenimine
***Mb*PylRS**	*Methanosarcina barkeri* PylRS
MOABK	methyl *ortho*-azidobenzyl lysine
MPABK	methyl *para*-azidobenzyl lysine
Ni-NTA	nickel(II)-nitrilotriacetic acid
NLS	nuclear localization sequence
NPC	nuclear pore complex
NTD	N-terminal domain
OABK	*ortho*-azidobenzyl lysine
P/S	penicillin/streptomycin
PABK	*para*-azidobenzyl lysine
PDK	poly-D-lysine
PylRS	pyrrolysyl-tRNA synthetase
PylT	$tRNA^{Pyl}$
RanGAP1	ran GTPase-activating protein 1
RanBP2	ran binding protein 2
RNΔN	RanGAP1, residues 400–589
SUMO	small ubiquitin-like modifier
TCO	*trans*-cyclooctene
UAA	unnatural amino acid

1. Introduction

1.1 Conditional control of protein function

A detailed understanding of how proteins function in living cells is essential to continued progress in the biomedical sciences. To gain this understanding, investigators need conditional tools that can quickly modulate protein function in living cells. While thousands of drugs have been developed that address this challenge, drugs only cover a limited subset of the proteome, may generate off-target effects, and usually can only deactivate protein function. To address this methodology gap, several techniques involving careful engineering of proteins and ligands have arisen. These methods include chemicals that induce dimerization (Fegan, White, Carlson, & Wagner, 2010) or dissociation (Rose et al., 2019), bump-and-hole approaches (Arora, Horne, & Islam, 2019), and designer receptors exclusively activated by designer drugs (DREADDs) (Dobrzanski & Kossut, 2017).

Several groups have developed unnatural amino acids (UAAs) with removable side-chain protecting groups. Reactions that achieved conditional deprotection include Pd-catalyzed removal of allenyl, allyl, and propargyl groups (Li et al., 2014; Wang et al., 2016) and tetrazine-mediated release of *trans*-cyclooctene (TCO)-masked lysine (Li, Jia, & Chen, 2014). These approaches face tradeoffs. Pd catalysis affords modest yields with slow kinetics (Li, Yu, et al., 2014; Wang, Zheng, et al., 2016). While tetrazine-mediated TCO deprotection is significantly faster, TCOs can convert to the unreactive *cis*-isomer in storage or during experiments in live cells or organisms (Li, Jia, & Chen, 2014), and tetrazines must sacrifice aqueous stability to obtain fast reaction rates (Karver, Weissleder, & Hilderbrand, 2011). To complement these methodologies, our group has developed several azido-UAAs utilizing the Staudinger reduction for protein activation (Luo, Liu, Morihiro, & Deiters, 2016; Wesalo, Luo, Morihiro, Liu, & Deiters, 2019). These UAAs quickly and quantitatively release lysine upon treatment with bioorthogonal phosphine reagents.

1.2 The Staudinger reduction for controlling biomolecular function

Several properties make the Staudinger reduction an excellent tool for controlling the function of biomolecules. The participating functional groups—azides and phosphines—are not found in nature (except for a single known example of an azide-containing natural product (Hossain, van der Helm, Sanduja, & Alam, 1985)). Because azides are soft electrophiles, they react chemoselectively with soft nucleophiles (e.g., phosphines). Thus, the azide-phosphine pair is bioorthogonal to biological nucleophiles and electrophiles, which are primarily hard in character (Sletten & Bertozzi, 2011). Azides have been extensively used in photocrosslinking experiments (Chin et al., 2002), carbohydrate labeling, and pharmaceuticals (e.g., zidovudine) (Sletten & Bertozzi, 2011), demonstrating their safety and stability in biological contexts. We have shown that phosphines typically used for the Staudinger reduction have insignificant cytotoxicity at micromolar concentration (Luo et al., 2016). However, phosphines are susceptible to oxidation under physiological conditions (Sletten & Bertozzi, 2011). Oxidation susceptibility complicates synthetic procedures. Degassed reaction, workup, and chromatography solvents are typically necessary for working with triarylphosphines (Chen et al., 2013). Since the solubility of oxygen in most organic solvents is an order of magnitude higher than its solubility in water (Sato, Hamada, Sumikawa, Araki, & Yamamoto, 2014), however, applications in biological systems may be feasible, even for phosphines that rapidly oxidize during synthesis.

The rate-determining step (and kinetic order) of the Staudinger reduction varies depending on the electronics of the reaction partners (Van Berkel, Van Eldijk, & Van Hest, 2011). We cannot increase phosphine nucleophilicity (Schilling, Jung, Biskup, Schepers, & Bräse, 2011) to hasten the (often) rate-determining bimolecular step (Van Berkel et al., 2011), as nucleophilicity promotes oxidation (Sletten & Bertozzi, 2011). The nature of P-substituents has been explored as well. While trialkylphosphines rapidly reduce disulfide functions, triaryl phosphines do not (Saxon & Bertozzi, 2000); this chemoselectivity, coupled with the increased steric bulk around the phosphorus center to prevent oxidation, makes triarylphosphines useful small molecule triggers. The most effective tuning for phosphines thus far is appending a carboxyl or amido group (particularly a primary amide) *ortho* to the phosphorus center on one of the phenyl groups (Lukasak, Morihiro, & Deiters, 2019; Luo et al., 2016; Saneyoshi et al., 2014). The phosphine 2-(diphenylphoshino)benzamide (**2DPBM**) (Saneyoshi et al., 2014) or commercially-available 2-(diphenylphosphino)benzoate are ideal, as the substituents accelerate hydrolysis of iminophosphorane intermediates via a neighboring group effect (Lukasak et al., 2019; Saneyoshi et al., 2014). Both substituents also increase water solubility and stability toward oxidation in our experience.

1.3 Unnatural amino acid (UAA) mutagenesis

Over the past two decades, genetic code expansion has enabled hundreds of UAAs to be encoded in both pro- and eukaryotic cells and animals (Brown, Liu, & Deiters, 2018). This technique has been leveraged for protein engineering, site-specific bioconjugation, therapeutics, probing protein structure and function, and controlling protein function (Young & Schultz, 2018).

In order to achieve improved small molecule activation of protein function, our group has focused on fine-tuning the azide-bearing protecting group rather than fine-tuning the phosphine reactivity, given the aforementioned constraints. Our starting point was *ortho*-azibobenzyl lysine (**OABK**),

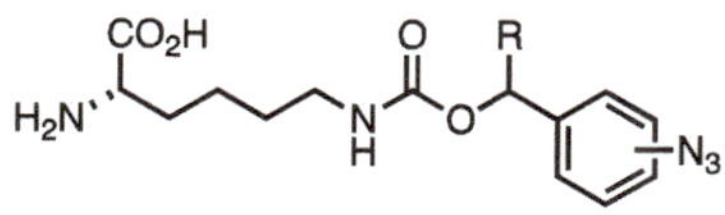

UAA	R	N_3
OABK	H	*ortho*
PABK	H	*para*
MOABK	Me	*ortho*
MPABK	Me	*para*

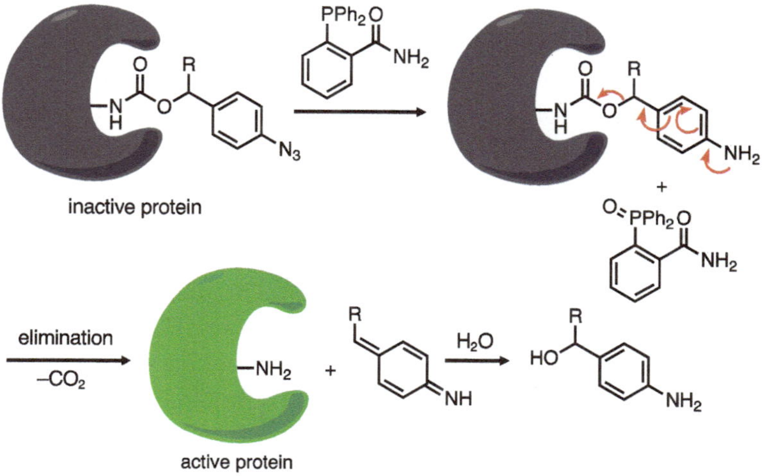

Fig. 1 UAAs structural variations and schematic illustrating 2DPBM-induced protein activation. *Adapted with permission from Wesalo, J. S., Luo, J., Morihiro, K., Liu, J., & Deiters, A. (2019). Phosphine-activated lysine analogues for fast chemical control of protein subcellular localization and protein SUMOylation,* ChemBioChem, *21, 1–9. https://doi.org/10.1002/cbic.201900464. Copyright 2019 John Wiley and Sons.*

which yields functional protein readouts in live cells with $t_{1/2}$ values from 98 to 118 min (Luo et al., 2016). We designed second-generation azido-lysines with modifications known to improve decaging kinetics (Fig. 1) (Alouane, Labruère, Le Saux, Schmidt, & Jullien, 2015). To this end, we (1) moved the azide from the *ortho* to the *para* position to remove steric congestion, and (2) methylated the benzylic position to stabilize the transient partial positive charge that develops there during 1,4- or 1,6-elimination. We found that both modifications accelerated deprotection significantly and improved decaging yield. *Para*-substitution was found to be more important than benzylic methylation, though the combined modifications showed an additive effect. Overall, **MPABK** had the fastest activation kinetics, followed by **PABK**, then **MOABK** and **OABK**. As **MOABK** was found to be surprisingly cytotoxic, we have omitted it from further testing in mammalian cells (Wesalo et al., 2019).

2. Methods

2.1 Preparation and handling of phosphines and azides

The main challenge in working with phosphorus(III) species is avoiding oxidation. Triarylphosphines should be handled using Schlenk techniques when possible, and the reader is referred to Shriver and Drezdzon's classic text, *The Manipulation of Air-Sensitive Compounds*, for detailed descriptions of airfree technique (Shriver & Drezdzon, 1986). Reaction yields increase with scale (Van Overschelde et al., 2009), presumably due to the decreased surface area:volume ratio for oxygen to diffuse into reaction mixtures with increasing scale. Indeed, many key syntheses reported in the literature have used multigram scales (Brauer et al., 2002). **2DPBM**, as phosphines go, is relatively stable. It may be prepared and isolated in modest yield without using degassed solvents as long as the work is completed quickly and the reaction mixture is kept under an inert gas atmosphere. To maximize yields, and as a general precaution for working with phosphines of unknown susceptibility, however, the following precautions should be observed:

1. Phosphines should be stored under inert gas at −20 °C.
2. Inert gasses always come with residual oxygen. While argon is more useful and easier to work with since it "blankets" samples more effectively, it typically has higher oxygen content than nitrogen unless very high-grade gas is purchased. If the residual oxygen content is high, phosphine solutions should not be exposed to high flow rates of inert gasses for long periods of time. If working with extremely sensitive primary/secondary phosphines, gas may need to be scrubbed of oxygen on-site (Busacca et al., 2009). We use Matheson Ultra High Purity argon (3 ppm O_2) for routine work.
3. Reaction solvents, liquid reagents, and NMR solvents should be degassed by at least 3 freeze-pump-thaw cycles on a Schlenk line before use.
4. Solvents and solutions used for workups and purifications may be partially degassed by sparging with nitrogen or argon for ≥30 min before use. Although the gas may be introduced through needles or Pasteur pipettes, gas dispersion tubes with glass frits are preferred for sparging >100 mL of liquids. Degassing requirements are less stringent for aqueous solutions due to lower oxygen solubility (Sato et al., 2014).
5. It is most important to exclude oxygen during and after purification to ensure that the final product is pure. While any phosphine oxides that

form before normal-phase chromatographic purification are easily removed due to their strong retention, oxidation during and after purification can compromise later experiments. Complex airfree column chromatography methods have been reported for highly sensitive compounds (Kremer & Helquist, 1984), but we find that a modified version of method reported by the Weil group (Chen et al., 2013) is satisfactory for triarylphosphines. We follow the Weil group's method with the following modifications:

a. As silica gel has a high capacity to adsorb gasses (Shriver & Drezdzon, 1986), silica gel is first degassed by heating ca. 100–150 g in a 500 mL round-bottom flask in a drying oven (≥130 °C) overnight, followed by cooling in vacuo on a Schlenk line and purging the flask with inert gas before use.

b. Flash chromatography is performed using a positive pressure of inert gas; setups that vent excess pressure (e.g., Schlenk lines) are only suitable for gravity columns.

c. If possible, a glassblower can add a sidearm with a stopcock near the top of the solvent reservoir to produce a "Schlenk" flash column. This greatly facilitates evacuation and purging with inert gas.

d. For sealing off test tubes after a fraction is collected, rubber septa are easier to apply and more durable then parafilm.

6. It is convenient to prepare stock solutions of phosphines (50 mM) for biological experiments. Phosphines (>5 mg) should be weighed into an inert gas-flushed vial or tube (note that argon will increase the mass in a container by ~0.5 mg mL^{-1} of headspace—this should be taken into account when taring vessels, or nitrogen may be used instead). After weighing, flush the headspace with inert gas (3 "vac-and-back" cycles on a Schlenk line are ideal). Cell-culture grade DMSO, added via syringe, is generally a suitable solvent if first degassed by at least 3 freeze-pump-thaw cycles. Once prepared, the stock solution should be drawn into a syringe and aliquoted into argon-flushed microcentrifuge or PCR tubes. A freshly-filled balloon containing argon that is attached to a 1 mL syringe fitted with a 22 ga. needle is convenient for purging the tubes' headspaces before closing them. Aliquots should be stored at −20 °C inside a larger argon-flushed centrifuge tube that is sealed with parafilm. Although we have found that ~1 mL aliquots of **2DPBM** stock solutions are >95% stable after 5 freeze-thaw cycles, we recommend using a new small (~5–20 μL) aliquot each time as a precaution.

7. To check phosphine integrity, $^{31}P\{^1H\}$-NMR is the method of choice, though the amount of phosphine oxide can be overestimated if the T_1 delay time is set too low (10 s is typically more than sufficient for accurate quantification (Lin et al., 2005)). No deuterated solvent is necessary (we often use degassed THF). If needed for locking the instrument, a few drops of deuterated solvent may be mixed in, or a flame-sealed melting point capillary tube containing D_2O or another deuterated solvent may be dropped into the NMR tube. When referenced to an 85% H_3PO_4 external standard, phosphines appear as singlets around $\delta = -6$ ppm, and phosphine oxides show a peak around $\delta = 30$ ppm (Kühl, 2008) (minor oxidation side-products, e.g., phosphinate esters, may appear as additional peaks in the 30–45 ppm range) (Barder & Buchwald, 2007). ESI-MS leads to overestimation of the extent of oxidation in our experience, as the $[M+H]^+$ and $[M+Na]^+$ adducts for most phosphine oxides ionize more efficiently the parent phosphines.

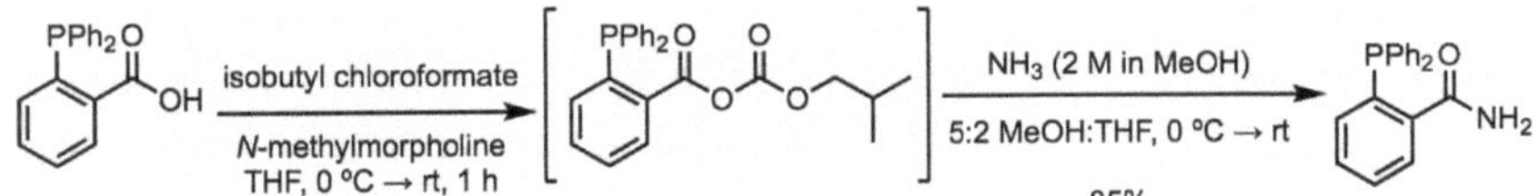

2DPBM

2DPBM was synthesized using a method modified from Saneyoshi et al. (2014) by adding 2-(diphenylphosphino)benzoic acid (246.0 mg, 0.803 mmol, 1 eq) in a flame-dried, argon-purged vial to a stirred solution of *N*-methylmorpholine (97 μL, 0.88 mmol, 1.1 eq) in anhydrous THF (2.0 mL) that had been degassed by 3 cycles of freeze-pump-thaw. The mixture was cooled to 0 °C on ice, and isobutyl chloroformate (115 μL, 0.88 mmol, 1.1 eq) was added dropwise over 2 min via syringe, resulting in opacification of the reaction mixture to a greenish-yellow suspension. The reaction mixture was allowed to warm to room temperature and was stirred for 1 h. At this time, TLC (1:1 hexanes:EtOAc) showed consumption of starting material and formation of a less polar intermediate (streak from R_f 0.2–0.95). The reaction mixture was again cooled on ice. Ammonia was added as an ice-cold 2 M solution in methanol (4.0 mL, 8.0 mmol, 10 eq) that had been briefly sparged with argon (2–3 min). The reaction mixture took on a white color, and a white precipitate formed. The reaction mixture was allowed to warm to room temperature and was stirred for 30 min. Additional ammonia (2 M in MeOH, 1.0 mL, 2.0 mmol, 2.5 eq) was added. After stirring for 30 more minutes, the reaction mixture was deposited onto degassed silica gel and purified by airfree flash chromatography

(55:45 EtOAc:hexanes, degassed by sparging with argon for 30 min before use), affording the product as a pale greenish-white powder in 85% yield (207 mg, 0.68 mmol). ^1H and ^{31}P NMR spectra matched reported values (Saneyoshi et al., 2014). TLC (1:1 hexanes:EtOAc, R_f 0.21).

Azides should be handled carefully. The same thermodynamic driving force that underlies their reactivity makes many of these compounds highly explosive (Keicher & Löbbecke, 2010). Heat and mechanical shock can result in rapid nitrogen extrusion and explosions. The following general precautions should be taken when working with azides:

1. Manipulate azides in fume hoods with sashes closed as much as possible.
2. Minimize scale within reason.
3. When using inorganic azide salts (e.g., NaN_3) for synthesis:
   a. Avoid contact with metals and with halogenated solvents to prevent explosive compounds from forming.
   b. Avoid exposure to acids. When acidic conditions are necessary, use a fume hood to prevent inhalation of toxic hydrazoic acid (HN_3) fumes.
   c. Remember that the azide ion is extremely toxic, with potency on par with cyanide. Avoid skin contact and ingestion.
4. Manipulate organic azides gently, and do not expose azides to strong heat sources. Distillation and sublimation are not appropriate purification techniques.
5. Store azides in the dark in a freezer.

As a general rule, the ratio of carbon and oxygen atoms to nitrogen atoms $(C+O)/N$ provides a general indicator of azide safety (Santa Barbara Environmental Health & Safety, 2006). A ratio ≥ 3 is relatively safe. In the 1–3 range, however, scale should be minimized, and compounds should be manipulated as dilute solutions (≤ 1 M). Compounds with a ratio ≤ 1 are quite dangerous. They are generally handled by specialist laboratories and require blast shields, leather or Kevlar gloves, and PVC- or adhesive film-coated glassware (Keicher & Löbbecke, 2010).

Note that aryl azides are competent photocrosslinking reagents (Sletten & Bertozzi, 2011). As such, they should not be exposed to strong light during their synthesis and use (Chin et al., 2002). For long-term storage, they should be stored in the dark (e.g., a box in a freezer).

Organic Azides: Syntheses and Applications (Keicher & Löbbecke, 2010) and UCSB's brief azide safety instructions (Santa Barbara Environmental Health & Safety, 2006) are excellent references that should be consulted before working with these compounds. We followed these precautions in our reported synthesis of azido-lysine derivatives (Wesalo et al., 2019) and in the ensuing biological experiments.

2.2 Site-specific incorporation of azido-lysine derivatives into proteins

We use the pyrrolysyl-tRNA synthetase/tRNA^{Pyl} (PylRS/PylT) pairs from *Methanosarcina mazei* and *M. barkeri* to enable UAA mutagenesis. This pair comprises a versatile aminoacyl-tRNA synthetase (aaRS) and tRNA with a CUA anticodon that natively binds the amber stop codon (TAG). It functions with the translational machinery in both prokaryotes and eukaryotes, and is orthogonal in both (Chin, 2017). In this work, we use the Y271A-Y349F double mutant (termed OABKRS), an efficient enzyme with an expanded catalytic site that accommodates bulky N$^{\epsilon}$-carboxybenzyl side chains (Yanagisawa et al., 2008), for **MPABK**. For **PABK**, we found that triple mutant (using *M. barkeri* numbering, L274A, C313A, Y349F; termed PABKRS) gives the highest incorporation efficiency. BHCKRS (Y271A, L274M) is also suitable for incorporating **PABK** and **MPABK** and can be used to facilitate comparisons to optical triggering (Luo et al., 2014).

2.2.1 *Protein expression in* Escherichia coli

We grow Top10 cells co-transformed two plasmids: one, e.g., pBAD-His$_6$-sfGFP-Y151TAG-PylT, containing the protein of interest and PylT, and another containing the proper PylRS mutant, e.g., pBK-PABKRS. UAAs should be added when the OD_{600} reaches 0.3–0.5 and arabinose (0.2% (w/w)) may be added 10–20 min later (to allow time for UAA uptake). Our group uses commercial LB-Miller media (Fisherbrand) supplemented with 1 mM $MgSO_4$ and a trace metal mix to support growth to a high density at saturation (Studier, 2005). When sfGFP-Y151UAA is expressed, fluorescence is proportional to the amount of unnatural protein produced. Further, the Y151 position is solvent-exposed, making it a convenient starting point for demonstrating small molecule-induced decaging by mass spectrometry (Luo et al., 2016) before moving on to other proteins.

Once the protein is expressed, it is purified using immobilized metal ion affinity chromatography (IMAC) with commercial nickel(II)-nitrilotriacetic acid (Ni-NTA) resin. Yields may then be determined densitometrically on an SDS-PAGE gel stained with Coomassie Brilliant Blue G dye. Three bovine serum albumin standards (0.5, 1, and 2 μg; diluted from a 2 mg mL^{-1} standard, Thermo Fisher) should be analyzed on the same gel.

We also use the Bradford protein concentration assay in a 96-well format, but this method tends to overestimate yields in our hands. In practice, we run Bradford assays before quantitative SDS-PAGE gels to estimate protein loadings that will fall within the dynamic range of the 0.5–2 μg standard curve. Once accurate concentration data is obtained, additional samples may be quantified accurately by measuring fluorescence intensity (e.g., on a plate reader), and comparing values to a standard curve. In addition to obtaining gels, MS analysis is essential to confirm the successful incorporation of the UAAs. Further, with caged UAAs, MS can be used to confirm that the caging group is stable.

2.2.1.1 Materials

- UAA stock solutions: 100 mM in DMSO is used for **MPABK**, **PABK**, and **MOABK**. Aqueous solutions are preferable when possible and the Barrick Lab maintains a good summary of effective solvents used for commercially-available UAAs, which often serve as useful positive controls: https://barricklab.org/twiki/bin/view/Lab/ProtocolsAntibioticStockSolutions. If strongly basic conditions are needed to prepare the stock solution (as with N^{ϵ}-boc-L-lysine), it should be prepared fresh to minimize racemization (Hammill, Miyake-Stoner, Hazen, Jackson, & Mehl, 2007).
- L-Arabinose: 20% (w/v) aqueous solution. We typically prepare 5–10 mL, then sterile filter into 1.7 mL tubes. Aliquots may be stored at −20 °C. When needed, one is thawed and stored at 4 °C.
- Antibiotic stock solutions: tetracycline hydrochloride: 5 mg mL^{-1} in 70% ethanol (200 ×; store in the absence of light at −20 °C); kanamycin sulfate: 50 mg mL^{-1} in water (1000 ×; sterile filter, aliquot, and store at −20 °C or 4 °C).
- LB-Miller media and supplements: sterile filtered or autoclaved 1 M (1,000 ×) magnesium sulfate and 5000 × trace metal mix (Hammill et al., 2007).

1. Inoculate a starter culture of LB supplemented with 1 mM $MgSO_4$ and trace metal mix containing kanamycin and tetracycline ("LB +Kan +Tet," 50 μg mL^{-1} and 25 μg mL^{-1}, respectively) with an authentic glycerol stock of the doubly-transformed expression host strain, and incubate overnight at 37 °C with shaking. (Note: PylRS plasmids may be recombinogenic based on our experience working with multiple synthetases, including pBK-*Mb*PylRS mutants and pTECH-chPylRS(IPYE) (Bryson et al., 2017), requiring periodic verification

of plasmid integrity and function and generation of new monoclonal glycerol stocks.)

2. Dilute saturated overnight cultures 1:200 into LB + Kan + Tet (5 mL) in a 15 mL centrifuge tube. Culture volumes may be scaled accordingly if larger amounts of protein are required. Vessels should not be more than ⅓ full and tubes should be incubated at an angle so that the cultures are well-aerated (Studier, 2005).
3. Grow cultures at 37 °C, 250 rpm until reaching an OD_{600} of 0.3–0.5. This typically takes 5–7 h; frequent checks should be performed before this time.
4. Add the UAA at a 1 mM final concentration. While this concentration is routinely used, less may be used for precious, toxic, or poorly-soluble UAAs. If it is difficult to find an efficient synthetase or if the UAA is only available as a racemic mixture, up to 5 mM concentrations can be used (Wang et al., 2017). For all experiments, a negative control culture should be treated with the vehicle (DMSO) so that background protein expression in the absence of UAA can be characterized.
5. Incubate for 10–20 min in the shaker to allow for UAA uptake into cells.
6. Add L-arabinose to a final concentration of 0.2% (w/v).
7. Incubate at 37 °C, 250 rpm for 16–20 h. For the highest possible protein yield, growth at lower temperature for longer times may improve EF-Tu's binding to $PylT^{UAA}$ and GTP (Wang, Kwiatkowski, & Forster, 2016), and will increase oxygen solubility, thereby supporting growth to higher densities (Studier, 2005). Protein expression may be followed by pipetting aliquots into multiwell plates and measuring sfGFP fluorescence/A_{600} with a plate reader or fluorimeter.
8. Pellet cells by centrifugation (5000 rcf, 10 min). Decant the supernatant and either freeze the cell pellets at −80 °C for later use, or proceed with lysis and protein purification (Wesalo et al., 2019). Retain a small (40–50 μL) sample of the crude lysate to visualize the amount of overexpressed protein relative to the rest of the proteome.
9. Purify the protein using IMAC. We recommend our reported protocol (Wesalo et al., 2019), with the optional addition of benzonase (Golden nuclease, EMD Millipore/Novagen, 2.5 U mL^{-1}) to the lysis buffer to lower the lysate's viscosity by degrading the genomic DNA.
10. Dilute 5 μL of the purified protein with PBS (25 μL) for analysis by ESI-MS. Alternatively, the whole protein may be analyzed by MALDI-TOF-MS after purification of a 4–5 μL sample admixed with

20% trifluoroacetic acid (1 μL) using a micro-C_{18} ZipTip (Millipore Sigma, eluting with 10 mg mL^{-1} sinapic acid in 3:2 MeCN:water containing 0.1% trifluoroacetic acid).

11. Determine approximate protein concentration by Bradford assay (Thermo Scientific Pierce Coomassie Plus (Bradford) Assay kit, 96 well format using the manufacturer's standard microplate protocol).

12. Analyze the protein by SDS-PAGE. We use 0.75 mm, 12% acrylamide gels. For quantification, 0.5, 1, and 2 μg BSA standards are analyzed on each gel. All +UAA sample loadings are based on the Bradford assay data so they fall within the standard curve (1.25 μg per lane is a good target). Each gel should include a molecular weight standard ("ladder"), crude sample (≥20 μL), −UAA elution, +UAA elutions, and BSA standards. BSA and sfGFP are easily resolved and can be analyzed in the same lane if necessary. After staining using Coomassie Brilliant Blue G, we perform densitometric quantification using Image Lab 6.0 software (Bio-Rad).

Using this methodology, we expressed sfGFP with Y151 changed to **PABK**, **MPABK**, or **MOABK** in yields of 8, 5.6, or 1.4 mg L^{-1} culture, respectively. This corresponds to 30%, 21%, or 5% of WT protein yields under identical conditions (Wesalo et al., 2019). Background incorporation of natural UAAs is negligible for OABKRS and PABKRS. ESI-MS analysis shows that all three UAAs are incorporated successfully into sfGFP, with no reduction to aniline intermediates or deprotection to lysine.

2.2.2 Protein expression in mammalian cells

It is essential that the mammalian cells are kept in excellent health and are used at a relatively low passage number (ideally ≤30 for HEK293T cells and ≤20 for NIH 3T3 cells) for efficient UAA mutagenesis. Black cell culture-treated 96-well plates are convenient for most imaging experiments. We find that 8-well chamber slides (Cellvis) offer greater image quality. Chamber slides, however, are more expensive, lower throughput, and require double the amount of all reagents compared to 96-well plates. For long-term imaging experiments (e.g., time courses), it is ideal to maintain the cells at 37 °C/5% CO_2 in an on-stage incubator. If CO_2 injection is not available during imaging, media can be supplemented with 25 mM HEPES and adjusted to maintain pH 7.4 in the absence of CO_2. We generally obtain excellent incorporation results using a two-plasmid system (Schmied, Elsässer, Uttamapinant, & Chin, 2014), modified with recently-discovered PylRS optimizations (Bryson et al., 2017). One plasmid contains

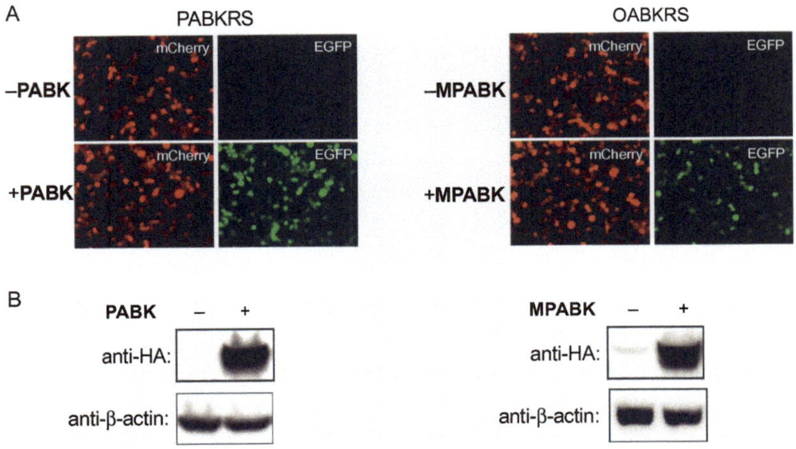

Fig. 2 Incorporation of **PABK** and **MPABK** (both 0.5 mM) into proteins in mammalian cells. (A) Widefield fluorescence micrographs show that these UAAs are incorporated into an mCherry-TAG-EGFP-HA reporter in a UAA-dependent fashion. (B) Anti-HA Western blots confirmed expression of the entire protein construct when UAAs were present. *Wesalo, J. S., Luo, J., Morihiro, K., Liu, J., & Deiters, A. (2019). Phosphine-activated lysine analogues for fast chemical control of protein subcellular localization and protein SUMOylation,* ChemBioChem, 21, *1–9. https://doi.org/10.1002/cbic.201900464. Copyright 2019 John Wiley and Sons.*

the PylRS and a PylT cassette containing four copies of the tRNA. The other plasmid contains the gene of interest, under control of the CMV promoter, with the desired codon mutated to TAG. For genes of interest terminating with a TAG stop codon, the stop codon must be mutated to TGA or TAA before use. To provide high transgene expression, two extra copies of PylT were added to the plasmid used for the imaging experiments in Fig. 2A.

In easy-to-transfect cell lines (e.g., HEK293T), our first-generation mammalian synthetase plasmid, pAG-*Mb*PylOABKRS-PylT_4 (Luo et al., 2016), may be used. Our second-generation PylRS-PylT plasmid, pPB 220PA-1-(U6-PylT*)$_4$/EF1α-PylRS, improves incorporation efficiency and enables UAA mutagenesis in more challenging cell lines (e.g., NIH 3T3). Recently, we optimized UAA incorporation efficiency by making four PylRS N-terminal domain mutations (V30I, T56P, H62Y, and A100E, or "IPYE") (Bryson et al., 2017) to this plasmid. The "IPYE" mutation set further increases unnatural protein expression approximately 40–60%.

Initial validation of UAA incorporation is performed with an mCherry-TAG-EGFP-HA construct. Using this construct, mCherry is always expressed in transfected cells. EGFP is expressed in proportion to amber suppression, as is the HA tag. Thus, the mCherry signal provides a transfection control, with GFP acting as a readout of amber suppression. A – UAA control is essential to evaluate potential background signal from misacylation of PylT with natural amino acids. The HA tag may be detected by Western blot to confirm successful expression of the full-length protein (Fig. 2A). We performed this analysis with **MPABK** and **PABK**. As a starting point, a 1:1 ratio of synthetase:reporter plasmid by mass works effectively; other groups have reported good results with a 1.5- to 2-fold excess of the reporter plasmid (Chen et al., 2009).

Unnatural proteins may be expressed in mammalian cells using the following protocol. All procedures must be performed in a biosafety cabinet using aseptic technique.

2.2.2.1 Materials

- Poly-D-lysine (PDK) hydrobromide, 70–150 kDa (MP Biomedicals). Coat plates according to the manufacturer's instructions. This coating helps HEK293T cells remain adhered to plates throughout transfections and washes. Though it is not necessary for 3T3 cells, it improves adhesion in our experience. It is cheaper and easier to use than protein-based alternatives (e.g., collagen and fibronectin).
- Black clear-bottom 96-well plates (Greiner) and/or 8-well chamber slides (CellVis).
- Linear polyethylenimine (LPEI, Polysciences) for transfecting HEK293T cells. Prepare a 1 $mg\,mL^{-1}$ solution according to *Cold Spring Harbor Protocols* (2008).
- Lipofectamine 3000 (Invitrogen) for transfecting NIH 3T3 cells. As a cheaper alternative, results are generally good but more variable with GenJet DNA Transfection Reagent for 3T3 cells (SignaGen Laboratories) if the "Advanced Protocol" is used.
- Opti-MEM medium (Gibco Life Technologies).
- DMEM medium, penicillin/streptomycin (P/S) solution, and fetal bovine serum (FBS).
- Sterile HEPES-buffered saline (HBS, 0.9% (w/v), pH 7.4) or commercial live cell imaging solution.
- Phenol red-free DMEM (high glucose). Note: FluoroBrite DMEM (Thermo Fisher) may be a suitable alternative with reduced background in the green channel, but we have not evaluated its effect on cell health during extended imaging experiments.

All volumes and masses are given on a per-well basis for a 96-well plate and should be doubled for 8-well chamber slides.

1. Trypsinize cells and seed at 20,000 per well in a black 96-well plate or 8-well chamber slide in DMEM supplemented with FBS (10%) and P/S.
2. Grow for 18–24 h to 70% confluency (HEK293T cells) or 80% confluency (NIH 3T3 cells).
3. Replace media with DMEM supplemented with 10% FBS and UAA or vehicle (antibiotic-free). **MPABK** and **PABK** are effectively used at 0.5 mM.
4. Transfections: prepare the reagents as described and add them dropwise by dispensing a small amount onto the pipette tip, and then gently touching the droplet to the media's surface. Gently tilt the plates at a ~30° angle and spin around to lower all sides of the plate for ~10 s. For sensitive azido-UAAs, it may be prudent to keep bright fluorescent lighting in cell culture hoods off.
   a. For LPEI (used with HEK293T cells):
      i. Prepare solutions of plasmid DNA (200 ng total, with equal masses of each plasmid for double- or triple-transfections) in Opti-MEM media (18 μL).
      ii. Dilute 1 $mg\,mL^{-1}$ LPEI to 0.33 $mg\,mL^{-1}$ with pre-warmed Opti-MEM media.
      iii. Add the diluted LPEI solution (2 μL) to the diluted DNA solution and immediately mix by pipetting up-and-down several times, while avoiding the formation of bubbles.
      iv. Incubate the solution for 10–15 min at room temperature.
      v. Carefully add the reagent (20 μL) to the cultured cells.
   b. For Lipofectamine 3000 (used with NIH 3T3 cells):
      i. Dilute DNA (200 ng total, equal masses of each plasmid) in Opti-MEM (5 μL). Add the P3000 reagent (0.2 μL).
      ii. Dilute Lipofectamine 3000 reagent (0.2 μL) in Opti-MEM (5 μL).
      iii. Add the diluted DNA/P3000 solution to the diluted Lipofectamine 3000. Pipette up-and-down several times to mix.
      iv. Incubate the solution for 10–15 min at room temperature.
      v. Carefully add the reagent (10 μL) to the cultured cells.
5. Incubate cells (37 °C, 5% CO_2) for 36–48 h before experimental readouts (Western blot, imaging, flow cytometry, etc.). We found that 24 h expressions are often insufficient.
6. For short-term imaging experiments, media should be changed to HBS or commercial live cell imaging solution. For time-course imaging experiments lasting a few hours, phenol red-free DMEM should be used.

2.3 Controlling protein translocation with phosphine triggers

Phosphine triggers must be chosen carefully, with attention to several properties. The ideal phosphine has good water solubility, resists off-target oxidation, penetrates cell membranes effectively, and reduces azides quickly without reducing disulfides. As described above, triarylphosphines with moderate electron density and a carboxylate or carboxamide substituent *ortho* to the phosphorus center are ideal. Despite lower solubility (0.1 mM) compared to the acid, we prefer the carboxamide **2DPBM** (Fig. 1) due to its faster kinetics (Lukasak et al., 2019).

To optimize phosphine-azide pairs for deployment in live cells, a fluorescent caged SATB1 nuclear translocation reporter provides a convenient readout. This reporter is validated for controlled activation of nuclear translocation in live cells (Engelke, Chou, Uprety, Jess, & Deiters, 2014; Luo et al., 2016; Wesalo et al., 2019). The reporter consists of EGFP-SATB1-K29TAG-mCherry remains trapped in the cytosol if a caged/protected lysine is incorporated, then translocates to the nucleus after decaging (Fig. 3B). The unusual NLS in SATB1 functions more slowly than canonical NLSs, with $t_{1/2}$ values on the order of several minutes rather than seconds (Engelke et al., 2014; Luo et al., 2016). The delay makes it easy to compare kinetics accurately in complex experiments requiring pipetting, microscope manipulations, and multichannel imaging of multiple wells. Imaging every 3–5 min is a sufficient sampling rate.

We compared the kinetics of **MPABK**, **PABK**, **OABK** side-by-side in the presence of **2DPBM** (100 μM) (Fig. 3C). Translocation $t_{1/2}$ values of 18, 27, and 94 min were measured for **MPABK**, **PABK**, and **OABK**, respectively. Nuclear translocation was quantitative in all 3 cases. Comparing **MPABK** and **PABK**, the former is significantly faster in the 15–50 min range. **PABK** and **MPABK** show a 61–81% improved $t_{1/2}$ relative to the first-generation UAA, **OABK**.

We have tested a range of concentrations (10–100 μM) for the **2DPBM** phosphine trigger. Varying the concentration allows fine-tuning of the rate and extent of protein translocation. Specifically, a full range of $t_{1/2}$ values ranging from 27 to 181 min was accessible by varying the phosphine concentration (Fig. 3D). Taken together, this methodology provides fast and orthogonal small molecule activation of proteins in live cells and its tunability represents finely-controllable rheostat rather than a simple OFF to ON switch.

These experiments were performed after transfecting cells as noted above. We performed these experiments with pAG-*Mb*PylOABKRS-PylT$_4$

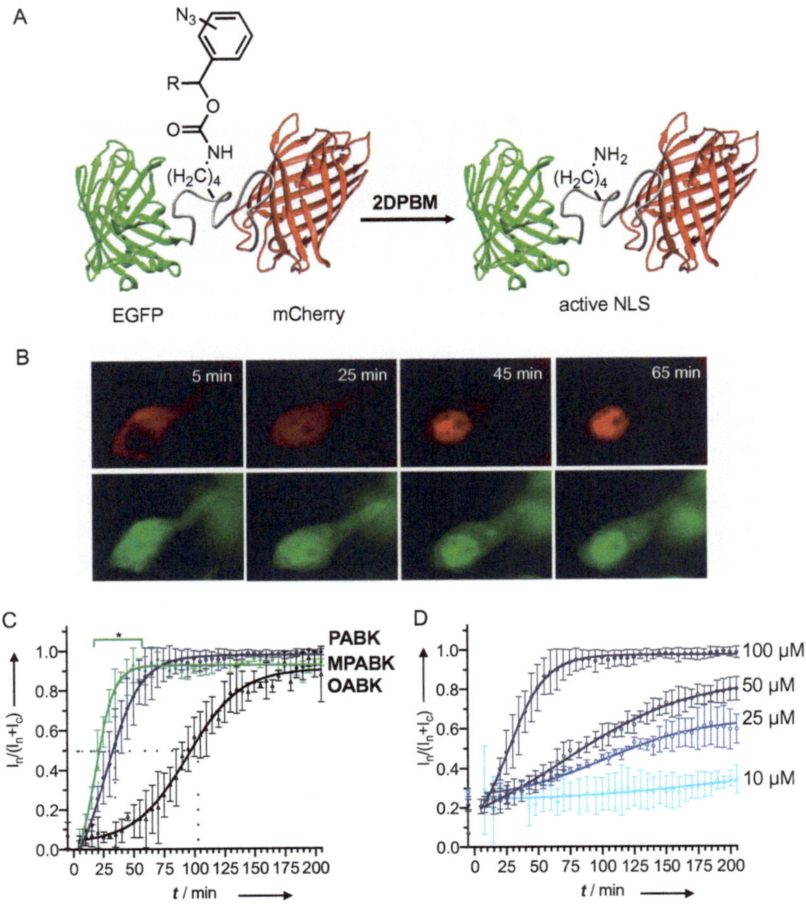

UAA	$t_{½}$ / min	95% CI / min	R^2
MPABK	18.0	14.0–20.5	0.97
PABK	26.7	18.0–31.9	0.96
OABK	94.4	89.9–98.7	0.95

[2DPBM] / μM	$t_{½}$ / min	95% CI / min	R^2
100	26.7	18.0–31.9	0.96
50	69.8	60.0–78.0	1.00
25	101.9	95.6–108.8	0.97
10	181.1	138.0–461.7	0.71

Fig. 3 (A) Phosphine-inducted turn-on SATB1 NLS in an EGFP-mCherry fusion construct. (B) Micrographs of mCherry-SATB1-K29**PABK**-EGFP in a representative cell. The full-length fusion construct begins completely in the cytosol, but rapidly and fully translocates to the nucleus after phosphine treatment. Truncated protein from termination at K29TAG spontaneously diffuses through nuclear pores, labeling the entire cell. (C) Kinetics for mCherry-SATB1-EGFP nuclear translocation with SATB1-K29 mutated to **OABK**, **PABK**, and **MPABK**. At $t=0$, after a 1 h UAA washout, media was replaced with phenol red-free DMEM supplemented with **2DPBM** (100 μM). Cells were imaged every 5 min (n = 3; from different fields of view), and nuclear mCherry signal intensity (I_n) as a fraction of all mCherry intensity (nuclear + cytosolic, $I_n + I_c$) was calculated. Mean ± SD is plotted, along with four-parameter logistic curve models. *, significant difference between **MPABK** and **PABK** by 2-way repeated measures ANOVA with Tukey's post hoc test, $\alpha = 0.05$. (D) Tunable control of SATB1-mediated translocation with multiple concentrations of **2DPBM** (10, 25, 50, and 100 μM). Mean ± SD (n = 3; from different fields of view) is plotted. *Wesalo, J. S., Luo, J., Morihiro, K., Liu, J., & Deiters, A. (2019). Phosphine-activated lysine analogues for fast chemical control of protein subcellular localization and protein SUMOylation*, ChemBioChem, 21, *1–9. https://doi.org/10.1002/cbic.201900464. Copyright 2019 John Wiley and Sons.*

(Luo et al., 2016) and pCDH-EGFP-OptoNLS-K29TAG-SATB1-mCherry ("eGFP-OptoNLS-SATB1-mCherry" from Engelke et al. (2014)). Since amber suppression is incomplete, cells express significant amounts of EGFP fused to the first 28 residues of SATB1 (EGFP-SATB1Δ). This acts as a useful marker of transfected cells. When most of the mCherry fluorescence (full-length construct) has translocated to the nucleus, remaining EGFP-SATB1Δ in the cytosol shows where the remaining cytosolic mCherry signal may be measured. Further, after complete translocation, the nuclear:cytosolic EGFP intensity ratio may be used to calculate suppression efficiency (Engelke et al., 2014).

1. If using an on-stage incubator, place an empty multiwell plate in it and equilibrate to 37 °C, 5% CO_2 for 10–15 min. We use a WSKM Stage Top Incubator (Tokai Hit). A 37 °C incubator is ideal for cell health and for faster decaging kinetics.
2. Wash out excess azido-UAAs by changing the media to phenol red-free DMEM for 1 h. During this time, place the cells on the microscope stage (ideally in an on-stage incubator) and remove the plate cover to prepare for imaging (sterile conditions are not needed for this short experiment). Select positions for each treatment condition and record the coordinates in the microscope software. Positions should include at least three fields of view with representative cells for each treatment group. If possible, set reference points in the microscope software to facilitate returning to the correct positions if the plate is moved during the media change (Note: $PylT^{UAA}$ conjugates will remain trapped in the cytosol after the wash, consuming some of the phosphine).
3. Thaw an aliquot of **2DPBM** (50 mM in degassed DMSO) and dilute to the desired concentration (10–100 μM) in prewarmed phenol red-free DMEM. Two serial dilutions are recommended for greater accuracy when preparing small volumes of media. Prepare a DMSO vehicle negative control as well (Note: adding DMSO solutions to media that is already in wells is not recommended, since the DMSO may "layer out" on the cells). To minimize oxidation, the **2DPBM** should be diluted with media immediately before use.
4. Start a timer and acquire baseline images (EGFP, mCherry, and brightfield channels). Suitable filter cubes are mCherry (excitation (ex); BP550/25, emission (em), BP605/70) and EGFP (ex, BP470/40; em, BP525/50). We use a Zeiss Axio Observer Z1, 20 × objective, numerical aperture 0.8 Plan-Apochromat M27 for these experiments.

5. Carefully open the on-stage incubator. Carefully remove the media (err toward leaving a few μL rather than risking dislodging cells). This operation is easiest with a single-channel pipette or multichannel pipette using two channels at a time. Add **2DPBM**-containing media (200 μL) or the negative control media (200 μL) slowly and carefully by pipetting it down the side of each well near the bottom.
6. Image the cells (all positions and all three channels) every 5 min until translocation has plateaued (~205 min is usually sufficient). Carefully monitor the focus and adjust as needed if not using software or hardware autofocus.
7. Open raw images simultaneously in Fiji. For each channel, use the Bio-Formats Import Options to open create a hyperstack with each time point. Use the "Synchronize Windows" options to sync the cursors, z-slices, scaling, and coordinates for the EGFP, mCherry, and brightfield channels, and move them for side-by-side view (easier using a large monitor or multiple displays). Save each stack as a .tif file with an appropriate name for each channel (mCherry, GFP, or brightfield).
8. Use the "Set Measurements" menu to include the mean gray value (area, median, and min/max may also be helpful for some analyses).
9. Draw a representative elliptical selection for the cytosol and measure fluorescence intensity for the same area in both channels at each time point. It is fastest to use keyboard shortcuts (for Macintosh, alternate between ⌘M and . or > to measure and advance to the next time point) and to get the measurements for all time points for one channel at a time. The GFP channel should be used as a guide to measure cytosolic fluorescence intensity in the mCherry channel late in the experiment, once most of the mCherry has entered the nucleus. Export each set of measurements as a table. Quantify fluorescence for at least three representative cells in separate fields of view. Use your preferred tool (Microsoft Excel, Python, R, etc.) to subtract background signal for mean intensity values (median may be used if outlier pixels are affecting the measurements, but the results are generally similar using either mean or median). Negative background-subtracted values may be assigned to zero as previously reported since negative fluorescence intensity is not possible (Engelke et al., 2014).
10. Calculate $\frac{I_n}{I_n + I_c}$ for each cell at each time point (I_n = nuclear intensity, I_c = cytosolic intensity). Determine the mean and standard deviation (SD) for this ratio for all of the cells at each time point. Plot mean ± SD over

time (GraphPad Prism is preferred for graphing and modeling tasks). Fit four-parameter logistic curves and use these models to determine $t_{1/2}$. To determine if statistically significant differences are present between $\frac{I_n}{I_n + I_c}$ ratios for different treatments at individual timepoints, perform a 2-way repeated measures ANOVA with an appropriate post-hoc test for multiple comparisons (Tukey's is recommended).

11. If desired, export the stacks as montages and/or videos. Apply false color using the Image > Lookup Table command if desired. Convert the image type to RGB color for greater compatibility with other programs.

3. Controlling protein SUMOylation with phosphine triggers

Azide-containing UAAs have been used to control a bacterial toxin (Ge, Fan, & Chen, 2016), EGFP fluorescence, luciferase activity, Cre recombinase, CRISPR/Cas9 gene editing, and nuclear translocation (Luo et al., 2016). Additionally, we recently demonstrated that second-generation **PABK** can turn on a post-translational modification. We decided to study Small Ubiquitin-like MOdifier (SUMO), as this is an important modification under active investigation for its roles in cancer, neurodegenerative disease, and responses to oxidative stress (Flotho & Melchior, 2013). Conditional turn-on of SUMOylation using the cell's natural SUMO protein and yielding a native isopeptide bond had not been reported.

To demonstrate control over SUMOylation with this methodology, we selected Ran GTPase Activating Protein 1 (RanGAP1) (Zhu et al., 2009) as our target. Once SUMOylated, RanGAP1 localizes from the cytosol to filaments radiating off of nuclear pore complexes (NPC) (Peters, 2006). **PABK** was incorporated at the acceptor lysine of a 20kDa C-terminal fragment of murine RanGAP1 (RGΔN). This domain, consisting of residues 400–589, is effectively SUMOylated then targeted to NPCs and is conveniently probed by Western blot. RGΔN-K526TAG expression was only observed in the presence of **PABK** (Fig. 4B, lanes 1 and 2). After treating with **2DPBM** (100 μM, 12 h), no more FLAG-RGΔN is detectable at 20 kDa (Fig. 4C, lane 3). Using HA-tagged SUMO1, we instead see RGΔN-HA-SUMO1 at 37 kDa. The 15–17 kDa mobility shift seen here typifies SUMOylation (Park-Sarge & Sarge, 2008). Moreover, a similar shift is observed when wild-type RGΔN and HA-SUMO1 are co-expressed

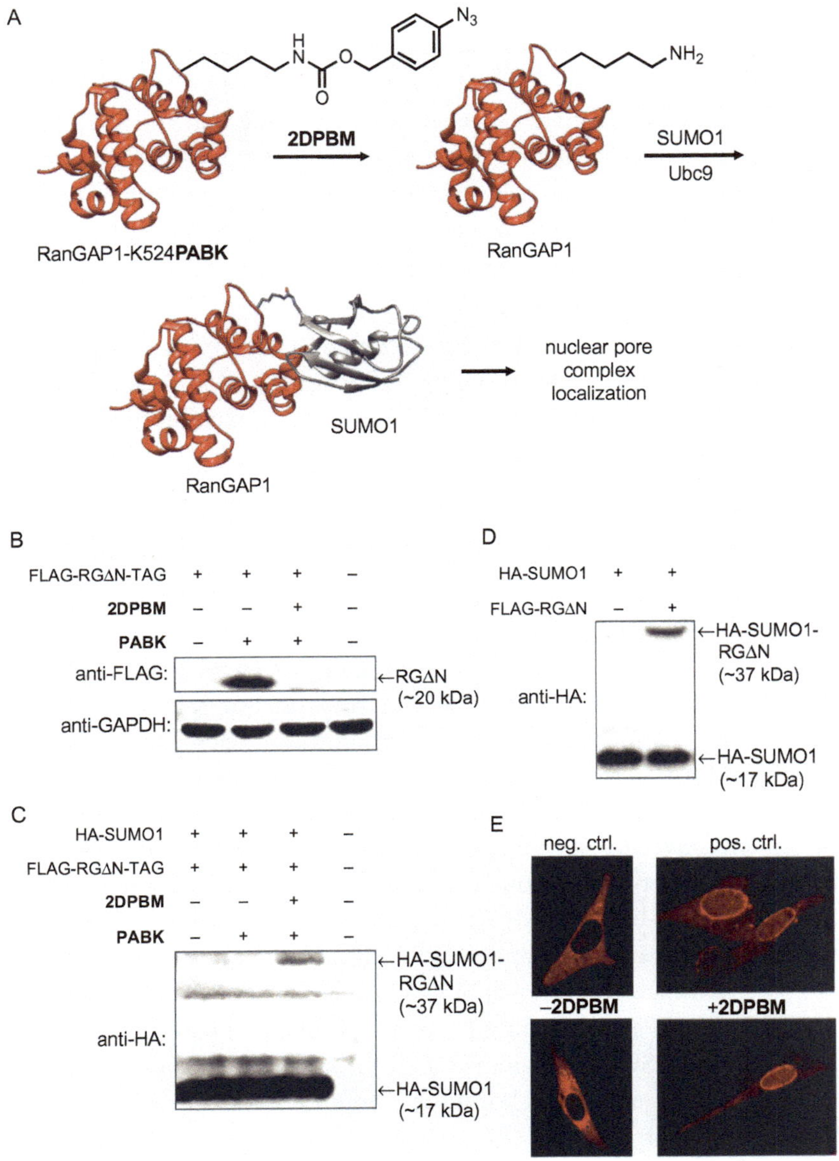

Fig. 4 (A) Schematic of **2DPBM**-triggered SUMOylation and subsequent translocation of RanGAP1 to the nuclear pore complex. (B) RanGAP1ΔN-K526**PABK** is seen at ~20 kDa, but vanishes after **2DPBM** treatment (100 μM, 12 h) following conjugation to endogenous SUMO. (C) The HA-SUMO1-RanGAPΔN complex is seen at 37 kDa using an anti-HA antibody exclusively after **2DPBM** exposure. (D) Similar RGΔN-HA-SUMO1 complex observed at 37 kDa when co-expressing wild-type RGΔN and HA-tagged SUMO1. (E) The protein is trapped in the cytosol in the RanGAP1-K524R (negative control mutant, top left). By contrast, wild-type RanGAP1-RFP (positive control, top right) is capable of accepting SUMO, and localizes to nuclear pore complexes with spillover into the cytosol when saturated. The caged K524**PABK** mutant (bottom) shows the negative control phenotype when DMSO is added, but recapitulates the positive control phenotype after **2DPBM** (100 μM, 16 h) is added. *Wesalo, J. S., Luo, J., Morihiro, K., Liu, J., & Deiters, A. (2019). Phosphine-activated lysine analogues for fast chemical control of protein subcellular localization and protein SUMOylation*, ChemBioChem, 21, *1–9. https://doi.org/10.1002/cbic.201900464. Copyright 2019 John Wiley and Sons.*

(Fig. 4D). Altogether, these results are consistent with cells SUMOylating FLAG-RGΔN-K526**PABK** after deprotection. No SUMO1 over-expression is required to observe disappearance of 20 kDa RGΔN-K526**PABK** after treating cells with **2DPBM**. This result is consistent with endogenous, wild-type SUMO1 as the substrate here. Similarly, wild-type FLAG-RGΔN is undetectable, and can only be seen when the K526R mutation (Mahajan, Gerace, Melchior, Delphin, & Guan, 1998) is made to block SUMOylation (Wesalo et al., 2019). Thus, we demonstrated chemical triggering to switch on natural SUMO1 modification in live cells with temporal control. This small molecule control was further confirmed by imaging RFP-tagged RanGAP1 translocation from the cytosol to the NPC in response to **2DPBM** treatment (Fig. 4E).

HEK293T cells were used to generate immunoblots, and NIH 3T3 cells were used for imaging. Transfections were performed according to the protocol above (see Section 2.2.1). Western blots were performed using standard procedures as detailed in Wesalo et al. (2019). The following protocol was used for the RanGAP imaging in NIH 3T3 cells:

1. Perform transfections with Lipofectamine 3000 in 8-well chamber slides as described in Section 2.2.1.
   a. pDsRed1-N1-RanGAP and the corresponding K524R mutant were used as the negative and positive controls to establish the phenotype.
   b. pDsRed1-N1-RanGAP-K524TAG with the terminal TAG stop codon mutated to TGA was co-transfected with pPB220PA-1-$(U6\text{-}PylT^*)_4$/EF1α-BhcKRS(IPYE)
2. At 48 h post-transfection, wash out excess UAAs by incubating cells in fresh phenol red-free DMEM for 1 h. An on-stage incubator may be used if available.
3. Acquire baseline images (mCherry and brightfield channels) of representative cells in the wells for each treatment condition. Note that cell-to-cell variability in transgene expression is known to be high (Mahajan et al., 1998); the exposure time may need to varied for some cells. Cells with an intermediate phenotype consistent with reported results should be imaged.
4. Freshly prepare phenol red-free DMEM containing **2DPBM** (100 μM) or vehicle (DMSO) as above (see Section 2.3, step 3).
5. Carefully replace the media, working with only 1 or 2 wells at a time, with **2DPBM** or vehicle-containing media (200 μL).
6. Incubate the cells for 16 h at 37 °C, 5% CO_2.
7. Image the cells once more and compare the observed phenotypes between the treatment groups.

4. Summary

The Staudinger reduction and its variations stand out as highly bioorthogonal reactions that are robust in demanding biological milieus. We made structural changes to first-generation azido-lysine **OABK** based on literature reports and chemical reasoning, thereby improving the rate of the Staudinger reduction/self-immolation pathway. Though precautions must be taken for the preparation and use of organic azides, the UAAs are easy to synthesize, readily incorporate into proteins, and rapidly decage upon treatment with **2DPBM**, the optimized phosphine reagent. Decaging is performed with a by simply changing to freshly-prepared media containing **2DPBM**. The methods reported here may be generalized to other proteins and can used in other cell lines and in animals. If desired, this methodology is compatible with photocaged groups in the visible range, and multiple biomolecules could be orthogonally deprotected in a single experiment with these combined techniques.

The UAAs developed here open up several possibilities for researchers. When the highest protein yields are needed, **PABK** may be used. On the other hand, **MPABK** can be used when a modest yield is acceptable, but the fastest kinetics for activation are desired. In either case, the kinetics may be tuned over a wide range by varying the phosphine concentration.

Acknowledgments

This work was supported by the National Institutes of Health (R01GM112728 and T32GM008208) and the National Science Foundation (CBET-1603930).

References

Alouane, A., Labruère, R., Le Saux, T., Schmidt, F., & Jullien, L. (2015). Self-immolative spacers: Kinetic aspects, structure-property relationships, and applications. *Angewandte Chemie, International Edition, 54*, 7492–7509.

Arora, S., Horne, W. S., & Islam, K. (2019). Engineering methyllysine writers and readers for allele-specific regulation of protein–protein interactions. *Journal of the American Chemical Society, 141*, 15466–15470. https://doi.org/10.1021/jacs.9b05725.

Barder, T. E., & Buchwald, S. L. (2007). Rationale behind the resistance of dialkylbiaryl phosphines toward oxidation by molecular oxygen. *Journal of the American Chemical Society, 129*, 5096–5101. https://doi.org/10.1021/ja0683180.

Brauer, D. J., Hingst, M., Kottsieper, K. W., Liek, C., Nickel, T., Tepper, M., et al. (2002). Water soluble phosphines Part XV. Syntheses of multiply functionalized and chiral phosphine ligands by Pd-catalyzed P C and C C coupling reactions. *Journal of Organometallic Chemistry, 645*, 14–26.

Brown, W., Liu, J., & Deiters, A. (2018). Genetic code expansion in animals. *ACS Chemical Biology, 13*, 2375–2386. https://doi.org/10.1021/acschembio.8b00520.

Bryson, D. I., Fan, C., Guo, L. T., Miller, C., Söll, D., & Liu, D. R. (2017). Continuous directed evolution of aminoacyl-tRNA synthetases. *Nature Chemical Biology*, *13*, 1253–1260. https://doi.org/10.1038/nchembio.2474.

Busacca, C. A., Farber, E., DeYoung, J., Campbell, S., Gonnella, N. C., Grinberg, N., et al. (2009). Ambient temperature hydrophosphination of internal, unactivated alkynes and allenyl phosphineoxides with phosphine borane complexes. *Organic Letters*, *11*, 5594–5597. https://doi.org/10.1021/ol9022547.

Chen, P. R., Groff, D., Guo, J., Ou, W., Cellitti, S., Geierstanger, B. H., et al. (2009). A facile system for encoding unnatural amino acids in mammalian cells. *Angewandte Chemie, International Edition*, *48*, 4052–4055. https://doi.org/10.1002/anie.200900683.

Chen, X., Henschke, L., Wu, Q., Muthoosamy, K., Neumann, B., & Weil, T. (2013). Site-selective azide incorporation into endogenous RNase A via a "chemistry" approach. *Organic & Biomolecular Chemistry*, *11*, 353–361. https://doi.org/10.1039/c2ob26561c.

Chin, J. W. (2017). Expanding and reprogramming the genetic code. *Nature*, *550*, 53–60. https://doi.org/10.1038/nature24031.

Chin, J. W., Santoro, S. W., Martin, A. B., King, D. S., Wang, L., & Schultz, P. G. (2002). Addition of p-azido-L-phenylalanine to the genetic code of Escherichia coli. *Journal of the American Chemical Society*, *124*, 9026–9027. https://doi.org/10.1021/ja027007w.

Cold Spring Harbor protocols. (2008). Polyethylenimine (PEI), linear (1 mg/mL). *Cold Spring Harbor Protocols*, *2008*, pdb.rec11323. https://doi.org/10.1101/pdb.rec11323.

Dobrzanski, G., & Kossut, M. (2017). Application of the DREADD technique in biomedical brain research. *Pharmacological Reports*, *69*, 213–221. https://doi.org/10.1016/j.pharep.2016.10.015.

Engelke, H., Chou, C., Uprety, R., Jess, P., & Deiters, A. (2014). Control of protein function through optochemical translocation. *ACS Synthetic Biology*, *3*, 731–736. https://doi.org/10.1021/sb400192a.

Fegan, A., White, B., Carlson, J. C. T., & Wagner, C. R. (2010). Chemically controlled protein assembly: Techniques and applications. *Chemical Reviews*, *110*, 3315–3336. https://doi.org/10.1021/cr8002888.

Flotho, A., & Melchior, F. (2013). Sumoylation: A regulatory protein modification in health and disease. *Annual Review of Biochemistry*, *82*, 357–385. https://doi.org/10.1146/annurev-biochem-061909-093311.

Ge, Y., Fan, X., & Chen, P. R. (2016). A genetically encoded multifunctional unnatural amino acid for versatile protein manipulations in living cells. *Chemical Science*, *7*, 7055–7060. https://doi.org/10.1039/c6sc02615j.

Hammill, J. T., Miyake-Stoner, S., Hazen, J. L., Jackson, J. C., & Mehl, R. A. (2007). Preparation of site-specifically labeled fluorinated proteins for 19F-NMR structural characterization. *Nature Protocols*, *2*, 2601–2607. https://doi.org/10.1038/nprot.2007.379.

Hossain, M. B., van der Helm, D., Sanduja, R., & Alam, M. (1985). Structure of 6-azidotetrazolo[5,1-a]phthalazine, C8H4N8, isolated from the toxic dinoflagellate Gymnodinium breve. *Acta Crystallographica Section C*, *41*, 1199–1202. https://doi.org/10.1107/S0108270185007132.

Karver, M. R., Weissleder, R., & Hilderbrand, S. A. (2011). Synthesis and evaluation of a series of 1, 2,4,5-tetrazines for bioorthogonal conjugation. *Bioconjugate Chemistry*, *22*, 2263–2270. https://doi.org/10.1021/bc200295y.

Keicher, T., & Löbbecke, S. (2010). Lab-scale synthesis of azido compounds: Safety measures and analysis. In S. Bräse & K. Banert (Eds.), *Organic Azides: Syntheses and Applications* (pp. 3–27). John Wiley & Sons, Ltd.

Kremer, K. A. M., & Helquist, P. (1984). Purification of air-sensitive organometallic compounds by a modified flash chromatography procedure. *Organometallics*, *3*, 1743–1745. https://doi.org/10.1021/om00089a025.

Kühl, O. (2008). Phosphorus-31 NMR spectroscopy: A concise introduction for the synthetic organic and organometallic chemist. Springer.

Li, J., Jia, S., & Chen, P. R. (2014). Diels-Alder reaction-triggered bioorthogonal protein decaging in living cells. *Nature Chemical Biology*, *10*, 1003–1007. https://doi.org/10.1038/nchembio.1656.

Li, J., Yu, J., Zhao, J., Wang, J., Zheng, S., Lin, S., et al. (2014). Palladium-triggered deprotection chemistry for protein activation in living cells. *Nature Chemistry*, *6*, 352–361. https://doi.org/10.1038/nchem.1887.

Lin, F. L., Hoyt, H. M., Van Halbeek, H., Bergman, R. G., Bertozzi, C. R., & Berkeley, L. (2005). Mechanistic investigation of the staudinger ligation. *Journal of the American Chemical Society*, *127*, 2686–2695.

Lukasak, B., Morihiro, K., & Deiters, A. (2019). Aryl azides as phosphine-activated switches for small molecule function. *Scientific Reports*, *9*, 1470. https://doi.org/10.1038/s41598-018-37023-6.

Luo, J., Liu, Q., Morihiro, K., & Deiters, A. (2016). Small-molecule control of protein function through Staudinger reduction. *Nature Chemistry*, *8*, 1027–1034. https://doi.org/10.1038/nchem.2573.

Luo, J., Uprety, R., Naro, Y., Chou, C., Nguyen, D. P., Chin, J. W., et al. (2014). Genetically encoded optochemical probes for simultaneous fluorescence reporting and light activation of protein function with two-photon excitation. *Journal of the American Chemical Society*, *136*, 15551–15558. https://doi.org/10.1021/ja5055862.

Mahajan, R., Gerace, L., Melchior, F., Delphin, C., & Guan, T. (1998). Molecular characterization of the SUMO-1 modification of RanGAP1 and its role in nuclear envelope association. A small ubiquitin-related polypeptide involved in targeting RanGAP1 to nuclear pore complex protein RanBP2. *The Journal of Cell Biology*, *140*, 259–270. https://doi.org/10.1083/jcb.140.2.259.

Park-Sarge, O.-K., & Sarge, K. D. (2008). Detection of sumoylated proteins. *Methods in Molecular Biology*, *464*, 255–265. https://doi.org/10.1007/978-1-60327-461-6-14.

Peters, R. (2006). Introduction to nucleocytoplasmic transport: Molecules and mechanisms. *Methods in Molecular Biology*, *322*, 235–258. https://doi.org/10.1007/978-1-59745-000-3_17.

Rose, J. C., Foight, G. W., Cunningham-Bryant, D., Loutey, D. E., Maly, D. J., & Dieter, E. M. (2019). A chemically-disrupted proximity system for controlling dynamic cellular processes. *Journal of the American Chemical Society*, *141*, 3352–3355. https://doi.org/10.1021/jacs.8b12382.

Saneyoshi, H., Ochikubo, T., Mashimo, T., Hatano, K., Ito, Y., & Abe, H. (2014). Triphenylphosphinecarboxamide: An effective reagent for the reduction of azides and its application to nucleic acid detection. *Organic Letters*, *16*, 30–33. https://doi.org/10.1021/ol402832w.

Santa Barbara Environmental Health, & Safety. (2006). Synthesizing, purifying, and handling organic azides. University of California. https://www.ehs.ucsb.edu/files/docs/ls/factsheets/Azides_FS26.pdf. (accessed December 13, 2019).

Sato, T., Hamada, Y., Sumikawa, M., Araki, S., & Yamamoto, H. (2014). Solubility of oxygen in organic solvents and calculation of the Hansen solubility parameters of oxygen. *Industrial and Engineering Chemistry Research*, *53*, 19331–19337. https://doi.org/10.1021/ie502386t.

Saxon, E., & Bertozzi, C. R. (2000). Cell surface engineering by a modified staudinger reaction. *Science*, *287*, 2007–2011.

Schilling, C. I., Jung, N., Biskup, M., Schepers, U., & Bräse, S. (2011). Bioconjugation via azide–Staudinger ligation: An overview. *Chemical Society Reviews*, *40*, 4840–4871. https://doi.org/10.1039/c0cs00123f.

Schmied, W. H., Elsässer, S. J., Uttamapinant, C., & Chin, J. W. (2014). Efficient multisite unnatural amino acid incorporation in mammalian cells via optimized pyrrolysyl tRNA synthetase/tRNA expression and engineered eRF1. *Journal of the American Chemical Society*, *136*, 15577–15583. https://doi.org/10.1021/ja5069728.

Shriver, D. F., & Drezdzon, M. A. (1986). *The manipulation of air-sensitive compounds* (2nd ed.). New York: John Wiley & Sons, Ltd.

Sletten, E. M., & Bertozzi, C. R. (2011). From mechanism to mouse: A tale of two bioorthogonal reactions. *Accounts of Chemical Research*, *44*, 666–676.

Studier, F. W. (2005). Protein production by auto-induction in high density shaking cultures. *Protein Expression and Purification*, *41*, 207–234. https://doi.org/10.1016/j.pep.2005.01.016.

Van Berkel, S. S., Van Eldijk, M. B., & Van Hest, J. C. M. (2011). Staudinger ligation as a method for bioconjugation. *Angewandte Chemie, International Edition*, *50*, 8806–8827. https://doi.org/10.1002/anie.201008102.

Van Overschelde, M., Vervecken, E., Modha, S. G., Cogen, S., Van der Eycken, E., & Van der Eycken, J. (2009). Catalyst-free alcoholysis of phosphane-boranes: A smooth, cheap, and efficient deprotection procedure. *Tetrahedron*, *65*, 6410–6415. https://doi.org/10.1016/j.tet.2009.05.063.

Wang, Z. A., Kurra, Y., Wang, X., Zeng, Y., Lee, Y. J., Sharma, V., et al. (2017). A versatile approach for site-specific lysine acylation in proteins. *Angewandte Chemie, International Edition*, *56*, 1643–1647. https://doi.org/10.1002/anie.201611415.

Wang, J., Kwiatkowski, M., & Forster, A. C. (2016). Kinetics of tRNAPyl-mediated amber suppression in Escherichia coli translation reveals unexpected limiting steps and competing reactions. *Biotechnology and Bioengineering*, *113*, 1552–1559. https://doi.org/10.1002/bit.25917.

Wang, J., Zheng, S., Liu, Y., Zhang, Z., Lin, Z., Li, J. J., et al. (2016). Palladium-triggered chemical rescue of intracellular proteins via genetically encoded allene-caged tyrosine. *Journal of the American Chemical Society*, *138*, 15118–15121. https://doi.org/10.1021/jacs.6b08933.

Wesalo, J. S., Luo, J., Morihiro, K., Liu, J., & Deiters, A. (2019). Phosphine-activated lysine analogues for fast chemical control of protein subcellular localization and protein SUMOylation. *ChemBioChem*, *21*, 1–9. https://doi.org/10.1002/cbic.201900464.

Yanagisawa, T., Ishii, R., Fukunaga, R., Kobayashi, T., Sakamoto, K., & Yokoyama, S. (2008). Multistep engineering of pyrrolysyl-tRNA synthetase to genetically encode Nε-(o-azidobenzyloxycarbonyl) lysine for site-specific protein modification. *Chemistry & Biology*, *15*, 1187–1197. https://doi.org/10.1016/j.chembiol.2008.10.004.

Young, D. D., & Schultz, P. G. (2018). Playing with the molecules of life. *ACS Chemical Biology*, *13*, 854–870. https://doi.org/10.1021/acschembio.7b00974.

Zhu, S., Goeres, J., Sixt, K. M., Békés, M., Zhang, X. D., Salvesen, G. S., et al. (2009). Protection from isopeptidase-mediated deconjugation regulates paralog-selective sumoylation of RanGAP1. *Molecular Cell*, *33*, 570–580. https://doi.org/10.1016/j.molcel.2009.02.008.

CHAPTER ELEVEN

Photopharmacological control of lipid function

Johannes Morstein, Dirk Trauner*
Department of Chemistry, New York University, New York, NY, United States
*Corresponding author: e-mail address: dirktrauner@nyu.edu

Contents

Abstract

Photopharmacology enables the optical control of biological processes through the use of molecular photoswitches, such as functionalized azobenzenes. In this protocol, we describe general considerations for working with bioactive azobenzene photoswitches. Our approach is exemplified with a photoswitchable sphingosine-1-phosphate analog (**PhotoS1P**).

1. Introduction

Functionalized synthetic photoswitches, such as azobenzenes, have enabled the optical control of an increasing number of biological targets. This approach, termed photopharmacology, is complementary to and in

Methods in Enzymology, Volume 638
ISSN 0076-6879
https://doi.org/10.1016/bs.mie.2020.04.025

some aspects congruent with optogenetics (Beharry & Woolley, 2011; Hüll, Morstein, & Trauner, 2018; Szymański, Beierle, Kistemaker, Velema, & Feringa, 2013). While optogenetics relies on genetically encoded photoreceptors, photopharmacology requires the addition of exogenous small molecules, which adds certain experimental challenges. Here, we seek to provide a practical guide for the application of photopharmacology. In the first part, we attempt to answer general questions about the use of photopharmacological tools, and in the second part we exemplify common assay methods with a recently published photoswitchable analog of sphingosine-1-phosphate, termed **PhotoS1P** (Morstein, Hill, et al., 2019).

2. General consideration when working with azobenzene-photoswitches

2.1 What types of azobenzenes are commonly used in photopharmacology?

Many azobenzene photoswitches used in photopharmacology behave like the unsubstituted parent azobenzene (Fig. 1A). Such "classical" azobenzenes (Fig. 1B) are typically isomerized with UV-A light (350–370 nm) to their *cis* form and with blue light (420–480 nm) to *trans*. The *trans*-form is thermodynamically favored and represents the dark-adapted state. By contrast, the *cis*-form represents a metastable state that will relax back to *trans* with a half-life dependent on the temperature and solvent. This half-life is generally in the range of several hours. Most photoswitchable lipids published to date are classical azobenzenes and closely adhere to this photophysical behavior.

Electron-rich substituents in *para* positions of azobenzenes red-shift the wavelength needed for maximum isomerization to *cis*. This also destabilizes the *cis*-form, leading to faster thermal relaxation (Fig. 1C). This effect can be further enhanced when combining an electron-rich substituent with an electron withdrawing group (Fig. 1D).

In addition, heterocyclic azobenzenes are becoming increasingly popular (Fig. 1E). Cyclic azobenzenes, such as diazocines (Fig. 1F), are interesting alternatives because they preferentially exist in their thermodynamically stable *cis* form in the dark and can be isomerized to the *trans* form with violet light (380–410 nm).

2.2 How do I dissolve an azobenzene photoswitch effectively?

The solubility of a biologically active azobenzene photoswitch often closely resembles that of the parent drug it is derived from. This is especially true for photoswitches designed through an azologization strategy (Morstein,

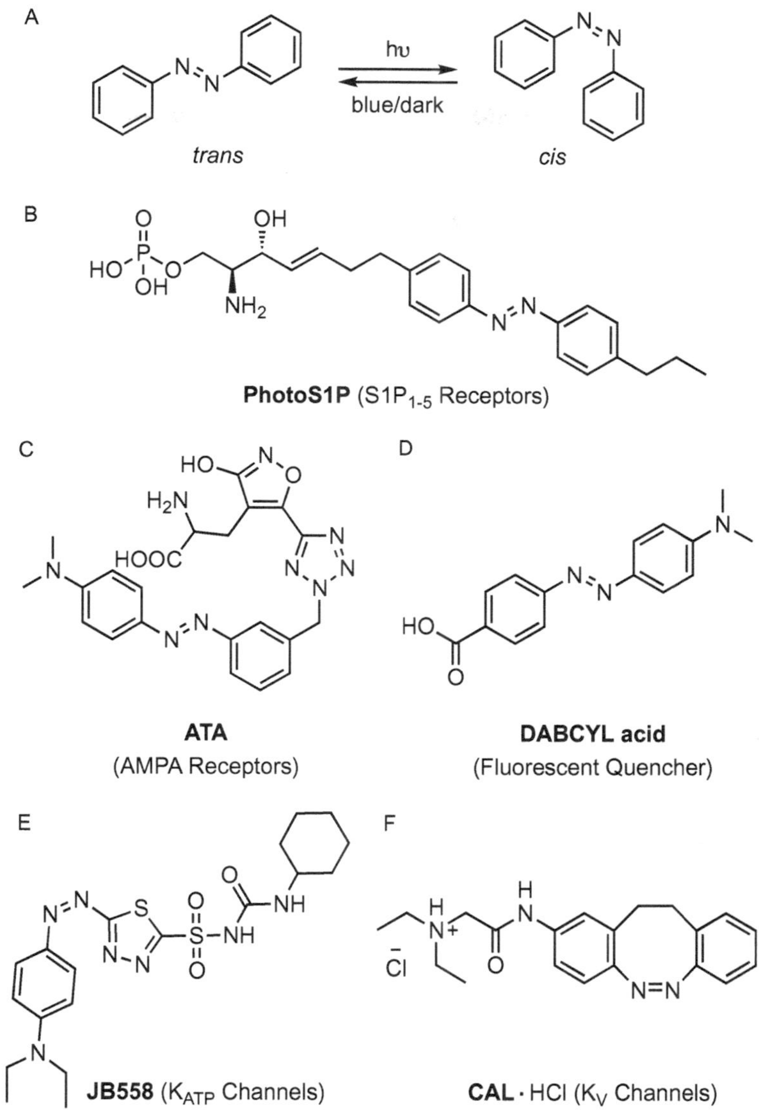

Fig. 1 (A) Photoisomerization of azobenzene. (B) Example of an azobenzene with classical photoswitching properties (Morstein, Hill, et al., 2019). (C and D) Examples of a red-shifted azobenzene (Stawski, Sumser, & Trauner, 2012) and a push-pull azobenzene. (E and F) Examples of a heterocyclic azobenzene (Broichhagen et al., 2015) and a diazocine photoswitch (Trads et al., 2019).

Awale, Reymond, & Trauner, 2019) and for photoswitchable lipids, wherein the lipophilic azobenzene moiety replaces parts of a saturated or (poly)unsaturated fatty acid chain (Frank et al., 2016; Frank, Moroni, Moshourab, Sumser, & Lewin, 2015; Morstein, Hill, et al., 2019).

Azobenzenes are inherently hydrophobic and exhibit limited water-solubility if not substituted with polar groups. Co-solvents are therefore often necessary to dissolve these compounds effectively. Due to the hydrophobicity and limited solubility of azobenzenes in aqueous solution, a 100 × stock solution leading to 1% final concentration of co-solvent (e.g., DMSO) is recommended to reduce the risk of precipitation in assays. With 1% aqueous DMSO, the solubility of many azobenzenes is limited to high micromolar concentrations. In some cases, complexation with bovine serum albumin (BSA) or cyclodextrins can be employed to increase the solubility of azobenzenes in aqueous solutions. Cyclodextrins have also been used to improve the biodistribution of photoswitches in vivo, e.g. in the context of vision restoration (Berry et al., 2017). A chemical strategy sometimes employed that increases solubility in aqueous solution is the attachment of polar groups, e.g., sulfonic acids, polyethylene glycols, or basic heterocycles, such as morpholine, that do not negatively affect the pharmacodynamics (Zhang, Timm, Arndt, & Woolley, 2010). Azobenzenes typically absorb light in the visible region of the spectrum, hence a stock solution will be colored (e.g., a stock solution at high micromolar or millimolar concentrations). If the solution is not tinted, chances are that the azobenzene is not properly dissolved.

2.3 Which light-sources should be used?

The required light-dose depends on the nature of the photoswitch, the employed wavelength, and the concentration of the molecule. Continuous irradiation can be phototoxic and is normally not required. Pulse-irradiation works well with photoswitches that relax relatively slowly, such as classical azobenzenes, and is preferred. Depending on the assay, different light-sources can be employed. In the Trauner group, we mostly rely on monochromators (e.g., OptoScan Monochromator 89 North), LEDs, or LED arrays. LED arrays are especially useful in experiments that require long-term incubation with the bioactive photoswitch. Our home-built "Cell DISCO" system (Fig. 2; (Borowiak et al., 2015)) enables to control the frequency and duration of light-flashes and can be optimized to deliver the minimal irradiation needed to keep photoswitches in the desired wavelengths-adapted state. In most assays, a 75 ms flash every 15 s has proven to be sufficient to keep classical azobenzenes (e.g., photolipids) in the *cis*-state at nanomolar or low micromolar concentrations in 24- or 96 well-plates with an LED array. The multiple well cell culture plates commonly employed are transparent at the wavelengths used and can be irradiated from below. Black bottom

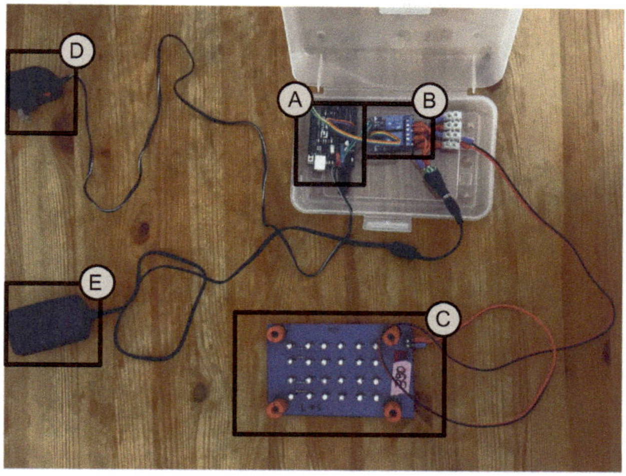

Fig. 2 The "Cell DISCO" system, as used in the Trauner group (Borowiak et al., 2015). (A) Single-board microcontroller (e.g., Arduino Uno SMD R3). (B) Power Relay Module Board. (C) 5mm LED plate. (D) 5V (DC) Power Supply. (E) 24V (DC) Power Supply.

or white bottom plates need to be irradiated from the top. The isomerization of photoswitches can be conveniently followed using an UV–Vis instrument or a well plate reader with absorption detection capabilities (see Section 2.4). Methods and assays for the in vivo light delivery are not discussed here but significant progress has been made to this end (Morstein & Trauner, 2019).

2.4 How do I establish if an azobenzene is sufficiently switched?

The ratio of azobenzene isomers is determined by the photostationary state (PSS) at the wavelength of light used. It takes some time to approach this PSS starting from a dark-adapted state. Whether the PSS is reached, can be conveniently determined using a standard UV–Vis spectrometer or a well-plate reader in absorption mode. Concentrations of the azobenzene photoswitch between 5 and 100 μM are typically used in these experiments. Switching of a classical azobenzene from *trans* to *cis* with UV-A light leads to a decrease in absorption at 340 nm (Fig. 3A, B). When no further changes in absorption are observed, one can safely assume that the PSS is fully established. Lower concentrations of photoswitches reach the PSS faster at a given intensity as a consequence of the Beer-Lambert law. To quantify the PSS, one can conduct a ^{1}H NMR experiment and measure peaks that can be assigned to either photoisomer (Fig. 3C). This is usually possible with azobenzenes since the aromatic signals of each isomer are well separated. If the UV–Vis spectra

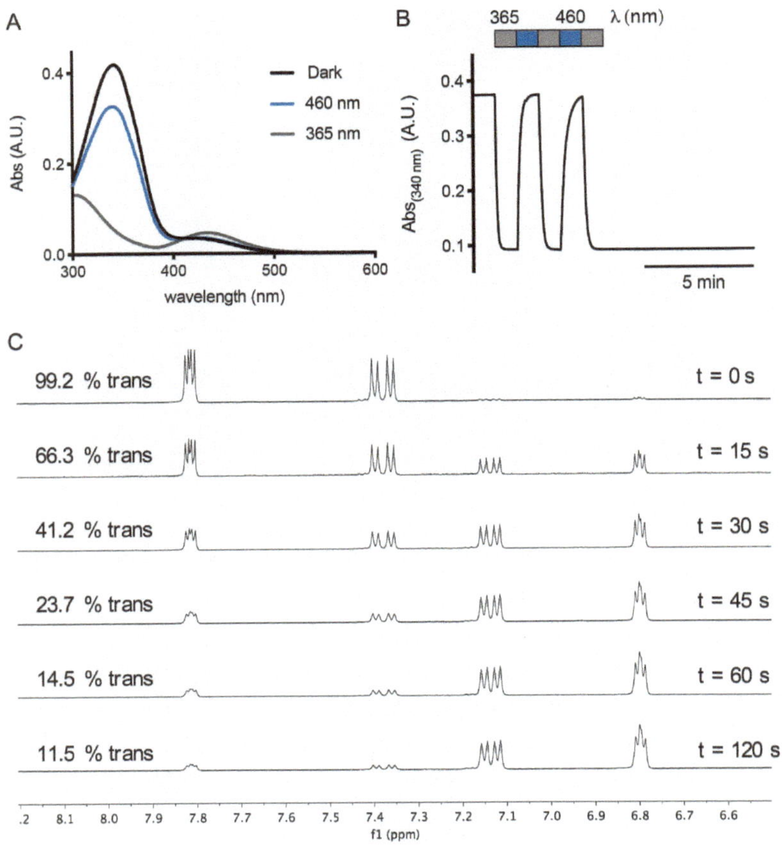

Fig. 3 Photophysical characterization of **PhotoS1P** (Morstein, Hill, et al., 2019). (A) UV–Vis spectrum of **PhotoS1P** in the dark-adapted state (black, *trans*), 365 nm adapted (gray, *cis*) and 460 nm adapted (blue, *trans)* photostationary states. (B) Reversible cycling between photoisomers with alternating illumination at 365/460 nm. This assay also established thermal relaxation rates in the dark. The example shown represents a "bistable" switch. (C) ^{1}H NMR of **PhotoS1P** after irradiation at 365 nm. **PhotoS1P** was dissolved in MeOD/ND_4OD and irradiated for varying durations at 365 nm. *Reprinted with permission from Morstein, J., Hill, R. Z., Novak, A. J. E., Feng, S., Norman, D. D., Donthamsetti, P. C., et al. (2019). Optical control of sphingosine-1-phosphate formation and function.* Nature Chemical Biology, 15, *623.*

of the individual isomers are known, e.g., from HPLC separations, the PSS can also be calculated from the UV-VIS spectra of the mixture.

2.5 Which colorimetric/fluorescent/luminescent assays are compatible?

Photopharmacology requires biological readouts. Numerous assays have been used to this end over the last 15 years. While non-optical assays have played a major role (e.g., using electrophysiology or mass spectrometry), photopharmacology is by no means limited to these. Most optical assays are compatible with photopharmacology, provided a proper dye, fluorophore, or photoluminescent agent is used. This is especially true if the assay is run as an endpoint assay, wherein the optical readout is done separately from the actuation. Caution only has to be taken if the photoswitch is still present when the optical readout is performed. Azobenzenes exhibit absorptivity that could interfere with colorimetric assays and can in principal act as fluorescent quencher. If an assay is run in a time-resolved manner, i.e., not as an endpoint assay, the compatibility of the switching wavelengths with the absorption and emission wavelengths of fluorophores or the emission wavelength of luciferins needs to be carefully evaluated. The best way to avoid undesired isomerization or activation is the use of chromophores that are spectrally orthogonal to the switches. In the case of classical azobenzenes, these would be chromophores that absorb and emit in the orange/red region of the spectrum. However, sometimes clever assay designs allow for the use wavelengths that are not orthogonal. An example is the calcium imaging of dorsal root ganglion neurons with **PhotoS1P** using Fura2 (Morstein, Hill, et al., 2019). While UV-A light is used in this assay to excite the employed fluorophore, the time between two imaging acquisitions is sufficient to apply blue light to isomerize the switch and illicit biological responses.

3. Photopharmacological assay methods—Exemplified with PhotoS1P

Recently, we developed photoswitchable analogs of the signaling lipids sphingosine and sphingosine-1-phosphate (S1P), respectively termed **PhotoSph** and **PhotoS1P** (Morstein, Hill, et al., 2019). **PhotoSph** is a photoswitchable substrate for sphingosine kinase and **PhotoS1P** enabled the optical control of S1P receptors in cell culture and in vivo.

3.1 Enzymatic assay using pre-irradiation mode—SphK assay with PhotoSph

Using pre-irradiation of two separate samples allows to directly compare the two isomers of a photoswitchable molecule side by side. In a sphingosine kinases assay, the two photoisomers *trans*-**PhotoSph** and *cis*-**PhotoSph** are compared as photoswitchable substrates for the enzyme sphingosine kinase I/II. This assay is based on the measurement of remaining ATP which is converted into a luminescent signal (Fig. 4).

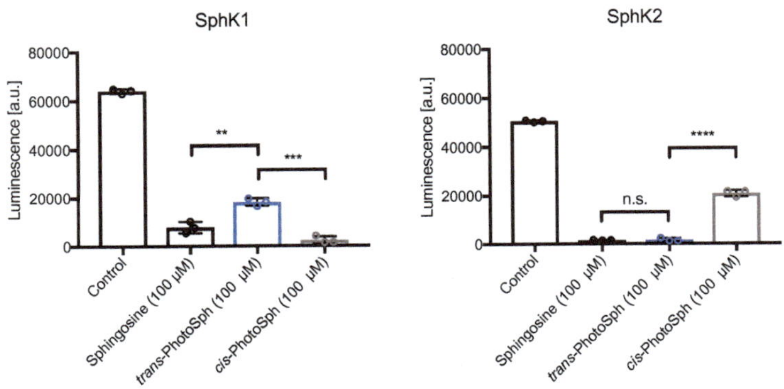

Fig. 4 In vitro SPHK Assay. 100 µM Sphingosine, *trans*-**PhotoSph** and *cis*-**PhotoSph** were incubated with SPHK1 (A) or SPHK2 (B) with ATP as the limiting reagent for 2 h. ATP was converted into a luminescence signal proportional to the amount of remaining ATP. Samples ($N = 3$) were run in triplicates. Error bars represent SEM. $^{**}P = 0.0031$; $^{***}P = 0.0002$; $^{****}P < 0.0001$; n.s. = 0.8662; two tailed student's *t*-test. *Reprinted with permission from Morstein, J., Hill, R. Z., Novak, A. J. E., Feng, S., Norman, D. D., Donthamsetti, P. C., et al. (2019). Optical control of sphingosine-1-phosphate formation and function.* Nature Chemical Biology, 15, *623.*

3.1.1 Equipment

1. Dark-Room equipped with 660 nm LED room light (ideal) or room with dim light.
2. Plate Reader for Luminescence (BMG Labtech FLUOstar Omega).
3. 365 nm light source (OptoScan Monochromator 89 North).
4. Sphingosine Kinase Activity Assay (Echelon Biosciences K-3500).
5. Ice-Container.

3.1.2 Materials

1. 4 mL Glass vials.
2. Sphingosine kinase 1 (Echelon Biosciences E-K068).
3. Sphingosine kinase 2 (Echelon Biosciences E-K069).
4. Sphingosine (18:0)—provided in assay kit (Echelon Biosciences K-3500).
5. **PhotoSph**—synthesized according to Morstein, Hill, et al. (2019).
6. ATP—provided in assay kit (Echelon Biosciences K-3500).
7. ATP detector mix—provided in assay kit (Echelon Biosciences K-3500).
8. Reaction Buffer—provided in assay kit (Echelon Biosciences K-3500).
9. DMSO.

3.1.3 Procedure

1. Prepare solutions as described in assay protocol (Echelon Biosciences K-3500).

2. Prepare 4× **PhotoSph** solution by diluting a 10 mM stock of **PhotoSph** in DMSO in reaction buffer to a concentration of 400 μM.
3. Split this solution into two glass vials (4 mL).
4. Immediately before pipetting the assay, keep one vial with 4× **PhotoSph** solution in the dark (*trans*-**PhotoSph** solution) and irradiate the other stock for 5 min with UV-A light (*cis*-**PhotoSph** solution).
5. Pipette the assay according to the assay protocol and replace sphingosine with *cis*-**PhotoSph** and *trans*-**PhotoSph** solutions.

3.2 Cell-based assay using pre-irradiation mode—Ca^{2+} mobilization assay with PhotoS1P

Activation of S1P-receptors can be assessed in HTC4 cells stably expressing $S1P_{1-5}$ receptors. A detailed protocol for this assay was previously described (Valentine & Tigyi, 2012). Herein we focus on the adaption of this protocol to study a photolipid using pre-irradiation mode. Similar to the enzymatic assay described in Section 3.1, this type of setup allows for the study of the pharmacology of both photoisomers of a photochromic ligand side by side. Here, the pharmacology of S1P (18:0), *trans*-**PhotoS1P**, and *cis*-**PhotoS1P** are compared. This assay is based on a fluorescent readout using the fluorescent calcium dye Fura2-AM (Fig. 5).

3.2.1 *Equipment*

1. General equipment for Ca^{2+} mobilization assay—see Valentine and Tigyi (2012).
2. 365 nm light source (e.g., OptoScan Monochromator 89 North, Cell DISCO (Borowiak et al., 2015), or other).

3.2.2 *Materials*

1. General equipment for Ca^{2+} mobilization assay—see Valentine and Tigyi (2012).
2. Dark-adapted **PhotoS1P** (stock solution prepared like S1P (18:0)) in Valentine and Tigyi (2012).

3.2.3 *Procedure*

1. The assay was run with LPA (18:0) and dark-adapted *trans*-**PhotoS1P** according to Valentine and Tigyi (2012).
2. The compound addition plate was removed from the FlexStation and irradiated for 2 min.

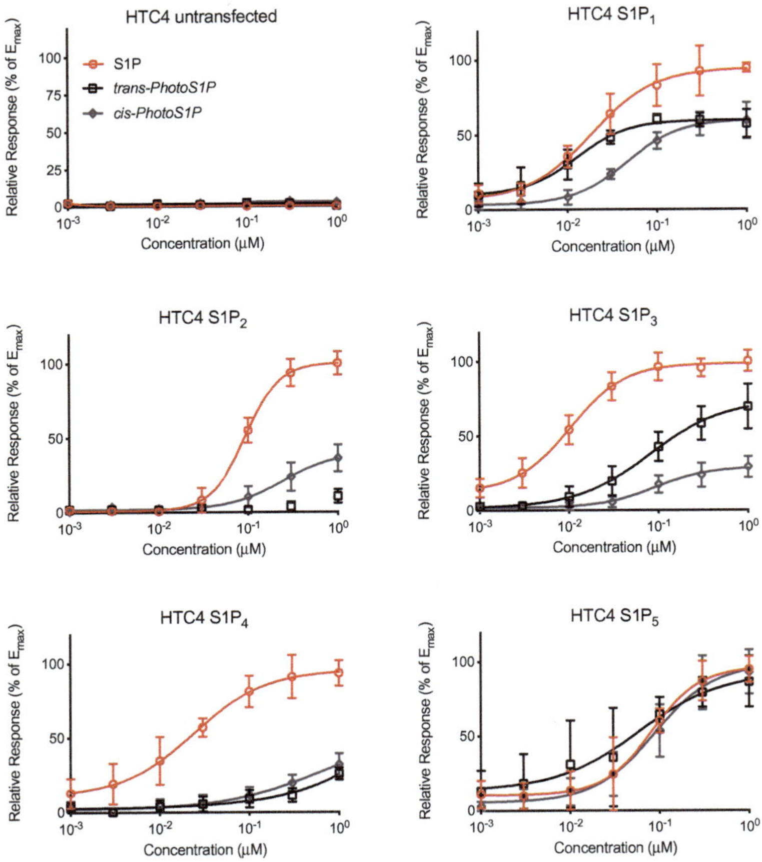

Fig. 5 Optical control of $S1P_{1-5}$ receptor-mediated calcium release. Fura2-AM calcium imaging in HTC4 cells stably transfected with $S1P_{1-5}$ receptors. Dose response of S1P, *trans*-**PhotoS1P** and *cis*-**PhotoS1P** in HTC4, HTC4 $S1P_1$, HTC4 $S1P_2$, HTC4 $S1P_3$, HTC4 $S1P_4$, and HTC4 $S1P_5$ cells. Samples ($N=8$) were run in quadruplicates and at two independent experiments. Data points were normalized to maximal S1P response for each receptor. Error bars represent mean ± S.D. *Reprinted with permission from Morstein, J., Hill, R. Z., Novak, A. J. E., Feng, S., Norman, D. D., Donthamsetti, P. C., et al. (2019). Optical control of sphingosine-1-phosphate formation and function.* Nature Chemical Biology, 15, *623.*

3. The compound addition plate was inserted into the FlexStation and the assay was repeated with 365 nm-adapted *cis*-**PhotoS1P** according to Valentine and Tigyi (2012).

The advantage of this format is that the same solution is used prior and after irradiation eliminating pipetting errors. Furthermore, cells are not irradiated with light eliminating potential phototoxicity effects.

3.3 Cell-based assay using live-switching mode—$S1P_1$-GIRK whole-cell patch clamp experiment with PhotoS1P

Electrophysiology experiments allow to assay a system with high temporal resolution. This type of experiment is therefore ideally suited to study a photochromic ligand in live-switching mode (Fig. 6).

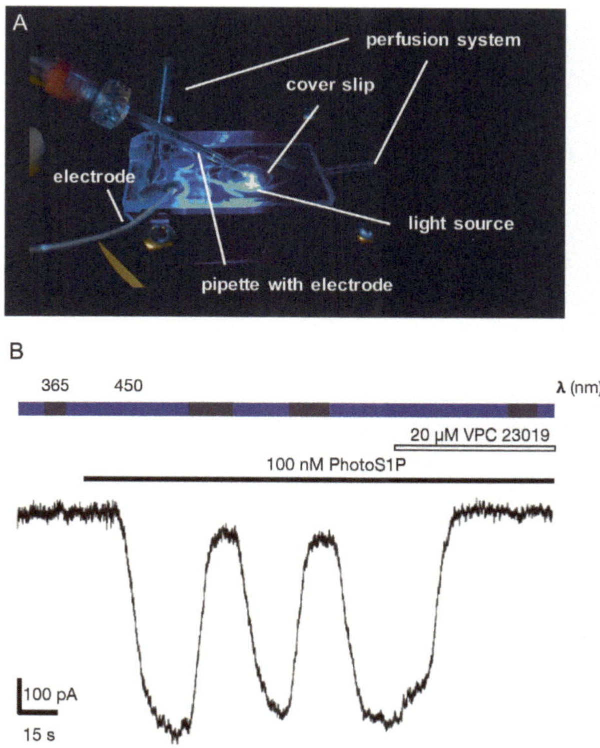

Fig. 6 Optical control of $S1P_1$-GIRK coupling in HEK293T cells. (A) Electrophysiology setup showing the electrodes, the cover slip and the perfusion system. Light is delivered from below through the objective of the inverted microscope. (B) A representative single current trace of **PhotoS1P** evaluated using whole-cell patch clamp recording in HEK293T cells transiently expressing $S1P_1$ receptors and GIRK channels. Switching illumination between 365 and 450 nm does not induce current in the absence of PhotoS1P. Application of PhotoS1P (100 nM) in *trans* state induces an inward current, which is reversed by illumination at 365 nm and re-activated by illumination at 450 nm. The current induced by **PhotoS1P** under 450 nm light is inhibited by the $S1P_1$ receptor antagonist VPC23019. Currents measured at constant membrane potential of −80 mV. *Reprinted with permission from Morstein, J., Hill, R. Z., Novak, A. J. E., Feng, S., Norman, D. D., Donthamsetti, P. C., et al. (2019). Optical control of sphingosine-1-phosphate formation and function.* Nature Chemical Biology, 15, *623.*

3.3.1 Equipment

1. Patch-Clamp SetUp with capability for whole cell patch clamp experiments.
2. Pipette Puller.
3. Polychrome V monochromator (TILL Photonics).
4. Biosafety Cabinet.
5. Incubator with CO_2 (for mammalian cell lines).
6. Pipettes.

3.3.2 Materials

1. HEK293T cell line.
2. 12 Well plate.
3. Cover Slips (18mm).
4. Ethanol.
5. Growth Medium: Dulbecco's Modified Eagle's Medium (DMEM), 10% Fetal Bovin Serum.
6. OptiMEM.
7. Lipofectamine 2000.
8. Poly-L-lysine.
9. GIRK1-F137S in mammalian expression vector.
10. tdTomato in mammalian expression vector.
11. PBS.
12. Trypsine.
13. **PhotoS1P**—synthesized according to Morstein, Hill, et al. (2019).

3.3.3 Procedure

1. Cover slips were prepared:
   Cover slips (18mm) were added to 12-well-plate.
   Poly-L-lysine was added to each well to cover slip.
   The plate was incubated in cell incubator at 37 degrees for 30min.
   Cover slips were washed with sterile water.
2. Growth medium was removed from HEK293T cells.
3. Cells were washed with PBS.
4. Trypsin (0.05%) was added (1mL in T25 flask) and the cells were incubated for 3min at 37 degrees.
5. After cells were detached, 5mL of growth medium was added to saturate trypsin.
6. Cells were seeded on cover slips (50 μL with approx. 30,000 cells).
7. After 4h, transfection mix was prepared.

8. For the transfection of cells in 6 wells, 6 μL lipofectamine 2000 and 150 μL of OptiMEM were added to microcentrifuge tube.
9. 4500 ng of GIRK1-F137S plasmid and 480 ng of tdTomato plasmid and 150 μL of OptiMEM were added (750 ng of GIRK1-F137S plasmid and 80 ng of tdTomato plasmid per well) and gently mixed in second microcentrifuge tube.
10. The two microcentrifuge tubes were mixed and incubated for 5 min.
11. 50 μL of transfection mix was added to each well.
12. Cells were incubated with transfection mixture for 16–24 h.
13. The cover slip was removed from growth medium with transfection mixture and transferred into a high-potassium solution (120 mM KCl, 25 mM NaCl, 10 mM HEPES, 2 mM $CaCl_2$ and 1 mM $MgCl_2$, pH 7.4).
14. Cover slip was mounted onto patch clamp setup.
15. Cell with intermediate expression of tdTomato was patched.
16. Compounds in high potassium solution were added through a gravity driven perfusion system.
17. Light was added with Polychrome V monochromator (TILL Photonics).

Acknowledgments

J.M. thanks the German Academic Scholarship Foundation for a PhD fellowship and New York University for a Margaret and Herman Sokol fellowship. The authors thank Tongil Ko for proofreading of the manuscript and Christopher Arp and Martin Reynders for helpful discussion.

References

Beharry, A. A., & Woolley, G. A. (2011). Azobenzene photoswitches for biomolecules. *Chemical Society Reviews*, *40*, 4422–4437.

Berry, M. H., Holt, A., Levitz, J., Broichhagen, J., Gaub, B. M., Visel, M., et al. (2017). Restoration of patterned vision with an engineered photoactivatable G protein-coupled receptor. *Nature Communications*, *8*, 1–12.

Borowiak, M., Nahaboo, W., Reynders, M., Nekolla, K., Jalinot, P., Hasserodt, J., et al. (2015). Photoswitchable inhibitors of microtubule dynamics optically control mitosis and cell death. *Cell*, *162*, 403–411.

Broichhagen, J., Frank, J. A., Johnston, N. R., Mitchell, R. K., Šmid, K., Marchetti, P., et al. (2015). A red-shifted photochromic sulfonylurea for the remote control of pancreatic beta cell function. *Chemical Communications*, *51*, 6018–6021.

Frank, J. A., Moroni, M., Moshourab, R., Sumser, M., & Lewin, G. R. (2015). Photoswitchable fatty acids enable optical control of TRPV1. *Nature Communications*, *6*, 7118.

Frank, J. A., Yushchenko, D. A., Hodson, D. J., Lipstein, N., Nagpal, J., Rutter, G. A., et al. (2016). Photoswitchable diacylglycerols enable optical control of protein kinase C. *Nature Chemical Biology*, *12*, 755–762.

Hüll, K., Morstein, J., & Trauner, D. (2018). In vivo photopharmacology. *Chemical Reviews*, *118*, 10710–10747.

Morstein, J., Awale, M., Reymond, J.-L., & Trauner, D. (2019). Mapping the azolog space enables the optical control of new biological targets. *ACS Central Science*, *5*, 607–618.

Morstein, J., Hill, R. Z., Novak, A. J. E., Feng, S., Norman, D. D., Donthamsetti, P. C., et al. (2019). Optical control of sphingosine-1-phosphate formation and function. *Nature Chemical Biology*, *15*, 623.

Morstein, J., & Trauner, D. (2019). New players in phototherapy: Photopharmacology and bio-integrated optoelectronics. *Current Opinion in Chemical Biology*, *50*, 145–151.

Stawski, P., Sumser, M., & Trauner, D. (2012). A photochromic agonist of AMPA receptors. *Angewandte Chemie, International Edition*, *51*, 5748–5751.

Szymański, W., Beierle, J. M., Kistemaker, H. A. V., Velema, W. A., & Feringa, B. L. (2013). Reversible photocontrol of biological systems by the incorporation of molecular photoswitches. *Chemical Reviews*, *113*, 6114–6178.

Trads, J. B., Hüll, K., Matsuura, B. S., Laprell, L., Fehrentz, T., Görldt, N., et al. (2019). Sign inversion in photopharmacology: Incorporation of cyclic azobenzenes in photoswitchable potassium channel blockers and openers. *Angewandte Chemie, International Edition*, *58*, 15421–15428.

Valentine, W. J., & Tigyi, G. (2012). High-throughput assays to measure intracellular Ca^{2+} mobilization in cells that express recombinant S1P receptor subtypes. In A. Pébay & K. Turksen (Eds.), *Vol. 874. Sphingosine-1-phosphate. Methods in molecular biology (methods and protocols)* (pp. 77–87). Humana Press.

Zhang, F., Timm, K. A., Arndt, K. M., & Woolley, G. A. (2010). Photocontrol of coiled-coil proteins in living cells. *Angewandte Chemie, International Edition*, *49*, 3943–3946.

CHAPTER TWELVE

Expanding the substrate selectivity of SNAP/CLIP-tagging of intracellular targets

Miguel Macias-Contreras, Kevin N. Little, Lei Zhu*
Department of Chemistry and Biochemistry, Florida State University, Tallahassee, FL, United States
*Corresponding author: e-mail address: lzhu@fsu.edu

Contents

Abstract

SNAP-tag belongs to a class of genetic tools of protein labeling that complements fluorescent proteins. This single-turnover enzyme is a mutant of human DNA repair protein O^6-alkylguanine-DNA alkyltransferase (hAGT). It accepts, in most cases, label-carrying O^6-benzylguanines or benzyl-2-chloro-6-aminopyrimidines as suitable substrates. In this article, strategies and methods to expand the scope of the labels for intracellular proteins of live cells via the actions of SNAP-tag are presented. CLIP-tag is another mutant of the hAGT that was engineered to have mutually exclusive substrate specificity from SNAP-tag. The use of complementary bioorthogonal chemical reactions in conjunction with orthogonal enzymatic SNAP/CLIP-tags for the purpose of dual-color intracellular labeling is also described.

Methods in Enzymology, Volume 638
ISSN 0076-6879
https://doi.org/10.1016/bs.mie.2020.02.021

1. Introduction

Selective and non-perturbative labeling of protein of interest (POI) in functioning cells or organisms allows biological processes to be visualized to the molecular details under a microscope (Zhang, Campbell, Ting, & Tsien, 2002). The discovery and subsequent expansion and refinement of fluorescent proteins (FPs) have provided such capacities (Fig. 1A). Therefore, FPs

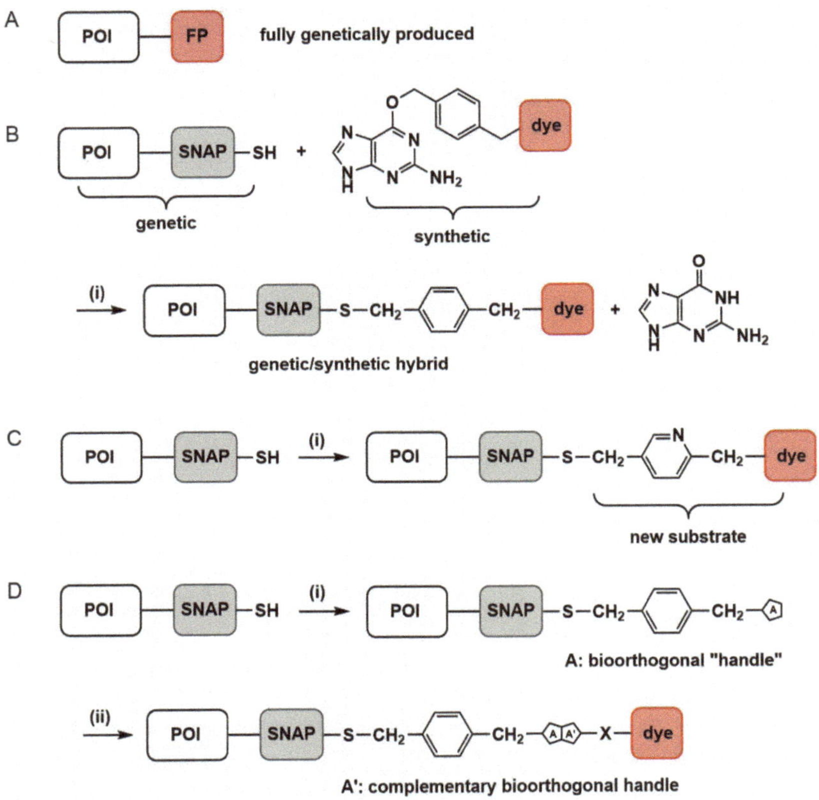

Fig. 1 (A) Fluorescent protein (FP)-labeled protein of interest (POI), which is fully genetically produced; (B) a POI is genetically coupled to SNAP-tag, which subsequently via the enzymatic step (i) reacts with an O^6-benzylguanine (BG) substrate to complete labeling and retains a phenyl linker. A cysteine residue (-SH) is the reactive locus; (C) 5-pyridylmethylguanine (**5PG**) is also accepted in the enzymatic step (i) with SNAP-tag to result in labeling via a pyridyl linker; (D) two-step labeling via the enzymatic step (i) to install a bioorthogonally reactive group ("handle" or "chemical reporter" A) (Sletten & Bertozzi, 2011), which is followed by the chemical step (ii) to conjugate with a dye that bears the complementary "handle" A′ to achieve labeling.

are universally recognized as indispensable tools for cell biologists and beyond (Day & Davidson, 2009; Tsien, 2009). Yet the quests for more discoveries would invariably expose the limits of existing tools and put forth challenges for innovations (Giepmans, Adams, Ellisman, & Tsien, 2006).

The light-emitting component of an FP is derived from a few amino acid residues in the posttranslational processing phase of protein production (Miyawaki, Nagai, & Mizuno, 2003; Remington, 2006; Yang, Moss, & Phillips, 1996). Consequently, only a limited set of photophysical (e.g., absorptivity, emission yield) and photochemical (e.g., photostability) attributes could be produced from FPs, which have become inadequate in applications that demand more (Fernández-Suárez & Ting, 2008; Heilemann, van de Linde, Mukherjee, & Sauer, 2009; Müller et al., 2017; Wombacher et al., 2010). Furthermore, developing FPs into probes that illuminate and track biological targets are tremendously challenging undertakings (Frommer, Davidson, & Campbell, 2009; Griesbeck, 2004; Zhang et al., 2002). Small molecular dyes or probes, although not selective in protein labeling and therefore free from spatial confinement, are not limited by the structural varieties of their building blocks, and therefore could satisfy increasingly demanding applications in terms of both photo-related attributes and sensitivity or signal dynamic range in response to targeted analyte molecules. The hybrid technologies that would combine the advantages of both FPs (genetic encoding and precise localization) and small molecular dyes (ample choices of optical properties and strategies for sensing) are poised to fill in the functional gap between FPs and synthetic fluorophores (Deo & Lavis, 2018).

Among the genetic/synthetic hybrid labeling technologies (Hinner & Johnsson, 2010; Jing & Cornish, 2011; Johnsson & Johnsson, 2003; Xue, Karpenko, Hiblot, & Johnsson, 2015; Zhang, Park, Suazo, & Distefano, 2018), SNAP- and (to a lesser extend) CLIP-tags are the subjects of this article. SNAP-tag is a mutant of human O^6-alkylguanine-DNA alkyltransferase (hAGT), and was produced via directed evolution to accept substituted O^6-benzylguanines (BGs) as substrates (Keppler et al., 2003). It was also found that substituted benzylchloropyrimidines are accepted as substrates (Correa et al., 2013). In the tagging reaction, the thiolate group of the cysteine residue in the active site of SNAP-tag displaces guanine in a BG substrate to rapidly and irreversibly capture the molecular cargo on the BG substrate (Fig. 1B). Yet with all the successes, the syntheses of label-conjugated BGs are not always trivial, and there is no guarantee that the molecular cargo would not adversely affect the labeling efficiencies

(Bosch et al., 2014; Correa et al., 2013; Srikun, Albers, Nam, Iavarone, & Chang, 2010). We have studied alternative substrates and strategies that aim to expand the utilities of SNAP-tag in intracellular protein labeling in both live and fixed cells. The structures and methods that we have used are reported in the next four sections (Macias-Contreras, He, & Zhu, 2018; Macias-Contreras et al., 2020).

2. Genetic construction

Four vectors were used in this study—SNAP-LaminA, SNAP-LifeAct, CLIP-LaminA, and CLIP-LifeAct—to target either nuclear envelope (LaminA) (Lutz, Trujillo, Denham, Wenger, & Sinensky, 1992) or filamentous actin (LifeAct) (Melak, Plessner, & Grosse, 2017). Similar plasmids have been reported in the literature in the past (Chen et al., 2018; Hong, Lam, & Tan, 2017; Liu et al., 2018; Müller et al., 2017; Zhang, Li, Zhang, Han, & Zhang, 2019). Same as other fusion vectors that were produced from the laboratory of the late Michael Davidson (e.g., mEmerald-LaminA-C-18 and mEmerald-LifeAct-7 as described below), the current vectors are based on either of the commercial Clontech-style cloning vectors C1 and N1 with the SNAP- or CLIP-tag replacing the FP (e.g., mEmerald) at the N-terminus of the fusion (LaminA) or the C-terminus (LifeAct). The primer sequences and the ligation strategies are the following.

2.1 SNAP-LaminA and SNAP-LifeAct

The SNAP-tag sequence (pSNAP$_f$ vector, NEB) was amplified using the forward primer AT CCG CTA GCG CTA CCG GTC GCC ACC ATG GAC AAA GAC TGC GAA ATG AAG CGC and the reverse primer GCT GCC AGA TCT GAG TCC GGA ACC CAG CCC AGG CTT GCC CAG TCT GT to introduce the restriction sites *Nhe*I and *Bgl*II, respectively. The amplified gene was digested with NheI and BglII and ligated into the NheI/BglII-digested vector mEmerald-LaminA-C-18 (Michael Davidson FP Collection, Addgene #54138) to produce SNAP-LaminA. The same strategy was used to obtain SNAP-LifeAct. In this case the SNAP-tag gene was amplified with the primers T CCA CCG GTC GCC ACC ATG GAC AAA GAC TGC GAA ATG AAG CGC and AGA GTC GCG GCC GCT TAA CCC AGC CCA GGC TTG CCC AGT CTG T with the purpose of introducing a stop codon and the restriction sites *Age*I and *Not*I. The amplified gene was digested at the mentioned sites and ligated into the AgeI/NotI-digested vector mEmerald-LifeAct-7 (Michael Davidson FP Collection, Addgene #54148).

2.2 CLIP-LaminA and CLIP-LifeAct

The primers CTA CCG GTC GCC ACC ATG GAC AAA GAC TGC GAA ATG AAG CGC and CAG ATC TGA GTC CGG AAC CCA GCC CAG GCT TGC C, forward and reverse respectively, were used to amplify the CLIP-tag gene (pCLIP$_f$ vector, NEB). The vector mEmerald-LaminA-C-18 was amplified with the forward primer TCC GGA CTC AGA TCT GGC AGC and reverse primer GGT GGC GAC CGG TAG CGC. The fragments were assembled using the Gibson Assembly (Gibson, 2011; Gibson et al., 2009) Cloning Kit (NEB) following the manufacturer's protocol to produce CLIP-LaminA. CLIP-LifeAct was assembled in the same way as above. The CLIP-tag sequence was amplified with the primers CCG GTC GAA TTC ACC ATG GAC AAA GAC TGC GAA ATG and TAG AGT CGC GGC CGC TCA TTA ATT AAC CTC GAG TTT AAA CG. Amplification of vector mEmerald-LifeAct-7 was done with the primers GCG GCC GCG ACT CTA GAT and GGT GAA TTC GAC CGG TGG.

3. SNAP-tag labeling of an O^6-(5-pyridylmethyl) guanine substrate

A major class of the substrates of SNAP-tag are substituted O^6-benzylguanines (BGs). The guanine component would depart as the result of the alkylative tagging reaction (Fig. 1B). The benzyl group, however, acts as a molecular linker but nothing more between the cargo and SNAP-tag in the completed product. We were wondering whether the benzyl group could be replaced by another moiety that would carry an active function beyond being a molecular linker, while preserving the reactivity with the SNAP-tag protein. The replacement we envisioned was pyridylmethyl group (Fig. 1C), which would act similarly to or better than benzyl as an electrophile in a S_N2 reaction, while assisting in sequestering metal ions in the context of, for example, a metal ion-sensitive fluorescent indicator (Macias-Contreras et al., 2018).

3.1 Substrate synthesis

The synthetic strategy of O^6-(5-pyridylmethyl) guanine (**5PG**) substrates is illustrated in Fig. 2 (Macias-Contreras et al., 2018). A **5PG** substrate has

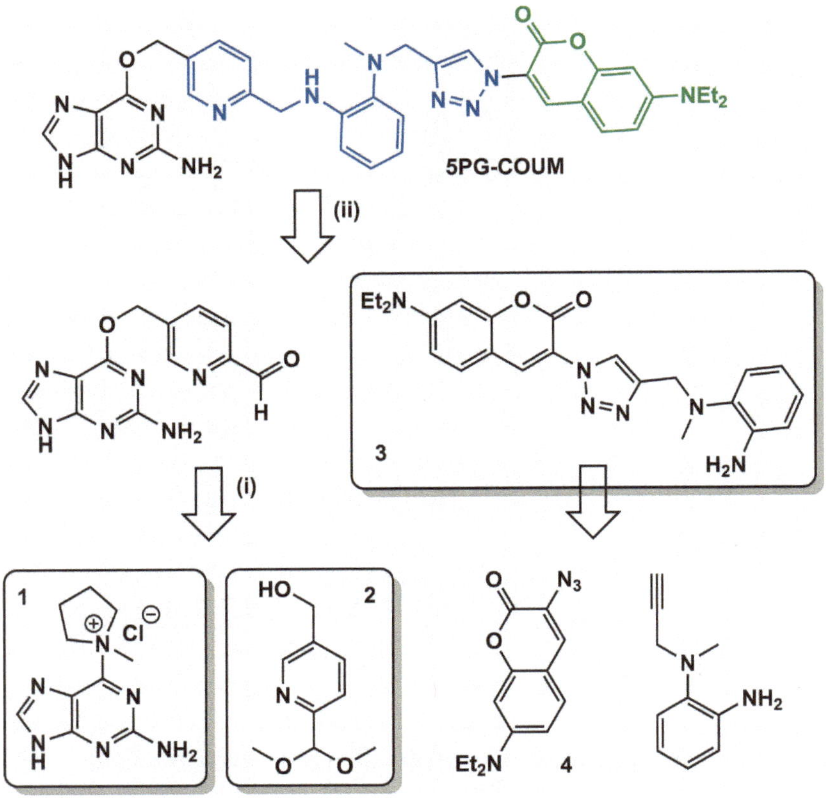

Fig. 2 Retrosynthetic analysis of **5PG-COUM**. Blue component: the zinc(II) binding site; green component: fluorophore.

three structural components that can be synthesized concurrently: the electrophilic guanine substrate (**1** in Fig. 2), the pyridylmethyl alcohol linker (**2**), and the cargo moiety (**3**). The connection between the linker and guanine is invariably a nucleophilic aromatic substitution (step i), while more conjugation reactions, such as amide bond formation (Keppler et al., 2004; Ramil et al., 2017; Srikun et al., 2010), copper(I)-catalyzed azide-alkyne cycloaddition (Song, Wang, Han, Xu, & Xiao, 2015), or in the current case reductive amination (step ii), have been employed in joining the linker and the cargo components. The major investment in the synthesis of **5PG** substrates is the preparation of the pyridylmethyl linker unit **2**, which takes four steps from a commercially available compound in a linear sequence. In this synthesis, compounds **3** and **4** could be any amino- or azido-substituted fluorescent dye, respectively. Therefore, **5PG** substrates with different colors regardless of metal coordination ability could be prepared.

3.2 Labeling and imaging

The intracellular labeling effectiveness of **5PG-COUM** is demonstrated in HeLa S3 cells. The culture dishes are seeded on Day 1, which are transfected with the chimera of SNAP-tag and the directing peptide on Day 2. The treatment of **5PG-COUM** to the transfected cells is carried out 24 h later on Day 3, followed by imaging on a confocal laser scanning fluorescence microscope. Two identically transfected dishes are usually prepared, one of which would react with SNAP-Cell® TMR-Star as a control for the transfection efficiency and the competency of the expressed SNAP-tag protein. Instead of being imaged on the FV1000 Confocal Laser Scanning Biological Microscope (Olympus) under controlled incubating conditions, the control dish is usually examined quickly from the eyepiece of a widefield fluorescence microscope using the green/orange fluorescence filter block that is suitable for Rhodamine B, to ensure that the transfection and the labeling steps have proceeded to the expectation. The detailed protocol is listed in the following two sections.

3.2.1 Materials

In addition to the equipment and supplies typically found in tissue culture and microscopy labs, the following items are specific to the experiments described in this article.

1. *RPMI 1640 culture medium* (Sigma-Aldrich) supplemented with 10% FBS (or 10% CCS) and penicillin, streptomycin, and amphotericin B (aka PSA) of 1 × each, stored at 4 °C.
2. *Cultured HeLa S3 cells* in the above medium at 37 °C, 95% humidity, and 5% CO_2 at least a week and two passages after being brought up from a frozen stock. Use it up to 2 months.
3. *ΔT culture dishes* (Bioptechs) mountable on an Olympus stage adapter. The ΔT culture dishes are placed inside Petri dishes (60 × 15 mm, polystyrene sterile) and sterilized by exposure to a UVP Multiple Ray Lamp equipped with a USHIO G8T5 germicidal UV bulb (254 nm) for 1–2 h.
4. *Vector DNAs* (Section 2) of SNAP-LaminA and SNAP-LifeAct at ~ 1 μg/μL.
5. *Effectene Transfection Reagent* (Qiagen) for transfection, which contains "bufferEC," "enhancer," and "Effectene."
6. *5PG-COUM*, 2 mM in DMSO. The DMSO stock solution of **5PG-COUM** is dispensed into 100-μL aliquots in Eppendorf tubes with screwcaps, and stored at −20 °C. An aliquot should be discarded after 10 freeze/thaw cycles.

7. *SNAP-Cell® TMR-Star* (NEB), 1 mM in DMSO. The DMSO stock solutions are stored at −20 °C.
8. FV1000 Confocal Laser Scanning Biological Microscope (Olympus) equipped with 405-, 488-, 543-, and 633-nm laser lines.
9. ΔT culture dish controller and Olympus stage adaptor (Bioptechs).

3.2.2 Protocols

The cell culture, transfection and labeling are done in a laminar flow hood equipped with Pipet Aid (Drummond) for both left and right hands. All samples (e.g., dyes and BG or **5PG** substrates), reagents (e.g., culture medium), and supplies need to be sterilized prior to being transferred into the hood. Sample and reagent containers are sterilized by rinsing with 70% ethanol. Pipet tips, glass and plastic containers were autoclaved, and the buffers/medium were either autoclaved or filtration-sterilized.

3.2.2.1 Transfection (Days 1 and 2)

1. Plate HeLa S3 cells on 35-mm ΔT culture dishes (Bioptechs) 16–18 h prior to transfection. They are allowed to grow to 50%–70% confluency at the time of transfection.
2. Transfer a vector DNA solution (1 μL, ~1 μg/μL) of either SNAP-LaminA or SNAP-LifeAct to a sterilized 1.5-mL Eppendorf tube.
3. Add BufferEC (50 μL) from the Effectene Transfection Reagent kit to the Eppendorf tube.
4. Add enhancer (3 μL) from the Effectene Transfection Reagent kit to the Eppendorf tube.
5. Vortex the Eppendorf tube (1–2 s), spin it down, and let it sit for 2 min.
6. Add Effectene (8 μL) from the Effectene Transfection Reagent kit, vortex (2–5 s) and spin, followed by a 10-min wait (up to 4 h).
7. Add warm (37 °C) culture medium (1 mL) in the Eppendorf tube, which is ready to replace the existing medium in the ΔT culture dish.
8. Aspirate off the existing medium from the ΔT dish, and pour in the Effectene/DNA-containing medium.
9. Incubate the dish for at least 12 h at 37 °C, 95% humidity, and 5% CO_2.
10. Replace the medium, and incubate for ~12 more hours (no more than 24 h post-transfection) before labeling.

3.2.2.2 Labeling (Day 3)

1. Warm the culture medium in a 37 °C water bath.
2. Take the aliquots of the DMSO stock solutions of **5PG-COUM** and SNAP-Cell® TMR-Star out of the −20 °C freezer. The tubes

containing the dyes were sterilized before transferred to a laminar flow hood, and the contents were thawed at rt covered with aluminum foil.

3. Transfer 1 mL of warmed medium to a 10-mL conical tube, and add 0.5 μL of the **5PG-COUM** solution in DMSO (2 mM) or 1 μL of SNAP-Cell® TMR-Star solution in DMSO (1 mM) to the medium to make a 1 μM of each dye in medium. Vortex to mix.
4. Aspirate off the existing medium that contains the transfection reagents, rinse the cells with fresh medium (×1). Replenish the dish with 1 mL of medium that contains a dye.
5. Incubate the dishes at 37 °C, 95% humidity, and 5% CO_2 for 30 min. The dye-containing medium was aspirated afterward, and the cells were washed with warm medium (×2), followed by another 30-min period of incubation.
6. Replace the medium after the incubation. The dish is now ready for imaging.

3.2.2.3 Imaging (Day 3)

1. Switch on the main control (Olympus IX2-UCB) of the FV1000 microscope.
2. Switch on the PSU unit, followed by turning the key to the "on" position.
3. Switch on the 405-nm diode laser (for **5PG-COUM**), followed by turning the key to the "on" position.
4. Switch on the 543-nm HeNe laser (for SNAP-Cell® TMR-Star).
5. Turn the computer on, and initialize the FV10-ASW4.2 program.
6. Place oil (Zeiss Immersol™ 518F) on the PlanAPO 60x/1.40 (Olympus) oil objective and on the heating contact on the Bioptechs stage adaptor.
7. Take the ΔT culture dish out of the petri dish, use an isopropanol-soaked tissue paper to wipe the bottom of the dish, and mount the dish on the stage adaptor. Tighten the screws evenly on the stage adaptor to secure the dish.
8. Fit the stage adaptor on the microscope stage, cover the dish with the heated lid, and turn the Delta T controller on with temperature set at 37 °C.
9. Turn the transmittance lamp on, and use eyepiece to bring the cells into focus, and then switch off the transmittance lamp.
10. In the Dye List, load the setting for "rhod-2" for imaging SNAP-Cell® TMR-Star (Channel 2, $\lambda_{ex} = 543$ nm, $\lambda_{em} = 560–660$ nm). In Channel 1, select 405-nm excitation, and set the emission window at 460–510 nm for imaging 7-aminocoumarin-based **5PG-COUM**. Check the light paths to make sure that correct dichroic mirrors and filters are selected.

11. In the live view, the scan rate is set at 4 μs/pixel and the aspect ratio is 512 × 512. When a region of interest (ROI) is identified, a proper zoom value is applied, and the scan rate is lowered to 10–20 μs/pixel. The aspect ratio is increased up to 1024 × 1024. The Kalman filter is turned on (5–10 lines), and the sequential imaging mode is used if multiple fluorescence channels are being imaged. The bright field image (i.e., the TD channel) is acquired with fluorescence images. The scale bar of 10 or 20 μm is placed at either lower-left of lower-right corner of the image.
12. Export images in 16-bit TIFF raw data format that include all the ROIs (e.g., scale bar). The figures are put together using Adobe Photoshop or ImageJ. Representative images from **5PG-COUM** labeling are shown in Fig. 3.

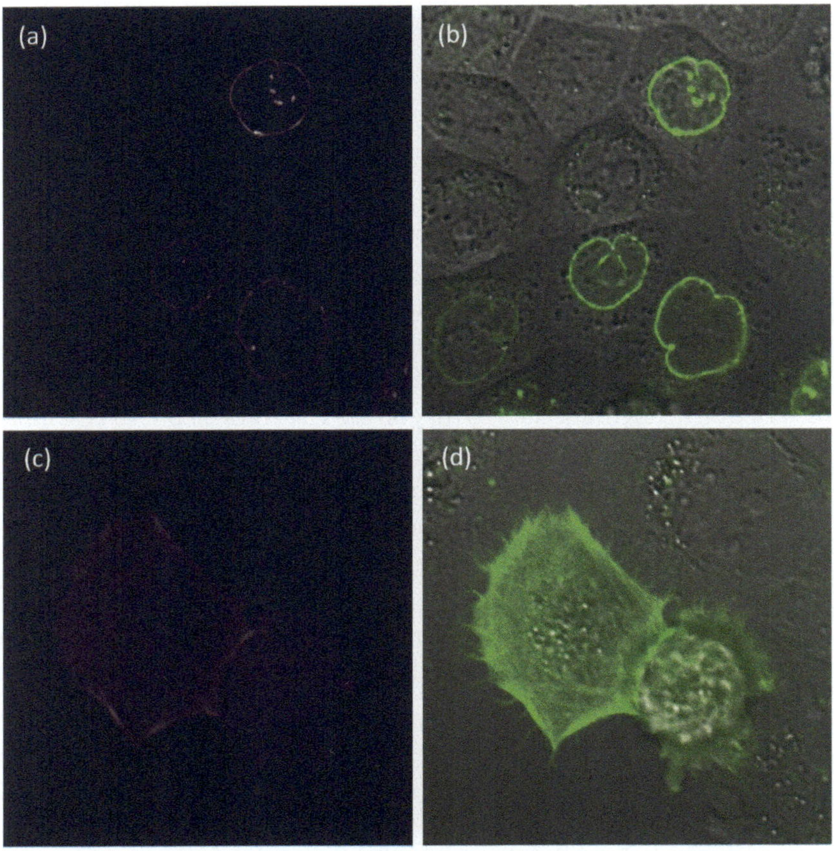

Fig. 3 Images of live HeLa S3 cells transfected with SNAP-LaminA (A, B) or SNAP-LifeAct (C, D) and labeled with **5PG-COUM**. (A, C): Fluorescence channels (λ_{ex} = 405 nm, λ_{em} = 460–510 nm); (B, D) merged image of both fluorescence and bright field channels.

The cells were also incubated in the medium that contained both SNAP-Cell® TMR-Star, the BG substrate that is available from NEB, and **5PG-COUM** at an equal molar concentration (1 μM). The purpose was to see how competitive, or the lack thereof, of **5PG-COUM** in the presence of a highly optimized substrate in labeling SNAP-tag. The presence of SNAP-Cell® TMR-Star did suppress the labeling by **5PG-COUM**. However, quite large heterogeneity among cells was observed in taking up **5PG-COUM** (Fig. 4). In each frame, there were only a few cells that were labeled by **5PG-COUM**, as seen in the green channel, while many

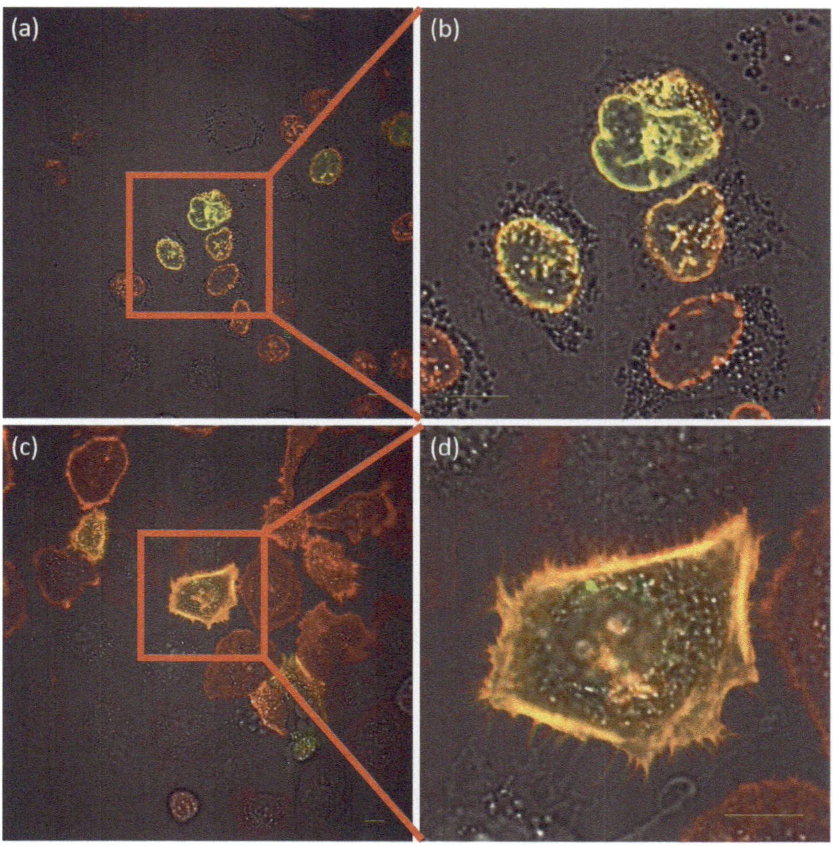

Fig. 4 Green/red/bright field-merged fluorescence images of live HeLa S3 cells transfected with SNAP-LaminA (A, B) or SNAP-LifeAct (C, D) labeled with both **5PG-COUM** (green channel) and SNAP-Cell® TMR-Star (red channel). Scale bar = 10 μm. The concentrations of both dyes during incubation were 0.5 μM. (A, C) Full frame; (B, D) zoomed section as indicated in (A) and (C).

more, including the **5PG-COUM**-labeled cells, were illuminated by SNAP-Cell® TMR-Star. It is likely that the heterogeneity originates from the varied levels of SNAP-tag expression that amplified the promiscuity of SNAP-tag in those cells that have produced overabundant quantities of SNAP-tag proteins.

4. Two-step bioorthogonal labeling

Assuming the size and function of the molecular cargo on a BG substrate might adversely affect the efficiency of the SNAP-tagging enzymatic reaction, the alternative strategy is to decouple BG from the molecular cargo to ensure that a BG substrate is efficiently taken up by SNAP-tag. This can be done by attaching a small sized bioorthogonal reactive group (aka, a "handle") on BG, which would be taken up by SNAP-tag efficiently due to the minimal nature of the functionalization of BG, while the molecular cargo that bears the complementary handle could be conjugated to the now bioorthogonally reactive SNAP-tag in the subsequent chemical step (Fig. 1D). Two options of such chemical steps are shown in Fig. 5A. This two-step labeling strategy involving SNAP-tag has been applied to decorate the cell surfaces (Ramil et al., 2017). In this article, results of concurrent labeling of two intracellular targets using the two-step approaches are described.

Due primarily to the consideration of size (Row & Prescher, 2018), azido and cyclopropenyl groups are attached to BG to create two compounds **BG-AZIDE** and **BG-CYP** (Fig. 5B). The former could undergo the "strain-promoted azide-alkyne cycloaddition" (SPAAC) (MacKenzie, Sherratt, Chigrinova, Cheung, & Pezacki, 2014; Sletten & Bertozzi, 2011), while the latter would engage in the "inverse-demand Diels-Alder reaction" (IEDDA) (Knall & Slugovc, 2013; Oliveira, Guo, & Bernardes, 2017), both of which are well-known and tested bioorthogonal reactions that can be made to not interfere with each other (Knall & Slugovc, 2013). The dyes that carry complementary bioorthogonal groups, namely dibenzoazacyclooctyne and 6-methyltetrazine (e.g., see Fig. 5C), are readily available from an increasing number of venders.

4.1 Substrate synthesis

The common intermediate between **BG-AZIDE** and **BG-CYP** is the amino-substituted BG (**BG-NH$_2$** in Fig. 5B), which would form peptide

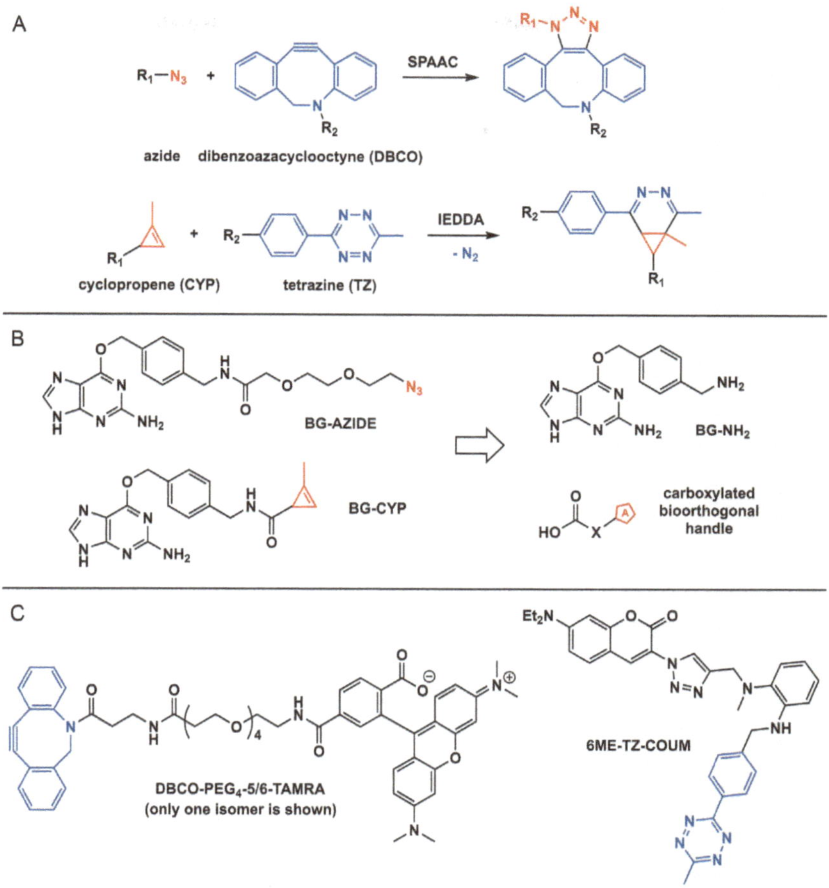

Fig. 5 (A) Two bioorthogonal reactions used in this work. SPAAC, strain-promoted azide-alkyne cycloaddition; IEDDA, inverse electron demand Diels-Alder reaction; (B) retrosynthetic dissection of **BG-AZIDE** and **BG-CYP**. "A": azido or cyclopropenyl; (C) representative DBCO- and TZ-bearing dyes.

bonds with either azido- or cyclopropenyl-substituted carboxylic acids to afford the target structures. The details of the synthesis are reported elsewhere (Macias-Contreras et al., 2020).

4.2 SPAAC and IEDDA

An organic azide and a dibenzoazacyclooctyl (DBCO)-functionalized molecule would undergo a SPAAC-type conjugation (see Fig. 5A) in an aqueous solution (Debets et al., 2010). The azido group therefore is attached to

BG to afford **BG-AZIDE** (Fig. 7B), which would (1) enzymatically conjugate with SNAP-tag from the BG end and (2) chemically couple with DBCO from the azido side. DBCO-conjugated dye molecules were purchased from Jena Bioscience that cover two spectral regions: DBCO-PEG_4-5/6-FAM emits in green (ex. 488/em. 517 nm), and DBCO-PEG_4-5/6-TAMRA in yellow (543/565 nm, Fig. 5C).

The IEDDA reaction applicable in bioconjugation involves an electron-deficient tetrazine as the diene, and a strained alkene as the dienophile. The high-lying HOMO of the dienophile interacts with the low-lying LUMO of the diene, which is the opposite arrangement of a typical Diels-Alder reaction (Oliveira et al., 2017). Several strained alkenes were examined as the functionalization on the BG substrate, and cyclopropenyl (CYP) group worked more consistently than norbornene (NB) or *trans*-cyclooctene (TCO) in live cells. Therefore, the results involving **BG-CYP** (Fig. 5B) are presented in this section. **6ME-TZ-COUM** was prepared as a tetrazine (TZ)-carrying green-emitting dye that would undergo IEDDA with the CYP group.

Three strategies were tried in the two-step intracellular labeling in live cells. (1) The cells were treated with a BG substrate prior to the introduction of DBCO or TZ-tagged dye under normal cultural conditions (complete medium, 37 °C, 5% CO_2, and 95% humidity). There is a washing step in between. (2) The electrophilicity of DBCO or TZ has been postulated to contribute to non-specific labeling via reaction with nucleophiles (e.g., thiols) (Murrey et al., 2015; van Geel, Pruijn, van Delft, & Boelens, 2012) available in the complete medium from serum and in the intracellular matrix. In order to reduce the extent of non-specific labeling from this origin, the second bioorthogonal chemical step was done at a lower temperature (e.g., rt or 4 °C), and/or in PBS with 1% of FBS to reduce the degree of non-specific consumption of the dyes (Lefebvre, Guetta, Poyer, Mahuteau-Betzer, & Teulade-Fichou, 2017; Li et al., 2018). (3) A BG substrate and complementarily tagged dye were mixed prior to the incubation with HeLa cells under normal conditions. In this case, BG substrate may react with SNAP-tag followed by the bioorthogonal conjugation, or choose to connect with the dye either in the medium or in the intracellular matrix prior to the enzymatic SNAP-tagging reaction. In our hands, option 3 was the least time-consuming that has achieved a consistent level of labeling efficiencies competitive with the first option. The second option that requires low-temperature incubation resulted more cell death than the others. Therefore, the protocols and results of option 3 are reported herein. Representative images acquired using the method of option 3 are shown in Fig. 6. Wary of the known thiol-yne reaction between SNAP-tag and DBCO-containing compounds (van Geel et al., 2012) that might result false positives of the programmed two-step reaction (SNAP-tagging and SPAAC), a control experiment was done in which SNAP-LaminA-expressing cells were treated with DBCO-PEG$_4$-5/6-TAMRA without **BG-AZIDE**. No specific lamina labeling was observed (Macias-Contreras et al., 2020).

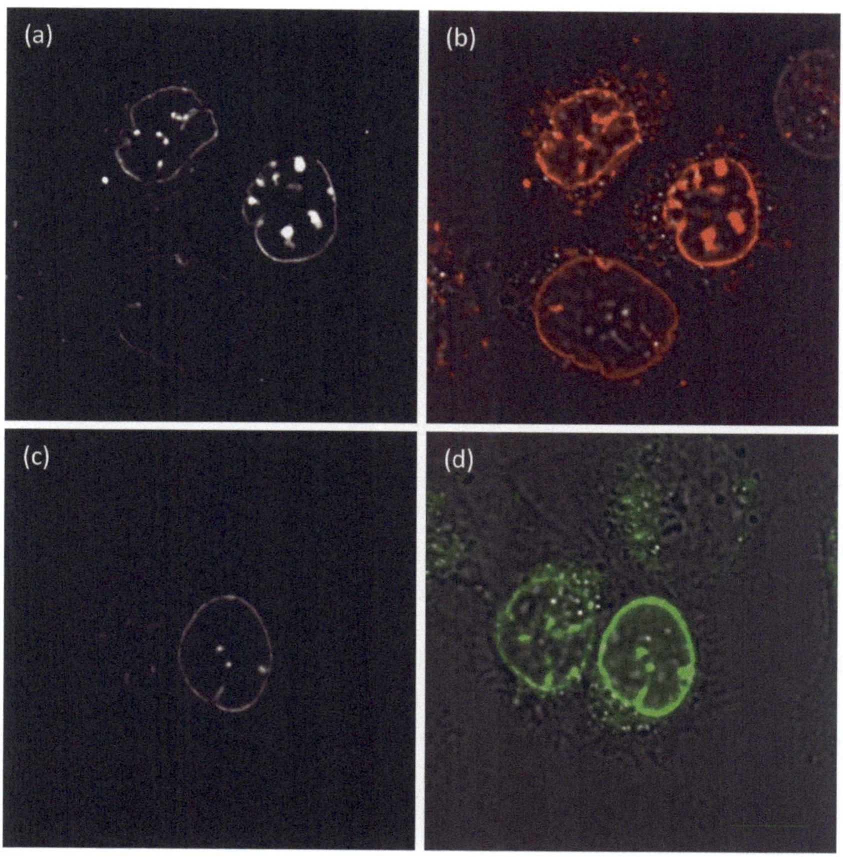

Fig. 6 Images of live HeLa S3 cells transfected with SNAP-LaminA and labeled with **BG-AZIDE**/DBCO-PEG$_4$-5/6-TAMRA (A, B) or **BG-CYP/6ME-TZ-COUM** (C, D) using the protocol described above. (A, C): Fluorescence channels, (A): $\lambda_{ex} = 543$ nm, $\lambda_{em} = 560–660$ nm, (C): $\lambda_{ex} = 405$ nm, $\lambda_{em} = 460–520$ nm; (B, D): merged image of both fluorescence and bright field channels.

4.2.1 Materials

1. *HeLa S3 cells* that are cultured in ΔT dishes to 50%–70% confluency at 37 °C, 95% humidity, and 5% CO_2.
2. *Vector DNAs* of SNAP-LaminA and SNAP-LifeAct (Section 2, ~0.5 μg/μL).
3. *BG-AZIDE and BG-CYP* stock solutions in DMSO (1–3 mM). The stock solutions are stored at −20 °C in 100-μL aliquots.
4. *DBCO-PEG$_4$-5/6-TAMRA* (Jena Bioscience) stock solutions in DMSO (1–3 mM). The stock solutions are stored at −20 °C in 100-μL aliquots.
5. *6ME-TZ-COUM* stock solutions in DMSO (1–3 mM). The stock solutions are stored at −20 °C in 100-μL aliquots.
6. *Glycine* (2 M) solution, autoclaved.

4.2.2 Protocols

A major drawback of the two-step labeling regardless the method is the significant non-specific labeling. By applying a glycine solution to coat the culture dish (Allen, Ross, & Davidson, 2013; Klein et al., 2011) prior to seeding the cells appears to reduce the degree of non-specific labeling that is attributable to the adhesion of dye molecules to the surface of the culture dish.

In the following steps, SNAP-LaminA, **BG-AZIDE**, and DBCO-PEG$_4$-5/6-TAMRA are used as an example for the two-step labeling and imaging.

1. (Day 1) Cover the bottom of a ΔT dish with a glycine solution (2 M) for at least 1 h, after which the glycine solution is aspirated. The dishes are then ready for plating cells.
2. (Day 2) Transfect the HeLa S3 cells adhering in a ΔT dish with SNAP-LaminA (Section 2) by following the steps described in Section 3.2.2.1.
3. (Day 3) Transfer the stock solutions of **BG-AZIDE** (1 mM in DMSO) and DBCO-PEG$_4$-5/6-TAMRA (1 mM in DMSO) in 1-mL screwcap Eppendorf tubes from a freezer and thawed at rt in a laminar flow hood. The tubes are vortexed if needed to accelerate defrosting. After the solutions are fully thawed, both tubes are centrifuged to consolidate the solutions.
4. Dispense the stock solution of **BG-AZIDE** (3 μL, 1 mM in DMSO) in warm medium (1 mL) in a 10-mL conical tube to a 3 μM solution of **BG-AZIDE** in the medium. The medium is vortexed for 2–5 s.

5. Add the stock solution of DBCO-PEG$_4$-5/6-TAMRA (3 μL, 1 mM in DMSO) in the **BG-AZIDE**-containing medium. The medium is vortexed for 2–5 s.
6. Aspirate the medium in the ΔT dish, which should be ~ 24 h post-transfection. Wash the cells with fresh warm medium (×1), and refill the dish with **BG-AZIDE** *and* DBCO-PEG$_4$-5/6-TAMRA-containing medium. Incubate the dish for 1 h at 37 °C, 95% humidity, and 5% CO_2.
7. Aspirate off the medium. Wash the cells with warm, fresh medium (×2). Refill the dish with 1 mL of medium, and incubate the dish for an additional 30 min at 37 °C, 95% humidity, and 5% CO_2.
8. Replace the medium. The dish is ready for imaging.
9. Set up the microscope and mount the dish as described in Section 3.2.2.3. Retrieve the spectral parameters available from the software for Rhodamine B (e.g., rhod-2), which are suitable for imaging TAMRA (λ_{ex} = 543 nm, λ_{em} = 560–660 nm). Follow the steps described in Section 3.2.2.3.

There are three modules in the two-step labeling process—the POI/SNAP-tag chimera, the BG substrate, and the functionalized dye. Combinations that involved actin fibers as the POI (LifeAct), BG substrate that functionalized by norbornene (**BG-NB**), and DBCO-PEG$_4$-5/6-FAM or TZ-BODIPY (both dyes were purchased from Jena Bioscience) as "clickable" dyes were all effective, albeit to different extent, in targeted labeling (Macias-Contreras et al., 2020). DBCO-PEG$_4$-5/6-FAM has not been as efficient in labeling as DBCO-PEG$_4$-5/6-TAMRA, probably for multiple reasons. FAM is easier to be photobleached than TAMRA, which creates technical difficulties. One report implies that FAM may have less than ideal cell membrane permeability (Yamagishi, Sawaki, Murata, & Takeoka, 2015), and that may have contributed to the lower degree of labeling. A dye-bioorthogonal handle conjugate has to have adequate cell membrane permeability for the two-step intracellular live-cell labeling to work. DBCO-sulfo-Cy5, which is cell impermeable, failed to engage in the two-step intracellular live-cell labeling as expected. Therefore, in addition to reducing background fluorescence, we also intend to study the disparity among the bioorthogonally tagged dyes, and to rank the effectiveness in the two-step intracellular live cell labeling in the near future. On the other hand, fixing and permeabilizing cells after the SNAP-tagging reaction (i.e., incubation with a BG-substrate) would allow more dyes carrying bioorthogonal handles to be effective. The results of the labeling of fixed cells are included in Macias-Contreras et al. (2020).

4.3 Two-color, two-step orthogonal/bioorthogonal labeling

4.3.1 CLIP-tag and orthogonal/bioorthogonal labeling

CLIP-tag is a mutant of SNAP-tag that accepts substrates that are orthogonal in specificity from those of SNAP-tag. Instead of O^6-benzylguanine (BG), O^2-benzylcytosine (BC, see one substrate in Fig. 7) was designed to be the selective substrate of CLIP-tag (Gautier et al., 2008). Therefore, when co-expressed in cells, the two tags are expected to accept their respective substrates, BG or BC, to concurrently deliver two molecular cargos to their respective destinations. If BG and BC substrates could be used to install two bioorthogonal handles that work in two different, non-interfering chemical ligation reactions, we may have a dual-color labeling strategy that could be referred as a type of orthogonal/bioorthogonal labeling. "Orthogonal" refers to the mutually exclusive substrate selectivity of SNAP- and CLIP-tags, while "bioorthogonal" refers to the faithfulness and specificity of the second, chemical conjugation step (Patterson & Prescher, 2015; Sletten & Bertozzi, 2011).

4.3.2 Protocols

Using the chemistry shown in Fig. 7B as an example, the protocols are listed as the following:

1. (Day 1) Coat the ΔT dishes with glycine (2 M) and seed with HeLa S3 cells as described in Section 4.2.2.
2. (Day 2) After ~24 h, transfect the cells with both vectors, SNAP-LifeAct and CLIP-LaminA (1–2 μg each in 1 mL of media). Follow the steps as described in Section 3.2.2.1.
3. (Day 3) Add **BC-AZIDE** (1.7 mM in DMSO, 1.8 μL) and **BG-CYP** (4 mM in DMSO, 0.8 μL) in warm medium (1 mL). Vortex 2–5 s.
4. Add DBCO-PEG_4-5/6-TAMRA (1 mM in DMSO, 3 μL) and **6ME-TZ-COUM** (2 mM in DMSO, 1.5 μL) to the medium prepared in Step 3. Vortex 2–5 s.
5. Replace the medium in the dish of transfected cells by the dye- and substrate-containing medium prepared in the previous step, followed by incubation at 37 °C for 1 h.
6. Aspirate off the medium, and wash the cells with fresh warm medium (×2), followed by incubation for an additional 30 min.
7. Replace the medium, and the dish is ready for imaging.
8. Switch on the 405-nm diode laser (for **6ME-TZ-COUM**) and 543-nm HeNe laser (for DBCO-PEG_4-5/6-TAMRA).

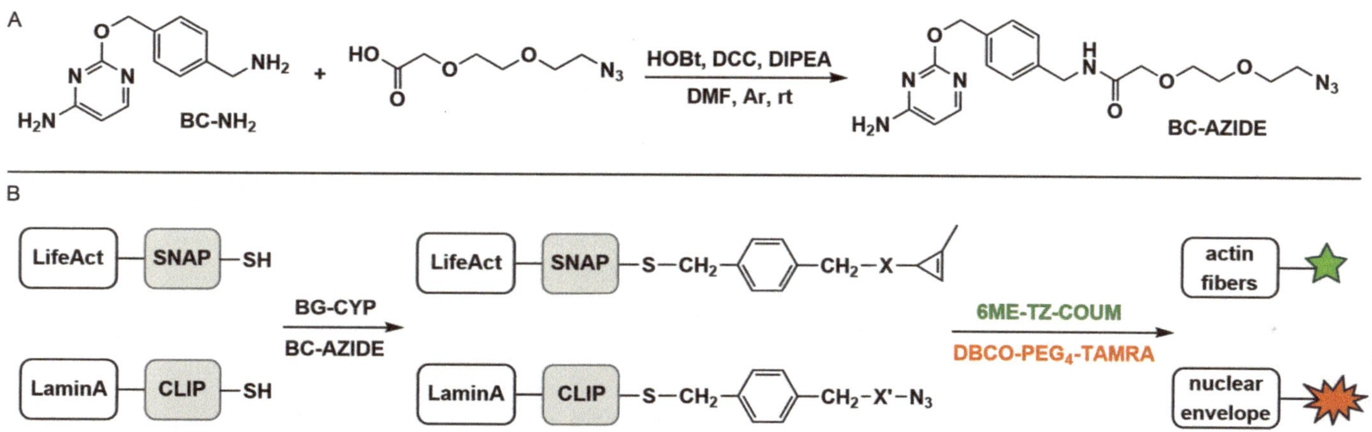

Fig. 7 (A) Synthesis of **BC-AZIDE**; (B) "orthogonal/bioorthogonal" two-color, two-step labeling. X and X′ are the linker components between bioorthogonal reaction handle and benzyl group.

9. Mount the dish and bring the cells to focus as described in Section 3.2.2.3. Use fluorescence channel 1 for detecting emission from **6ME-TZ-COUM** ($\lambda_{ex} = 405$ nm, $\lambda_{em} = 460$–520 nm). Use channel 2 for detecting emission from TAMRA ($\lambda_{ex} = 543$ nm, $\lambda_{em} = 560$–660 nm). Make sure the correct mirrors are selected on the light path.
10. In Live View, designate one window as the overlay of the two fluorescence channels from which the dually labeled cells will be identified. After an appropriate zoom value is applied, acquire an image from each channel individually because the focal planes are often different in two separate channels. The merged fluorescence images are the composites from the two separate acquisitions (Fig. 8).

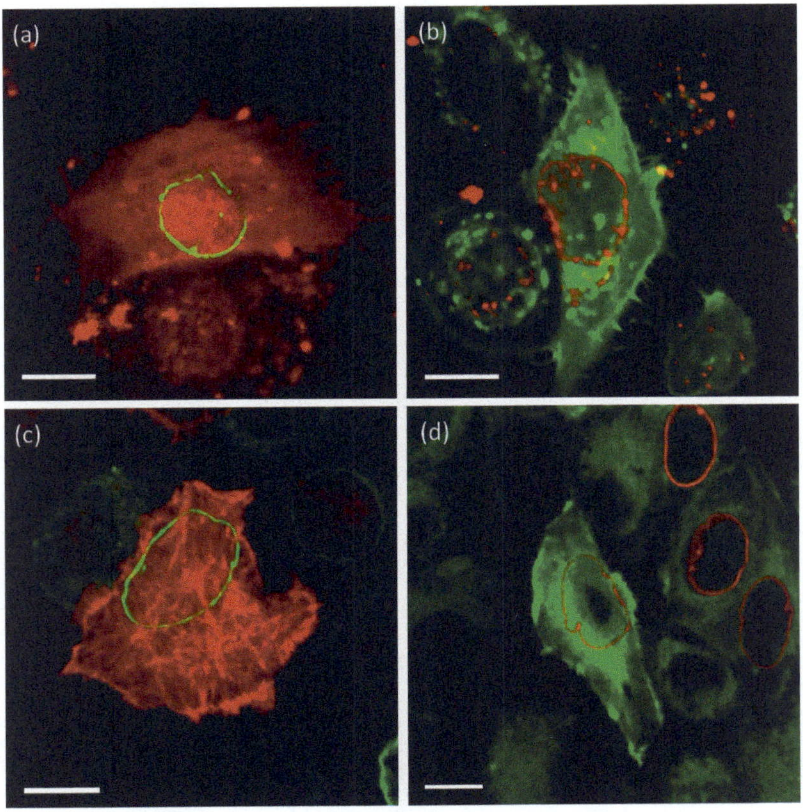

Fig. 8 Dually labeled images using the orthogonal/bioorthogonal strategy. HeLa cells transfected with CLIP-LaminA and SNAP-LifeAct (A, C), and CLIP-LifeAct and SNAP-LaminA (B, D). In (A) and (B), the cells were treated with **BC-AZIDE**, **BG-CYP**, DBCO-PEG$_4$-5/6-TAMRA and **6ME-TZ-COUM**, and imaged live; in (C) and (D), the cells were treated with **BG-TCO**, **BC-AZIDE**, DBCO-PEG$_4$-5/6-FAM, and TZ-Cy5. The cells were fixed and permeabilized after the enzymatic tagging step. Green channel: $\lambda_{ex} = 405$ nm, $\lambda_{em} = 460$–520 nm; red channel: $\lambda_{ex} = 543$ nm, $\lambda_{em} = 560$–660 nm.

The major limitations of the two-step intracellular labeling of live cells are (1) the relatively heavy background attributable to the non-specific adherence of dyes in the intracellular matrix, and (2) the limited scope of applicable handle-conjugated dyes due to, among other issues, inadequate ability to penetrate the cell membrane. If cells were fixed after the enzymatic tagging step, i.e., after the reaction with BG- and/or BC-substrates is over, the cell membrane could be permeabilized by a detergent such as Triton to allow more bioorthogonally equipped dyes to be effective. Included in Fig. 8C-D are two images of dually labeled fixed cells, in which **BG-TCO** was used to pair with TZ-Cy5 in the red channel. Detailed descriptions of the dual-labeling of fixed cells are included in Macias-Contreras et al. (2020). The problem caused by non-specific dye adhesion (limitation #1) affects cytoskeletal protein structures, such as actin fibers, disproportionally, rending them difficult to distinguish from the background should the latter be high.

5. Conclusion

The purpose of this work is to expand the scope of the molecular cargos that could be delivered to a protein of interest (POI) in live cells via the SNAP/CLIP-tag technology. In addition to the canonical benzylguanine (BG) substrates, 5-pyridylmethyl guanines (**5PG**) could also be accepted by SNAP-tag, albeit with a lower efficiency. The pyridyl group could engage in metal coordination, an additional function beyond being the molecular linker between the cargo and the POI. The connection between the POI and molecular cargo via SNAP-tagging can be interpolated by a bioorthogonal reaction, which in addition to providing flexibility of the labels to given targets, could be a remedy to those BG substrates that might be rejected by SNAP/CLIP-tags. The detailed procedures of labeling and imaging that involve these new tools in SNAP/CLIP-tags are described in this article, which is, we would admit a work in progress. We are actively working to improve the issues we have encountered so far, which include reducing background labeling, addressing the challenges in benzylcytosine (the CLIP-tag substrates, e.g., **BC-AZIDE**) synthesis, and expanding the limited set of bioorthogonally tagged dyes effective in the two-step labeling procedures that are workable in live cells. Those are efforts worth investing considering the added utilities to the already powerful SNAP/CLIP-tag technologies and more generally to the area of bioconjugate chemistry.

Acknowledgments

This work was supported by the National Science Foundation (CHE1566011). The authors are grateful to the late Michael Davidson (National High Magnetic Field Laboratory) for the encouragement in working in the areas of fluorescence microscopy and genetically based labeling, the technical guidance, and the material support. We also thank John Allen (Nikon), Eric Clark (NHMFL), and John Marchlenski (Hunt Optics & Imaging) for setting up the cell culture and microscopy labs, and Cheryl Pye (FSU Biology) for molecular cloning assistance.

References

Allen, J. R., Ross, S. T., & Davidson, M. W. (2013). Sample preparation for single molecule localization microscopy. *Physical Chemistry Chemical Physics*, *15*, 18771–18783.

Bosch, P. J., Corrêa, I. R., Sonntag, M. H., Ibach, J., Brunsveld, L., Kanger, J. S., et al. (2014). Evaluation of fluorophores to label SNAP-tag fused proteins for multicolor single-molecule tracking microscopy in live cells. *Biophysical Journal*, *107*(4), 803–814. https://doi.org/10.1016/j.bpj.2014.06.040.

Chen, H.-J., Chew, C. Y., Chang, E.-H., Tu, Y.-W., Wei, L.-Y., Wu, B.-H., et al. (2018). *S-cis* diene conformation: A new bathochromic shift strategy for near-infrared fluorescence switchable dye and the imaging applications. *Journal of the American Chemical Society*, *140*(15), 5224–5234. https://doi.org/10.1021/jacs.8b01159.

Correa, I. R., Baker, B., Zhang, A., Sun, L., Provost, C. R., Lukinavicius, G., et al. (2013). Substrates for improved live-cell fluorescence labeling of SNAP-tag. *Current Pharmaceutical Design*, *19*(30), 5414–5420. Retrieved from https://www.ingentaconnect.com/content/ben/cpd/2013/00000019/00000030/art00011.

Day, R. N., & Davidson, M. W. (2009). The fluorescent protein palette: Tools for cellular imaging. *Chemical Society Reviews*, *38*, 2887–2921.

Debets, M. F., van Berkel, S. S., Schoffelen, S., Rutjes, F. P. J. T., van Hest, J. C. M., & van Delft, F. L. (2010). Aza-dibenzocyclooctynes for fast and efficient enzyme PEGylation via copper-free (3+2) cycloaddition. *Chemical Communications*, *46*(1), 97–99. https://doi.org/10.1039/B917797C.

Deo, C., & Lavis, L. D. (2018). Synthetic and genetically encoded fluorescent neural activity indicators. *Current Opinion in Neurobiology*, *50*, 101–108. https://doi.org/10.1016/j.conb.2018.01.003.

Fernández-Suárez, M., & Ting, A. Y. (2008). Fluorescent probes for super-resolution imaging in living cells. *Nature Reviews. Molecular Cell Biology*, *9*, 929–943.

Frommer, W. B., Davidson, M. W., & Campbell, R. E. (2009). Genetically encoded biosensors based on engineered fluorescent proteins. *Chemical Society Reviews*, *38*, 2833–2841.

Gautier, A., Juillerat, A., Heinis, C., Ivan Reis Corrêa, J., Kindermann, M., Beaufils, F., et al. (2008). An engineered protein tag for multiprotein labeling in living cells. *Chemistry & Biology*, *15*, 128–136.

Gibson, D. G. (2011). Chapter fifteen—Enzymatic assembly of overlapping DNA fragments. In C. Voigt (Ed.), *Vol. 498. Methods in enzymology* (pp. 349–361). Academic Press.

Gibson, D. G., Young, L., Chuang, R.-Y., Venter, J. C., Hutchison, C. A., & Smith, H. O. (2009). Enzymatic assembly of DNA molecules up to several hundred kilobases. *Nature Methods*, *6*(5), 343–345. https://doi.org/10.1038/nmeth.1318.

Giepmans, B. N. G., Adams, S. R., Ellisman, M. H., & Tsien, R. Y. (2006). The fluorescent toolbox for assessing protein location and function. *Science*, *312*(5771), 217–224. https://doi.org/10.1126/science.1124618.

Griesbeck, O. (2004). Fluorescent proteins as sensors for cellular functions. *Current Opinion in Neurobiology, 14*(5), 636–641. https://doi.org/10.1016/j.conb.2004.08.002.

Heilemann, M., van de Linde, S., Mukherjee, A., & Sauer, M. (2009). Super-resolution imaging with small organic fluorophores. *Angewandte Chemie International Edition, 48*(37), 6903–6908. https://doi.org/10.1002/anie.200902073.

Hinner, M. J., & Johnsson, K. (2010). How to obtain labeled proteins and what to do with them. *Current Opinion in Chemical Biology, 21*, 766–776.

Hong, Y.-R., Lam, C. H., & Tan, K.-T. (2017). Fluorogenic protein labeling probes to study the morphological interplay between PreLamin and Mature Lamin. *Bioconjugate Chemistry, 28*(11), 2895–2902. https://doi.org/10.1021/acs.bioconjchem.7b00611.

Jing, C., & Cornish, V. W. (2011). Chemical tags for labeling proteins inside living cells. *Accounts of Chemical Research, 44*, 784–792.

Johnsson, N., & Johnsson, K. (2003). A fusion of disciplines: Chemical approaches to exploit fusion proteins for functional genomics. *Chembiochem, 4*(9), 803–810. https://doi.org/10.1002/cbic.200200603.

Keppler, A., Gendreizig, S., Gronemeyer, T., Pick, H., Vogel, H., & Johnsson, K. (2003). A general method for the covalent labeling of fusion proteins with small molecules in vivo. *Nature Biotechnology, 21*, 86–89.

Keppler, A., Kindermann, M., Gendreizig, S., Pick, H., Vogel, H., & Johnsson, K. (2004). Labeling of fusion proteins of O^6-alkylguanine-DNA alkyltransferase with small molecules in vivo and in vitro. *Methods, 32*, 437–444.

Klein, T., Löschberger, A., Proppert, S., Wolter, S., van de Linde, S., & Sauer, M. (2011). Live-cell dSTORM with SNAP-tag fusion proteins. *Nature Methods, 8*(1), 7–9. https://doi.org/10.1038/nmeth0111-7b.

Knall, A.-C., & Slugovc, C. (2013). Inverse electron demand Diels–Alder (iEDDA)-initiated conjugation: A (high) potential click chemistry scheme. *Chemical Society Reviews, 42*(12), 5131–5142. https://doi.org/10.1039/C3CS60049A.

Lefebvre, J., Guetta, C., Poyer, F., Mahuteau-Betzer, F., & Teulade-Fichou, M.-P. (2017). Copper–alkyne complexation responsible for the nucleolar localization of quadruplex nucleic acid drugs labeled by click reactions. *Angewandte Chemie International Edition, 56*(38), 11365–11369. https://doi.org/10.1002/anie.201703783.

Li, S., Liu, Y., Liu, L., Feng, Y., Ding, L., & Ju, H. (2018). A hierarchical coding strategy for live cell imaging of protein-specific glycoform. *Angewandte Chemie, 130*(37), 12183–12187. https://doi.org/10.1002/ange.201807054.

Liu, J., Xiong, R., Brans, T., Lippens, S., Parthoens, E., Zanacchi, F. C., et al. (2018). Repeated photoporation with graphene quantum dots enables homogeneous labeling of live cells with extrinsic markers for fluorescence microscopy. *Light: Science & Applications*, 7(1), 47. https://doi.org/10.1038/s41377-018-0048-3.

Lutz, R. J., Trujillo, M. A., Denham, K. S., Wenger, L., & Sinensky, M. (1992). Nucleoplasmic localization of prelamin A: Implications for prenylation-dependent lamin A assembly into the nuclear lamina. *Proceedings of the National Academy of Sciences of the United States of America, 89*(7), 3000–3004. https://doi.org/10.1073/pnas.89.7.3000.

Macias-Contreras, M., He, H., & Zhu, L. (2018). Beyond O^6-benzylguanine: O^6-(5-pyridylmethyl)guanine as a substrate for the self-labelling enzyme SNAP-tag. *Bioconjugate Chemistry, 29*, 4104–4109. https://doi.org/10.1021/acs.bioconjchem.8b00703.

Macias-Contreras, M., He, H., Little, K. N., Lee, J. P., Campbell, R. P., Royzen, M., et al. (2020). SNAP/CLIP-Tags and Strain-Promoted Azide-Alkyne Cycloaddition (SPAAC)/Inverse Electron Demand Diels-Alder (IEDDA) for intracellular orthogonal/bioorthogonal labeling, *Bioconjugate Chemistry 31*, Accepted. https://doi.org/10.1021/acs.bioconjchem.0c00107.

MacKenzie, D. A., Sherratt, A. R., Chigrinova, M., Cheung, L. L. W., & Pezacki, J. P. (2014). Strain-promoted cycloadditions involving nitrones and alkynes—Rapid tunable reactions for bioorthogonal labeling. *Current Opinion in Chemical Biology*, *21*, 81–88. https://doi.org/10.1016/j.cbpa.2014.05.023.

Melak, M., Plessner, M., & Grosse, R. (2017). Actin visualization at a glance. *Journal of Cell Science*, *130*(3), 525–530. https://doi.org/10.1242/jcs.189068.

Miyawaki, A., Nagai, T., & Mizuno, H. (2003). Mechanisms of protein fluorophore formation and engineering. *Current Opinion in Chemical Biology*, 7(5), 557–562. https://doi.org/10.1016/S1367-5931(03)00097-8.

Müller, A., Neukam, M., Ivanova, A., Sönmez, A., Münster, C., Kretschmar, S., et al. (2017). A global approach for quantitative super resolution and electron microscopy on cryo and epoxy sections using self-labeling protein tags. *Scientific Reports*, 7(1), 23. https://doi.org/10.1038/s41598-017-00033-x.

Murrey, H. E., Judkins, J. C., am Ende, C. W., Ballard, T. E., Fang, Y., Riccardi, K., et al. (2015). Systematic evaluation of bioorthogonal reactions in live cells with clickable HaloTag ligands: Implications for intracellular imaging. *Journal of the American Chemical Society*, *137*(35), 11461–11475. https://doi.org/10.1021/jacs.5b06847.

Oliveira, B. L., Guo, Z., & Bernardes, G. J. L. (2017). Inverse electron demand Diels–Alder reactions in chemical biology. *Chemical Society Reviews*, *46*(16), 4895–4950. https://doi.org/10.1039/C7CS00184C.

Patterson, D. M., & Prescher, J. A. (2015). Orthogonal bioorthogonal chemistries. *Current Opinion in Chemical Biology*, *28*, 141–149.

Ramil, C. P., Dong, M., An, P., Lewandowski, T. M., Yu, Z., Miller, L. J., et al. (2017). Spirohexene-tetrazine ligation enables bioorthogonal labeling of class B G protein-coupled receptors in live cells. *Journal of the American Chemical Society*, *139*(38), 13376–13386. https://doi.org/10.1021/jacs.7b05674.

Remington, S. J. (2006). Fluorescent proteins: Maturation, photochemistry and photophysics. *Current Opinion in Structural Biology*, *16*(6), 714–721. https://doi.org/10.1016/j.sbi.2006.10.001.

Row, R. D., & Prescher, J. A. (2018). Constructing new bioorthogonal reagents and reactions. *Accounts of Chemical Research*, *51*(5), 1073–1081. https://doi.org/10.1021/acs.accounts.7b00606.

Sletten, E. M., & Bertozzi, C. R. (2011). From mechanism to mouse: A tale of two bioorthogonal reactions. *Accounts of Chemical Research*, *44*, 666–676.

Song, X., Wang, C., Han, Z., Xu, Y., & Xiao, Y. (2015). Terminal alkyne substituted O^6-benzylguanine for versatile and effective syntheses of fluorescent labels to genetically encoded SNAP-tags. *RSC Advances*, *5*(30), 23646–23649. https://doi.org/10.1039/C4RA17072E.

Srikun, D., Albers, A. E., Nam, C. I., Iavarone, A. T., & Chang, C. J. (2010). Organelle-targetable fluorescent probes for imaging hydrogen peroxide in living cells via SNAP-tag protein labeling. *Journal of the American Chemical Society*, *132*, 4455.

Tsien, R. Y. (2009). Constructing and exploiting the fluorescent protein paintbox (nobel lecture). *Angewandte Chemie International Edition*, *48*(31), 5612–5626. https://doi.org/10.1002/anie.200901916.

van Geel, R., Pruijn, G. J. M., van Delft, F. L., & Boelens, W. C. (2012). Preventing thiol-yne addition improves the specificity of strain-promoted azide–alkyne cycloaddition. *Bioconjugate Chemistry*, *23*(3), 392–398. https://doi.org/10.1021/bc200365k.

Wombacher, R., Heidbreder, M., van de Linde, S., Sheetz, M. P., Heilemann, M., Cornish, V. W., et al. (2010). Live-cell super-resolution imaging with trimethoprim conjugates. *Nature Methods*, 7(9), 717–719. https://doi.org/10.1038/nmeth.1489.

Xue, L., Karpenko, I. A., Hiblot, J., & Johnsson, K. (2015). Imaging and manipulating proteins in live cells through covalent labeling. *Nature Chemical Biology*, *11*(12), 917–923. https://doi.org/10.1038/nchembio.1959.

Yamagishi, K., Sawaki, K., Murata, A., & Takeoka, S. (2015). A Cu-free clickable fluorescent probe for intracellular targeting of small biomolecules. *Chemical Communications*, *51*(37), 7879–7882. https://doi.org/10.1039/C4CC09947H.

Yang, F., Moss, L. G., & Phillips, G. N. (1996). The molecular structure of green fluorescent protein. *Nature Biotechnology*, *14*(10), 1246–1251. https://doi.org/10.1038/nbt1096-1246.

Zhang, J., Campbell, R. E., Ting, A. Y., & Tsien, R. Y. (2002). Creating new fluorescent probes for cell biology. *Nature Reviews Molecular Cell Biology*, *3*(12), 906–918. https://doi.org/10.1038/nrm976.

Zhang, M., Li, M., Zhang, W., Han, Y., & Zhang, Y.-H. (2019). Simple and efficient delivery of cell-impermeable organic fluorescent probes into live cells for live-cell superresolution imaging. *Light: Science & Applications*, *8*(1), 73. https://doi.org/10.1038/s41377-019-0188-0.

Zhang, Y., Park, K.-Y., Suazo, K. F., & Distefano, M. D. (2018). Recent progress in enzymatic protein labelling techniques and their applications. *Chemical Society Reviews*, *47*(24), 9106–9136. https://doi.org/10.1039/C8CS00537K.

CHAPTER THIRTEEN

Light-induced protein proximity by activation of gibberellic acid derivatives in living cells

Michael J. Ziegler, Richard Wombacher*

Institute of Pharmacy and Molecular Biology, Ruprecht-Karls-Universität Heidelberg, Heidelberg, Germany

*Corresponding author: e-mail address: wombacher@uni-heidelberg.de

Contents

Abstract

Light controlled tools are highly attractive for the modulation and manipulation of biological processes. As an external trigger light can be applied with high temporal and special control to various samples. In the recent years a number of optochemical and -genetic tools have been developed to translate the input of light into molecular changes that result in specific biological responses. Here we present a highly efficient system for light-induced protein dimerization in live cells using photocaged derivatives of the plant hormone gibberellic acid (GA3). We provide a precise protocol for a simple one-step synthesis of the photocaged CIP and its application for protein translocation in living cells.

Abbreviations

CAR chimeric antigen receptor
CIP chemical inducer of proximity
CRISPR clustered regularly interspaced short palindromic repeats
DMEM Dulbecco's modified Eagle's medium
DNA deoxyribonucleic acid
DPBS Dulbecco's phosphate-buffered saline

Methods in Enzymology, Volume 638
ISSN 0076-6879
https://doi.org/10.1016/bs.mie.2020.02.022

ESI	electrospray ionization
GFP	green fluorescent protein
HR	high resolution
MS	mass spectrometry
NMR	nuclear magnetic resonance
PC	photocage
POI	protein of interest
R_f	retardation factor
RNA	ribonucleic acid
RNAi	RNA interference
TLC	thin layer chromatography
TOM20	target of outer mitochondrial membrane
UV	ultra violet (light)

Chemical abbreviations

ABA	abscisic acid
DAPI	4′,6-diamidin-2-phenylindol
DCM	dichloromethane
DMSO	dimethyl sulfoxide
EDTA	ethylenediaminetetraacetic acid
EtOAc	ethyl acetate
GA_3	gibberellic acid
THF	tetrahydrofuran

1. Introduction

The ability to alter protein localization, activity or concentration is essential to the study of protein function *in vivo* and allows for investigation of the role of individual proteins in cellular processes. While genetic knock-outs of individual proteins have been invaluable for cell biology, chemical biology has the potential to provide new tools allowing for reversible control of protein function and activity with fine temporal resolution. Nature itself regulates a vast number of cellular events by proximity of different biomolecules such as proteins, RNA or DNA. These interactions underlie high spatiotemporal control. Current biological methods used for protein regulation control protein concentration at the genetic or post-transcriptional level. While these are powerful molecular biology tools, knock-out and knock-down methods can have several drawbacks. Knocking out a protein of interest may result in lethality or cause other unanticipated phenotypes.

Protein knock-down using CRISPR or RNAi techniques can result in several "off-target" (Jackson et al., 2003; Kim et al., 2015) effects or unwanted interferon response (Sioud, 2005). Moreover, techniques that operate on the level of DNA or RNA do not enable a fine temporal control over protein expression levels. An elegant way to induce proximity between two proteins is the use chemical inducers of proximity (CIP). These small, drug like molecules bind (either covalently or non-covalently) specifically two identical (homodimerization) or different (heterodimerization) protein domains and thereby induce proximity (Fegan, White, Carlson, & Wagner, 2010). Proteins of interest (POI) can be fused to the dimerization domains by genetic manipulation and thereby be brought into proximity upon addition of the appropriate CIP. One major advantage of CIPs is that the manipulation does not take place at the level DNA or RNA but at the level of protein. As a result, the delay between perturbation and effect is reduced tremendously (from 24–36 h to 1–60 min) (Fig. 1A) (Lampson & Kapoor, 2006). However, one major limitation of the perturbation speed is the rate of cellular uptake of the CIP molecule (k_1, Fig. 1B), which depends on a number of factors such as the CIP molecule itself and cell/tissue type, just to name a few. One strategy to increase the perturbation speed is to use light as an external trigger. A caged CIP molecule is used to make the perturbation independent of cellular uptake. After crossing the membrane, the inactive caged CIP can be converted to the active CIP upon irradiation with light. Thus, an immediate intracellular pool of active CIP can be generated at any time point within any region of choice using light as a trigger with high spatiotemporal control.

In the last few years CIP have found wide application in biology, *e.g.*, induced signal transduction, transcription control, protein translocation, protein degradation or regulation of chromatin. More recent studies emphasize their application in medicine as therapeutics or for CAR-T cells therapy.

1.1 Chemical inducers of proximity

The most widely used CIP system is the heterodimerizer rapamycin (Fig. 1C). Rapamycin is macrolide isolated from *Streptomyces hygrosopicus* and an immunosuppressant in mammalians. Rapamycin shows strong interactions with its primary protein target FKBP and forms a complex, which then binds to the FRB "FKBP12-rapamycin-binding" domain of the target of rapamycin (TOR) kinase. This interaction interrupts the signal cascade of the immune response (Choi, Chen, Schreiber, & Clardy, 1996; Mangelsdorf et al., 1995). Fusion of arbitrary POIs to the two dimerizing domains allows

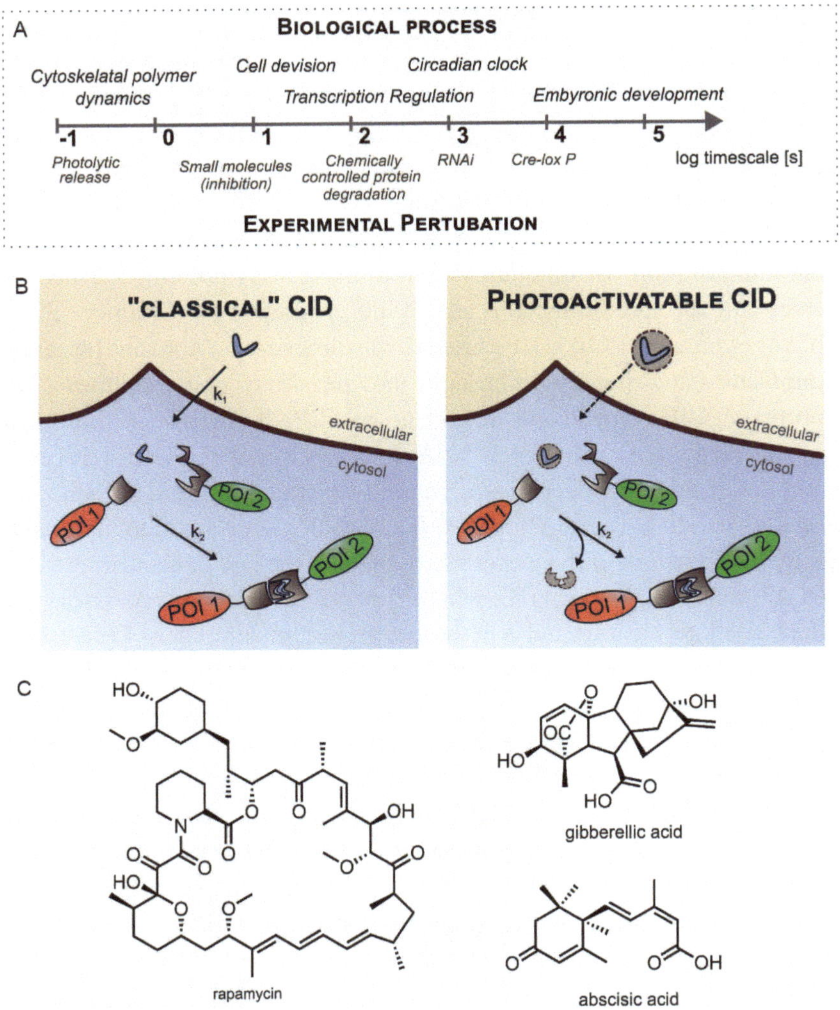

Fig. 1 (A) Timescales of biological processes and appropriate experimental perturbation. (B) General scheme showing the difference between "classical" and photoactivatable CIPs, k_1 rate of cellular uptake, k_2 rate of dimerization process within the cell. (C) Chemical structure of rapamycin, gibberellic acid (GA_3) and abscisic acid (ABA). Photo protectable moiety depicted in blue. *Panel (A) adapted and modified from Lampson, M.A.; Kapoor, T.M. (2006). Unraveling cell division mechanisms with small-molecule inhibitors. Nature Chemical Biology, 2(1), 19–27.*

the use of rapamycin as a tool to specifically create artificial proximity between the POIs. However, the mammalian target of rapamycin is always addressed, and therefore, there is no complete orthogonality to the biological system.

In the past decade different plant-based CIP systems were reported. Compared to earlier CIPs they are derived from other another kingdom (plants) and are orthogonal in mammalians, hence have no unspecific target.

Recently a new class of CIP based on plant hormone receptor interactions has been reported (Liang, Ho, & Crabtree, 2011; Miyamoto et al., 2012). The diterpenoid gibberellic acid (GA_3, Fig. 1C) is a plant hormone that controls growth and development in plants (Davière & Achard, 2013). The bioactive gibberellic acid GA_3 is binding to the receptor "gibberellin insensitive dwarf" (GID1) and thereby induces a conformational change that results in the strong binding of the GID1-GA_3 complex to the protein GAI ("gibberellin insensitive") (Murase, Hirano, Sun, & Hakoshima, 2008). Inoue et al. could demonstrate that this plant hormone receptor complex formation can be used as a CIP system in mammalian cells (Miyamoto et al., 2012). As the carboxylic acid is negatively charged at physiological pH, gibberellin is not able to pass cell membranes. For intracellular application as CIP they used a cell permeable acetoxymethyl derivative of gibberellic acid (GA3-AM), which gets converted into active GA3 upon de-esterification by endogenous esterases in the cytosol of mammalian cells and successfully induced protein dimerization.

A second plant-based CIP system is based on the phytohormone abscisic acid (ABA, Fig. 1C). ABA regulates plant water consumption and drought tolerance by controlling the activity of different protein kinases (Miyazono et al., 2009). The formation of a protein-phytohormone complex with the protein domains PYR "pyrabactin resistance" (PYR) or "pyrabactin resistance like" (PYL) exposes a binding side for "abscisic acid insensitive" (ABI) resulting in protein dimerization and downstream signaling transduction (Melcher et al., 2009).

Both plant hormone-based systems have been successfully adopted for the use in mammalian cells to induce protein dimerization in an orthogonal and specific manner. Until now no side effects or endogenous targets in mammalian cells have been identified until now.

1.2 Photoactivatable CIP

A major challenge is the spatial and temporal control of the protein dimerization process. The dimerization process is temporally not only controlled by diffusion of the small molecule within the cell but mainly by cellular dependent uptake. Also, precise activation of single cells is not possible. To circumvent these limitations, so-called photo removable protecting groups can be attached to the CIPs. The use of light as external trigger is non-invasive and provides a high level of spatiotemporal control. One major challenge of the development of new photoactivatable CIP is to identify positions, that modified with photo removable protecting groups, suppress the activity of the CIP.

Photoactivatable derivatives of the CIP working horse rapamycin were not available for a long-time. A reason for that is the complex structure of rapamycin which is synthetically challenging to access and identification of prone positions for photocaging groups is not trivial. Three different systems based on rapamycin were reported so far Umeda et al. developed a photocaged rapamycin derivative, which was tethered to extracellular avidin *via* biotin moiety. However, photorelease is conducted in extracellular environment, hence poor membrane permeability is not bypassed (Umeda, Ueno, Pohlmeyer, Nagano, & Inoue, 2010). Deiters et al. reported a photocaged rapamycin derivative developed by structured guided design, however, binding affinity to FKBP is not completely suppressed by caging, which makes the use of a modified FKBP protein necessary (Karginov et al., 2010). The same group developed a photocleavable homodimer of rapamycin, which binds two FKBP domains prior to uncaging. Once irradiated heterodimerization with FRB (Brown et al., 2015). Due to complex synthesis and rather specific working conditions, photoactivatable rapamycin derivatives have not found widespread application in biology laboratories.

Recent studies showed that plant hormone-based CIPs can be chemically converted to photoactivatable CIPs with little synthetic effort. Plant hormones possess complex structures and often have a very narrow structure-activity relationship, which is why slight modifications prevent them from binding to the receptor and often greatly reduce their biological activity. For both abscisic acid and gibberellic acid photoactivatable derivatives have been reported (Schelkle et al., 2015; Wright, Guo, & Liang, 2015). They both possess a carboxyl acid moiety, which are crucial for receptor binding and synthetic modification to an ester renders both plant hormones inactive. Simple conversion of the carboxylic acid to a photocleavable ester therefore deactivates the dimerization activity, while irradiation with light releases the active CIP, restoring its biological activity. The photocaging group can be chosen depending on experimental requirements (in terms of decaging wavelength, molecular properties) and installed in a modular principle.

2. Methods

2.1 Synthesis of GA_3-PC

In a previous work we demonstrated that gibberellic acid can be photocaged by simple esterification of the carboxylic acid moiety (Schelkle et al., 2015). We installed several different photo protective groups which allow to tune photorelease conditions in terms of decaging wavelengths. To provide a protocol that can be easily followed by scientists with less experience in chemical synthesis, we here describe the one-step synthesis of *o*-nitrobenzyl protected gibberellins. Both starting materials, gibberellic acid and photocage precursor, are commercially available at low price and can be directly used in a one-step protocol to synthesize the simplest photoactivatable CIP (Fig. 2). In this way we hope to make this attractive method for live cell protein manipulation accessible to a more biologically oriented community.

2.1.1 Materials, equipment, reagents

1. Solvents: tetrahydrofuran (THF, CAS-# 109-99-9), ethyl acetate (EtOAc, CAS-# 141-78-6), dichloromethane (DCM, CAS-# 75-09-2)
2. Reactants/reagents: Gibberellin A_3 (100 mg, CAS-# 77-06-5), cesium carbonate (150 mg, CAS-# 534-17-8), 4,5-dimethoxy-2-nitrobenzylbromid (100 mg, 53413-67-5), Na_2SO_4 anhydrous (10 g, CAS-# 7757-82-6)
3. Solutions for aqueous workup: saturated aqueous $NaHCO_3$ solution, saturated aqueous NaCl solution
4. Glassware and supplies: round bottom flasks, Erlenmeyer flasks, separatory funnel, glass chromatography columns, TLC separation chamber
5. Disposables: syringes, cannulas, Pasteur pipettes
6. Magnetic stirrer and stir bars

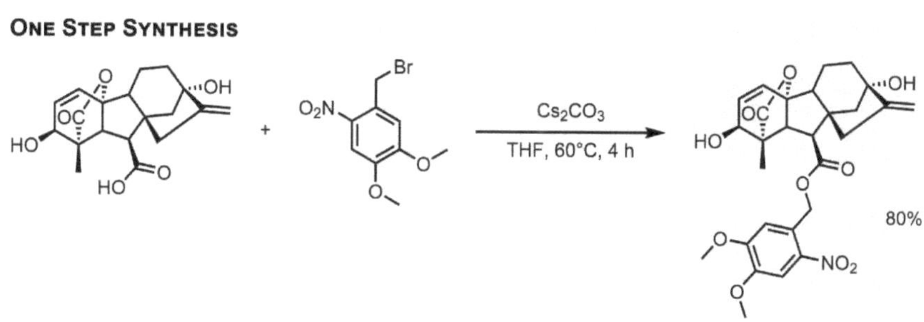

Fig. 2 Synthesis of photocaged gibberellic acid GA_3-PC.

7. Column purification and thin layer chromatography: silica gel for column chromatography (10g, 60Å pore size, 230–500 mesh participle size), TLC plates with fluorescence indicator (*e.g.*, POLYGRAM, SIL/G UV_{254}), Hanessian's staining solution (12.0g ammonium molybdate, 0.5g ceric ammonium molybdate, 15mL concentrated sulfuric acid acid in 235mL H_2O)
8. Equipment: rotary evaporator, UV lamp for TLC, Schlenk line for high vacuum
9. Analytics: NMR spectrometer, NMR tube, $CDCl_3$, mass spectrometer

2.1.2 Protocols

2.1.2.1 Chemical synthesis

Photocaged gibberellin GA_3-PG can be synthesized in a one-step procedure (Fig. 2)

1. Charge a round bottom flask (25mL) with a stirring bar and 100mg gibberellin GA_3
2. Dissolve gibberellin in 6mL THF
3. Suspend $CsCO_3$ (141mg, 1.5 eq) and 96mg 4,5-dimethoxy-2-nitrobenzylbromide in reaction mixture
4. Seal reaction flask and reflux mixture for 4h at 67°C, then allow mixture to reach room temperature
5. Remove solvent with rotary evaporator
6. Suspend dry reaction mixture in 5mL DCM
7. Wash organic layer with 5mL saturated aqueous solution of $NaHCO_3$, then with 5mL brine
8. Dry organic layer over Na_2SO_4
9. Remove organic solvent using a rotary evaporator (800mbar, 40°C)
10. Purify the crude product by column chromatography (10g silica gel). Start elution with 200mL DCM, continue with 200mL DCM:EA 10%, then 400mL DCM:EA 20%. Use TLC to determine product fractions ($R_f = 0.18$ DCM:EA 20%)
11. Remove solvent under reduced pressure using rotary evaporator (start with 800mbar, 40°C, then reduce to 200mbar). Remove traces of residual solvent at <1 mbar using high vacuum
12. The final compound is isolated as a yellow powder with a yield about 80%

2.1.3 Safety considerations

Conduct all experiments in a well-ventilated fume hood. Prior to synthesis, make sure to check hazards of all chemicals and solvents used.

Wear appropriate protective gear (safety goggles, lab coat, gloves). Before applying vacuum always check glassware for cracks as precaution of implosions.

2.1.4 Analysis

2.1.4.1 NMR characterization

^{1}H NMR (500 MHz, $CDCl_3$) δ 7.68 (s, 1H), 6.98 (s, 1H), 6.30 (dd, J=9.3, 0.9 Hz, 1H), 5.90 (dd, J=9.3, 3.7 Hz, 1H), 5.52 (d, J=1.5 Hz, 2H), 5.26 (dd, J=3.0, 1.9 Hz, 1H), 4.93 (d, J=2.1 Hz, 1H), 4.14 (d, J=3.7 Hz, 1H), 3.98 (s, 3H), 3.96 (s, 3H), 3.22 (d, J=10.8 Hz, 1H), 2.85 (d, J= 10.8 Hz, 1H), 1.65–2.2 (m, 8H), 1.24 (s, 3H).

^{13}C NMR (126 MHz, $CDCl_3$) δ 178.43, 156.84, 153.56, 148.83, 140.60, 132.94, 132.62, 125.68, 111.60, 108.55, 107.92, 90.50, 78.23, 77.16, 69.88, 64.33, 56.79, 56.61, 53.60, 52.97, 51.34, 50.86, 50.71, 44.97, 43.14, 38.34, 17.15, 14.70.

2.1.4.2 Mass spectrometry

HR-ESI, calculated for $[C_{28}H_{31}NO_{10}+Na]^{+}$: 564.1840, found: 564.1851.

2.1.4.3 Thin layer chromatography

R_f(DCM:EA 80:20) = 0.18.

2.2 Photoactivation of gibberellic acids in living cells

In this section we describe a proof of principle experiment to visualize protein dimerization (Fig. 3). GID1 dimerization domain is expressed as a fusion of green fluorescence protein (eGFP) and expressed in the cytosol. GAI dimerization domain is fused to red fluorescence protein (mCherry) targeted to mitochondria by TOM20 protein. Both plasmids have been provided by Takanari Inoue and are available at Addgene (Miyamoto et al., 2012).

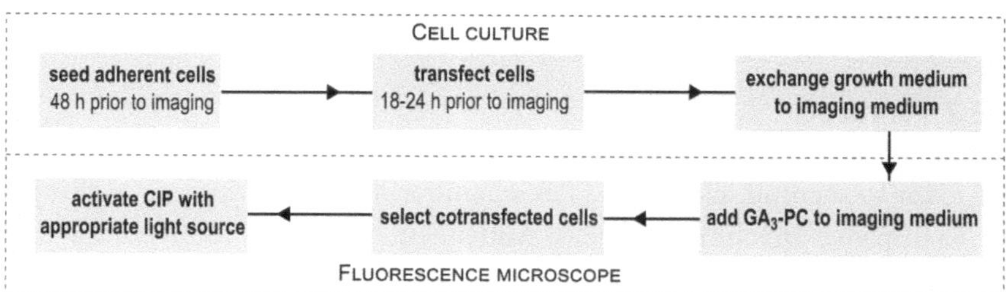

Fig. 3 Workflow of cell preparation and fluorescence microscopy for photoactivatable chemical induced proximity.

Depending on the experimental setup a number of different light sources can be used for the activation. Active GA3 can be either released by a mounted light-emitting diode, a 405 nm laser line or a mercury-vapor lamp with DAPI excitation filter. Gibberellin can be released with high spatiotemporal precision by irradiation of single cells. This proof of principle experiment can be adapted to any proteins of interest by replacing eGFP and tom20-mCherry in the plasmids by using common cloning techniques.

2.2.1 Materials, equipment, reagents

1. Adherent cell lines, with appropriate transfection protocols (*e.g.*, COS7, HeLa, HEK 293)
2. Equipment: Laminar flow cabinet, water bath (with autoclaved, deionized water, sterilized), cell incubator (37 °C, 5% CO_2), temperature-controlled centrifuge, Neubauer counting chamber, light microscope, vortex mixer, Eppendorf Pipettes (different volumes)
3. Disposables: cell culture flasks for adherent cells (surface area of 25 cm^2 are sufficient), conical centrifuge tubes (15 mL), Eppendorf tubes (1.5 mL), pipette tips
4. Cell media: DMEM supplemented with 2 mM L-glutamine, 1 mM sodium pyruvate and 10% (v/v) fetal bovine serum, DPBS, Trypsin-EDTA (0.05%) solution, Opti-MEM I reduced serum media, Leibowitz's L15 medium. All stored at 4 °C
5. Chemicals: DMSO, transfection reagent (*e.g.*, TransIT-LT1), HF (0.1 M) in water, GA_3-PC
6. Cell chamber: Lab-Tek chamber slide
7. GFP-GID1 (from Takanari Inoue, Addgene plasmid #37306) and Tom20-mCherry-GAI (from Takanari Inoue, Addgene plasmid #37316) as a 200 ng/μL solution in H_2O
8. Epifluorescence microscope with a 40–100 × objective, equipped either with 405 activation laser line or mercury-vapor lamp and appropriate DAPI filters

2.2.2 Protocols

2.2.2.1 Sample preparation

1. Prior to seeding, clean Lab-Tek chamber slides with hydrogen fluoride to improve attachment of cells. Add 300 μL 0.1 M HF solution, incubate for 1 min, remove solution and wash twice with 300 μL MilliQ water. Repeat washing procedure twice
2. Thaw and cultivate cells for at least two passages

3. Upon reaching 80% confluency, detach cells with 1 mL Trypsin-EDTA solution (for T25 cell culture flask) add 5 mL DMEM medium after complete detachment
4. Pelletier cells, remove supernatant and resuspend in 1 mL DMEM medium. Determine number of cells with Neubauer counting chamber
5. Adjust number of cells to 7000 cells per well in 200 μL DMEM medium
6. Incubate cells overnight
7. Transfect cells ~18–24 h prior to imaging. Dilute 0.1 μg GFP-GID1 and 0.1 μg Tom20-mCherry-GAI plasmids in 9 μL Opti-MEM medium. Add 0.3 μL TransIT and mix carefully by frequent pipetting. Incubate for 22.5 min at room temperature. Add transfection mix to seeded cells. Mix growth medium carefully to guarantee even distribution of transfection mix
8. Incubate cells for 18–24 h

2.2.2.2 Imaging in living cells

1. Prepare following stock solutions:
   100 μM solution of GA_3-PC in DMSO
   Use DMSO main stock to prepare 2 μM GA_3-PC in L15 (98:2)

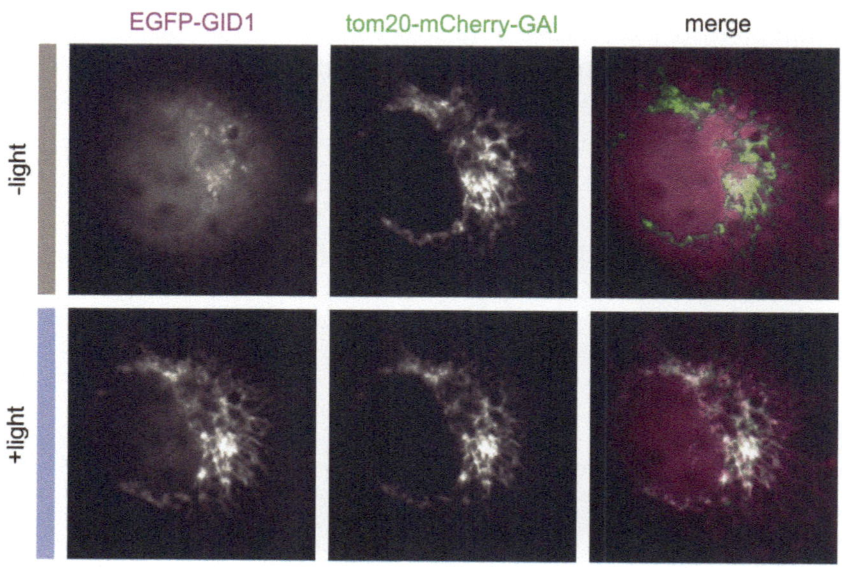

Fig. 4 Epifluorescence microscopy images visualize localization of GID1 and GAI protein before and after irradiation with 405 nm laser (3–7 s). Dimerization completed within 20 s after irradiation. A video of the dimerization process is available Supplementary video 1 in the online version at https://doi.org/10.1016/bs.mie.2020.02.022.

2. Exchange DMEM medium on cells with L15 medium
3. Perform life cell imaging on widefield epifluorescence microscope with lasers emitting at 488 and 561 nm
4. Identify and focus cotransfected cells (signal in GFP and mCherry channel)
5. Add 200 μL of 2 μM GA_3-PC solution to imaging medium
6. Perform uncaging with laser emitting at 405 nm for 3–7 s
7. Process and analyze data with appropriate image processing program (*e.g.*, ImageJ, Fig. 4; Supplementary video 1 in the online version at https://doi.org/10.1016/bs.mie.2020.02.022)

3. Summary

With this protocol we provide simple and rapid access to a technology for light-induced protein dimerization within living cells. It can be easily carried out by experimentalists with little experience in chemical synthesis. The used photoactivatable CIP can be chemically synthesized in a simple one-step procedure and directly used for light-induced protein manipulation in live cells with high spatial and temporal precision. The plasmids used for this protocol are available from Addgene and the introduction of other proteins of interest is easily possible by using standard cloning technologies. Fluorescence microscopy analysis and light activation can be performed on a simple epifluorescence microscope as well as a confocal microscope.

Acknowledgments

This work was supported by funding from the Deutsche Forschungsgemeinschaft DFG (SPP 1623). Michael J. Ziegler acknowledges funding from the Carl-Zeiss-Stiftung. The authors thank Klaus Yserentant for scientific and technical advice. We thank Dirk-Peter Herten for access to microscopes.

References

Brown, K. A., Zou, Y., Shirvanyants, D., Zhang, J., Samanta, S., Mantravadi, P. K., et al. (2015). Light-cleavable rapamycin dimer as an optical trigger for protein dimerization. *Chemical Communications*, *51*(26), 5702–5705.

Choi, J., Chen, J., Schreiber, S. L., & Clardy, J. (1996). Structure of the FKBP12-rapamycin complex interacting with the binding domain of human FRAP. *Science*, *273*, 239–242.

Davière, J.-M., & Achard, P. (2013). Gibberellin signaling in plants. *Development*, *140*(6), 1147–1151.

Fegan, A., White, B., Carlson, J. C., & Wagner, C. R. (2010). Chemically controlled protein assembly: Techniques and applications. *Chemical Reviews*, *110*(6), 3315–3336.

Jackson, A. L., Bartz, S. R., Schelter, J., Kobayashi, S. V., Burchard, J., Mao, M., et al. (2003). Expression profiling reveals off-target gene regulation by RNAi. *Nature Biotechnology*, *21*(6), 635–637.

Karginov, A. V., Zou, Y., Shirvanyants, D., Kota, P., Dokholyan, N. V., Young, D. D., et al. (2010). Light regulation of protein dimerization and kinase activity in living cells using photocaged rapamycin and engineered FKBP. *Journal of the American Chemical Society*, *133*(3), 420–423.

Kim, D., Bae, S., Park, J., Kim, E., Kim, S., Yu, H. R., et al. (2015). Digenome-seq: Genome-wide profiling of CRISPR-Cas9 off-target effects in human cells. *Nature Methods*, *12*(3), 237–243.

Lampson, M. A., & Kapoor, T. M. (2006). Unraveling cell division mechanisms with small-molecule inhibitors. *Nature Chemical Biology*, *2*(1), 19–27.

Liang, F.-S., Ho, W. Q., & Crabtree, G. R. (2011). Engineering the ABA plant stress pathway for regulation of induced proximity. *Science Signaling*, *4*(164), rs2.

Mangelsdorf, D. J., Thummel, C., Beato, M., Herrlich, P., Schütz, G., Umesono, K., et al. (1995). The nuclear receptor superfamily: The second decade. *Cell*, *83*(6), 835.

Melcher, K., Ng, L.-M., Zhou, X. E., Soon, F.-F., Xu, Y., Suino-Powell, K. M., et al. (2009). A gate–latch–lock mechanism for hormone signalling by abscisic acid receptors. *Nature*, *462*(7273), 602.

Miyamoto, T., DeRose, R., Suarez, A., Ueno, T., Chen, M., Sun, T.-P., et al. (2012). Rapid and orthogonal logic gating with a gibberellin-induced dimerization system. *Nature Chemical Biology*, *8*(5), 465–470.

Miyazono, K.-I., Miyakawa, T., Sawano, Y., Kubota, K., Kang, H.-J., Asano, A., et al. (2009). Structural basis of abscisic acid signalling. *Nature*, *462*(7273), 609–614.

Murase, K., Hirano, Y., Sun, T.-P., & Hakoshima, T. (2008). Gibberellin-induced DELLA recognition by the gibberellin receptor GID1. *Nature*, *456*(7221), 459.

Schelkle, K. M., Griesbaum, T., Ollech, D., Becht, S., Buckup, T., Hamburger, M., et al. (2015). Light-induced protein dimerization by one-and two-photon activation of gibberellic acid derivatives in living cells. *Angewandte Chemie International Edition*, *54*(9), 2825–2829.

Sioud, M. (2005). Induction of inflammatory cytokines and interferon responses by double-stranded and single-stranded siRNAs is sequence-dependent and requires endosomal localization. *Journal of Molecular Biology*, *348*(5), 1079–1090.

Umeda, N., Ueno, T., Pohlmeyer, C., Nagano, T., & Inoue, T. (2010). A photocleavable rapamycin conjugate for spatiotemporal control of small GTPase activity. *Journal of the American Chemical Society*, *133*(1), 12–14.

Wright, C. W., Guo, Z. F., & Liang, F. S. (2015). Light control of cellular processes by using photocaged abscisic acid. *Chembiochem*, *16*(2), 254–261.

CHAPTER FOURTEEN

Photoactivatable trimethoprim-based probes for spatiotemporal control of biological processes

Daniel Z. Wu[a], Michael A. Lampson[b], David M. Chenoweth[a,]*

[a]Roy and Diana Vagelos Laboratories, Department of Chemistry, University of Pennsylvania, Philadelphia, PA, United States
[b]Lynch Laboratories, Department of Biology, University of Pennsylvania, Philadelphia, PA, United States
*Corresponding author: e-mail address: dcheno@sas.upenn.edu

Contents

Abstract

Optogenetic tools allow regulation of cellular processes with light, which can be delivered with spatiotemporal resolution. By combining the chemical versatility of photoremovable protecting groups with the biological specificity of self-labeling tags, we developed a series of chemi-optogenetic tools that enable protein recruitment with precise spatiotemporal control. To this end, we created a modular platform for chemically inducible proximity (CIP), a technique in which two proteins of interest are brought together by the presence of a small molecule to induce a biological effect. The local proximity of a protein and its substrate has been shown to be sufficient to

Methods in Enzymology, Volume 638
ISSN 0076-6879
https://doi.org/10.1016/bs.mie.2020.03.015

initiate a desired biological effect, making CIP a valuable technique towards probing cellular processes. The high affinity and specificity of these tags result in rapid initiation of dimerization, allowing biochemical processes to be studied on a biologically relevant timescale. In this chapter, we describe the synthesis and application of chemi-optogenetic probes for spatiotemporal control of protein proximity.

Abbreviations

CIP	chemically inducible proximity (also chemically inducible dimerization)
CTH	coumarin-trimethoprim-haloligand
eDHFR	*E. coli* dihydrofolate reductase
FKBP	FK506 binding protein
NTH	NVOC-trimethoprim-haloligand
PPI	protein-protein interactions
SLF	synthetic ligand for FKBP
TNH	trimethoprim-NVOC-haloligand

Chemical abbreviations

CDI	1,1′-carbonyldiimidazole
DBU	1,8-diazabicyclo[5.4.0]undec-7-ene
DCM	dichloromethane
DIEA	*N,N*-diisopropylethylamine or *N*-Ethyldiisopropylamine or Hünig's base
DMAP	4-(dimethylamino)pyridine
DMF	*N,N*-dimethylformamide
DMSO	dimethyl sulfoxide
HATU	1-[bis(dimethylamino)methylene]-1*H*-1,2,3-triazolo[4,5-*b*]pyridinium-3-oxid hexafluorophosphate or *N*-[(dimethylamino)-1*H*-1,2,3-triazolo-[4,5-*b*]pyridin-1-ylmethylene]-*N*-methylmethanaminium hexafluorophosphate *N*-oxide
NVOC or NV	6-nitroveratryloxycarbonyl
NVOC-Cl	6-nitroveratryloxycarbonyl chloride or 4,5-dimethoxy-2-nitrobenzyl chloroformate
TEA	triethylamine
TFA	trifluoroacetic acid
TMP	trimethoprim

1. Introduction

1.1 Chemically induced proximity

Biological processes are often regulated by the distance of a protein and its interacting partner. Proximity is a regulatory mechanism that has been shown to promote acetylation, methylation, phosphorylation, and a handful of other protein modifications. Recruitment of proteins are also capable of influencing gene regulation and protein degradation. As explained by Stanton et al., the proximity of proteins can produce a biological effect because the probability of an effective collision between two proteins is a function of distance (Stanton et al., 2018). In contrast, organelles can also sequester components to decrease the effective concentration of relevant substrates (An et al., 2008; Robinson et al., 2010).

Chemically inducible proximity (or chemically inducible dimerization) is a method in which a small molecule promotes the heterodimerization between two proteins (often a chemical tag or a complimentary protein pair) that are individually fused to proteins of interest (POIs). This dimerization event brings the two POIs within proximity, triggering potential downstream biological effects. To clarify the terminology: two protein tags are dimerized bringing their respective fused POI within proximity. The major advantages of this technique over traditional genetic manipulations (i.e., knockout, transfection, silencing) are its temporal control and chemical specificity. In these systems, the protein fusions are already expressed and remain diffuse until the small molecule is introduced. Small molecule diffusion occurs on the order of μm/s (Mastro et al., 1984), initiating chemically induced proximity within seconds (Miyamoto et al., 2012). In addition, the heterodimerizing protein pair generally do not interact with endogenous mammalian protein.

Chemical tags (also known as self-labeling tags) are proteins that can be fused to other proteins of interest and covalently tagged with a complementary ligand. Halotag (Promega) and the SNAP/CLIP system (New England Biolabs) are three commercial choices that are commonly utilized. Noncommercial chemical tags, such as BL-tags, FAPs, PYP-tag and TMP/eDHFR, can be found on Addgene, the global plasmid repository. In general, chemical tags possess high molecular specificity towards their respective ligands and can be used in conjunction with one another. These desirable bioorthogonal trails are attributed to a combination of their exogenous nature and directed evolution. Chemical tag ligands can also be tethered to various chemical probes, adding another dimension of versatility that cannot be achieved through genetic encoding alone.

Chemical inducers of proximity employ a small molecule that can recruit protein pairs to the same position of a biological system. These proteins bind to different portions of the same small molecule. For example, FKBP (FK506-rapamycin binding protein) and FRB (FKBP-rapamycin binding domain of mTOR) bind to different sections of rapamycin (Benjamin et al., 2011; Li et al., 2014; Spencer et al., 1993). Drug hybrids can also be synthesized where two different drugs are tethered to heterodimerize their respective proteins. One such example is the TMP-SLF (trimethoprim—synthetic ligand for FKBP) conjugate that is capable of dimerizing bacterial dihydrofolate reductase (eDHFR) with FKBP, which was used to conditionally activate glycosyltransferase (Czlapinski et al., 2008). Proteolytic targeting chimeras (PROTAC) also fall within this category (Coleman & Crews, 2018).

1.2 Photoactivatable systems

To understand the spatiotemporal organization of biological processes, we designed a chemi-optogenetic platform utilizing self-labeling protein tags and photocages (DEACM and NVOC). One potential drawback of this approach is that uncaging is irreversible, unlike optogenetic protein or chemical photoswitches (e.g., azobenzene), which can move between the "OFF" and "ON" states multiple times (Hüll et al., 2018). However, this trait works to our advantage when used to study long-term localization or highly dynamic proteins, in which constant pinpoint illumination through three-dimensional space is unfeasible. Moreover, one round of recruitment and release is sufficient to study many biological systems. Photocages are commercially available and can be typically installed onto small molecules with minimal chemical manipulation (Klán et al., 2013). By installing a photocage directly onto pertinent functional groups of a drug, the modification serves to directly block protein binding interactions, which can then be removed after illumination to initiate binding with spatiotemporal control. In addition, direct obstruction of the drug-protein interaction allows recruitment to be fully "OFF" in the dark state, whereas photoswitches and optogenetic proteins exist at an equilibrium between "ON" and "OFF" states even in the absence of light. Regardless of the photoactivation modality, each system offers unique advantages and should be chosen based on the design of the experiment.

In this chapter, we describe the chemi-optogenetic platforms we have developed that provide control over dimerization of genetically tagged proteins using orthogonal wavelengths of light. The spatiotemporal and reversible limitations of existing chemical inducers of proximity (CIP) systems inspired us to develop this optogenetic platform by incorporating photoactivation. Our system consists of three components: a Halotag fusion protein that localizes to an intracellular structure, an eDHFR fusion construct, and a photocaged chemical dimerizer (Fig. 1). By using a Halotag fusion as an anchor for our

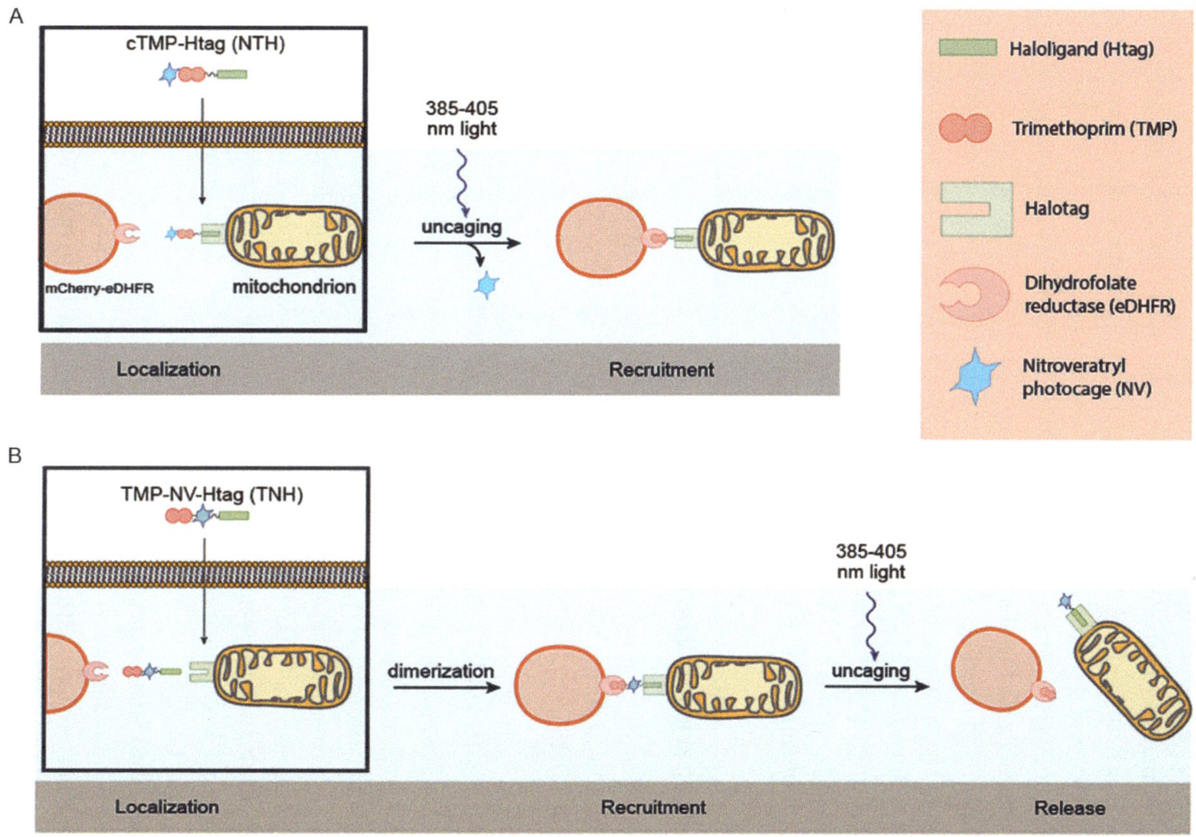

Fig. 1 *General scheme of photoactivatable or photoreversible dimerization events.* (A) The general scheme of the photocaged dimerizer is shown. NTH (shown as cartoon) enters the cell and undergoes covalent bond formation with the Halotag protein, excess probe is washed out. Upon exposure to 405 nm light, NVOC is released allowing the eDHFR fusion to bind to trimethoprim. (B) Photoreversible dimerizer TNH (shown as cartoon) enters the cell and recruits Halotag fusion and eDHFR fusion together. Upon exposure to 405 nm light, NVOC is released allowing the eDHFR fusion to diffuse from the Halotag fusion.

probe, we can saturate the binding sites and sufficiently washout excess probe to enhance spatial resolution upon photoactivation. The eDHFR fusion binds non-covalently to the trimethoprim portion of the dimerizer (after uncaging) and can be reversed by addition of free trimethoprim.

We show that [7-(diethylamino)coumarin-4-yl] methyl (DEACM) can be uncaged with 444 nm light or 405 nm light at 10% intensity of the laser source (28 mW) and the same laser conditions do not uncage 6-nitroveratryl oxycarbonyl (NVOC) (Zhang et al., 2017). NVOC is only sensitive to light up to 405 nm because the absorption spectra is more blue-shifted than that of DEACM. Additionally, DEACM has a significantly higher photon absorption and high quantum efficiency at 440 nm, whereas NVOC exhibits its highest absorption at 365 nm with almost no absorption at >400 nm. Lastly, we showcase the modularity of the system, by assembling a probe with the NVOC embedded in-between haloligand and trimethoprim, thus creating a photoreversible probe.

2. Preparation and synthesis

A detailed description of the synthetic procedure and characterization can be found in the supplementary section of (Ballister et al., 2014; Zhang et al., 2017).

2.1 Equipment, reagents, and general procedures

Reagents used in the chemical synthesis are listed below. Low resolution electrospray ionization mass spectra (LRMS) were obtained on a Waters Acquity Ultra Performance LC connected to a single quadrupole detector (SQD) mass spectrometer (Milford, MA, USA). High-resolution electrospray ionization mass spectra (HRMS) were collected with a Waters LCT Premier XE liquid chromatograph/mass spectrometer. Nuclear magnetic resonance (NMR) spectra were obtained on a Bruker 400 MHz NMR instrument (Billerica, MA, USA).

Equipment and instrumentation

1. Fume hood (SafeAire 6' Chemical Fume Hood)
2. Rotovapor system (Buchi R-200 System with V-800 Vacuum Controller)
3. Stirring hot plate (IKA RCT basic; No: 003810001)
4. Nitrogen or argon gas (Airgas NI R200SS or AR 200)

5. Round-bottom flasks (24/40 bottle neck used)
6. Stir bars
7. Separatory funnel
8. Rubber septum or plastic caps (for 24/40 bottle neck)
9. TLC plates (Sorbent Technologies silica plates, 250 μm thickness w/UV254 + 366 glass backed; No: 1615226)
10. TLC chamber
11. Method of purification:
    Automated flash chromatography (Teledyne ISCO CombiFlash Rf+) or
    Manual chromatography silica gel (Silicycle silica gel 55–65 Å pore diameter)
12. NMR for analysis (Bruker 400 MHz NMR processed by TopSpin software)
13. LCMS/HRMS for characterization

2.2 Materials

Building blocks

1. 1-Bromo-6-chlorohexane (CAS: 6294-17-3; TCI America B199825G).
2. Trimethoprim (CAS: 738-70-5; Chem-Impex Int'L Inc. 01634100G).
3. 3-Bromopropylamine hydrobromide (CAS: 5003-71-4; TCI America B126325G).
4. Succinic anhydride (CAS: 108-30-5; Alfa Aesar A12245).
5. 4,5-Dimethoxy-2-nitrobenzyl chloroformate (CAS: 42855-00-5; Sigma-Aldrich 420069-1G).
6. 2-(2-Aminoethoxy)ethanol (CAS: 929-06-6; ACROS 103631000).
7. Tetraethyleneglycol (CAS: 112-60-7; ACROS AC149590010).
8. 7-Diethylamino-4-methylcoumarin (CAS: 91-44-1; Chem Impex Int'L Inc. 27087100G).
9. Carbonyldiimidazole (CAS: 530-62-1; Sigma-Aldrich 115533-10G).
10. Vanillin (CAS: 121-33-5; TCI America H026425).

Reagents

1. Carbontetrabromide (CAS: 558-13-4; TCI America T0038100G).
2. Triphenylphosphine (CAS: 603-35-0; ACROS 140422500).
3. Sodium hydride, 57–63% oil dispersion *[Pyrophoric]* (CAS: 7646-69-7; 1343118).

4. Sodium meta-periodate (CAS: 7790-28-5; Fisher S398-100).
5. *N,N*-Dimethylformamide diethyl acetal (CAS: 4637-24-5; TCI D1293100ML).
6. Sodium borohydride (CAS: 16940-66-2; ACROS 189300100).
7. Hydrobromic acid (48%) *[Highly Corrosive]* (CAS: 10035-10-6; Oakwood Products 0447071KG).
8. Boc anhydride (CAS: 24424-99-5; Chem Impex Int'L Inc. 00128250G).
9. Potassium carbonate (CAS: 584-08-7; ACROS 424081000).
10. DBU (CAS: 6674-22-2; ACROS 160611000).
11. Benzyl bromide (CAS: 100-39-0; ACROS 105871000).
12. Fuming nitric acid *[Highly Corrosive]* (CAS: 7697-37-2; Fisher A202400).
13. Acetic anhydride (CAS: 108-24-7; Fisher A10-100).
14. 4-Nitrophenyl chloroformate (CAS: 769-46-1; ACROS 170800050).
15. DMAP (CAS: 1122-58-3; ACROS 14820050).
16. DIEA (CAS: 7087-68-5; ACROS 115220250).
17. Trifluoroacetic acid (CAS: 76-05-1; Fisher O4901-500).
18. HATU (CAS: 148893-10-1; Oakwood 02392625G).
19. Triethylamine (CAS: 121-44-8; Alfa Aesar A12646K).
20. Sodium hydroxide, pellets *[Caustic]* (CAS: 1310-73-2; Fisher S318-100).
21. Potassium permanganate (CAS: 7722-64-7; Fisher P279-500).

Solvents

1. THF, anhydrous (CAS: 109-99-9; Alfa Aesar 44608K2).
2. DCM, reagent grade (CAS: 75-09-2; Fisher D37-500).
3. DMF, anhydrous (CAS: 68-12-2; ACROS 610941000).
4. DMSO, anhydrous (CAS: 67-68-5; Alfa Aesar 43998AE).
5. EtOH, absolute [200 Proof] (CAS: 64-17-5; ACROS 615095000).
6. Diethyl ether, anhydrous (CAS: 60-29-7; Fisher E19804).
7. Hexanes, reagent grade (CAS: 92112-69-1; Fisher H292-20).

2.3 Synthesis of chemical dimerizers

2.3.1 Synthesis of uncaged dimerizer (TH)

Design strategy

Trimethoprim-Halotag (TH) was synthesized by forming an amide linkage between trimethoprim and the chlorohexane ligand (Scheme 1). Linker length and composition were considered based on literature

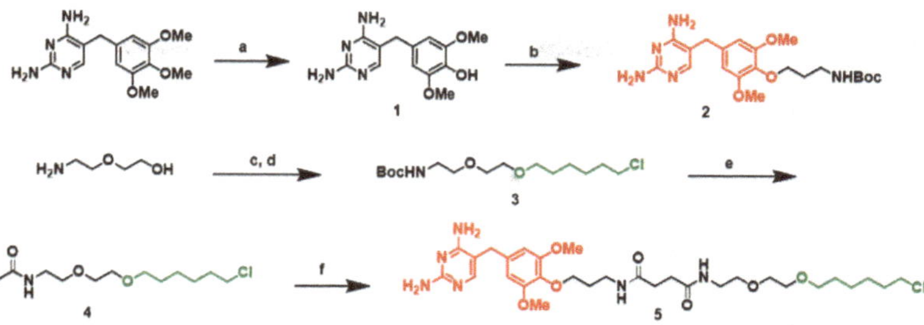

Scheme 1 Chemical synthesis of TH (*Trimethoprim–Haloligand*) Reagents and conditions: (a) 48% HBr_{aq}, 100 °C, 15 min, 75% (b) *tert*-butyl-3-bromopropylcarbamate, DBU, DMSO, 25 °C, 16 h, 42% (c) di-tert-butyl dicarbonate, ethyl ether, EtOH, 0 to 25 °C, 2 h, quantitative (d) NaH, 1-bromo-6-chlorohexane, THF/DMF (2:1), 0 °C to rt., 16 h, 45% (e) compound **3** was deprotected with TFA:DCM (1:1), rt., 2 h to yield (Halo-PEG_2-NH_2 TFA salt), dried in vacuo, followed by addition of succinic anhydride, DMAP, DCM, 25 °C, 16 h, 90% (2 steps) (f) compound **2** was deprotected with TFA:DCM (1:1), rt., 2 h to yield (TMP-NH_2 TFA salt), dried in vacuo, followed by addition of compound **4,** HATU, DIEA, DMF, 25 °C, 2 h, 67% (2 steps).

precedent (Encell et al., 2012; Los et al., 2008). The amide bond placed after the two ethylene glycol units was predicted to hydrogen bond Thr148 and Phe144 due to the increased binding kinetics of Halotag (Ohana et al., 2015). The optimal linker length beyond the 13-atom spacer (chlorohexane with two PEG units) was accessed by analyzing crystal structures of Halotag (PDB: 1BN7) and eDHFR (PDB: 1RG7).

Synthetic notes

(a) Care should be taken when using both 48% HBr and 50% (w/v) NaOH (Caustic). The timing of the methoxy cleavage should be precise to avoid over-demethoxylation. HBr should be heated precisely to temperature before the addition of trimethoprim, and the progress of the reaction can be monitored by TLC (DCM/MeOH 10:1). The reaction is quenched with 50% (w/v) NaOH in water. The addition of NaOH is highly exothermic and done slowly. We found that a 25 g scale reaction with 200 mL of 48% HBr for 15 min afforded good yields. However, we have also noticed variations in yields (40–80%) between commercial batches of trimethoprim and the quality of HBr.

(b) *tert*-Butyl-3-bromopropylcarbamate was synthesized from 3-bromopropylamine hydrobromide using traditional protecting conditions.

(c) A potassium permanganate stain was used to visualize the product on TLC. The starting material is water-soluble and removed thorough separation.

(d) 1-Bromo-6-chlorohexane was found to be a comparable substitute for 1-iodo-6-chlorohexane. This compound is a more economical substitute and provides similar yields.

Safety Hazard: Sodium hydride is corrosive and pyrophoric. Care should be taken when quenching the reaction, such as slow addition of ice water.

(e) TFA:DCM can be removed by roto-evaporation to yield the TFA salt. Water should be placed in the solvent trap to catch condensed TFA. The system should be rinsed with acetone to remove residual TFA.

(f) Compound **4** was first activated with DIPEA and HATU and then added to a solution of deprotected Compound **2** in DCM, DIPEA.

2.3.2 Synthesis of caged dimerizers (CTH and NTH)

Design strategy

CTH and NTH were synthesized by building off trimethoprim-NHBoc (Compound **2**). We note that the yields obtained from photocaging with NVOC (41%) are much higher than that of Coumarin (DEACM) caging (12%). We separated a single isomer of caged TMP for characterization. However, both photoisomers prevent binding to eDHFR. Additionally, the NVOC chloroformate (CAS: 42855-00-5) is commercially available. Although DEACM-OH (Geißler et al. 2005) (Compound **7**; CAS: 54711-38-5) has recently become commercially available through TCI (Catalog D4970), it is more economical to synthesize the compound. We report a gram-scale synthesis for preparing Coumarin-OH that does not utilize selenium dioxide based on a literature procedure (Weinrich et al., 2017) (Scheme 2).

Synthetic notes

(g) The major isomers (shown) are more nonpolar.

(h) The reaction should reach ~140 °C otherwise the chemistry does not proceed. The imine product precipitates out in cold DMF and is filtered. The filter cake was washed with cold acetone and used without further purification. Addition of sodium meta-periodate forms a slurry after some time, which is filtered at the end of the reaction and washed with ethyl acetate. The filtrate is collected and separated to yield an aldehyde. The aldehyde formed after step ii is a nonpolar red product.

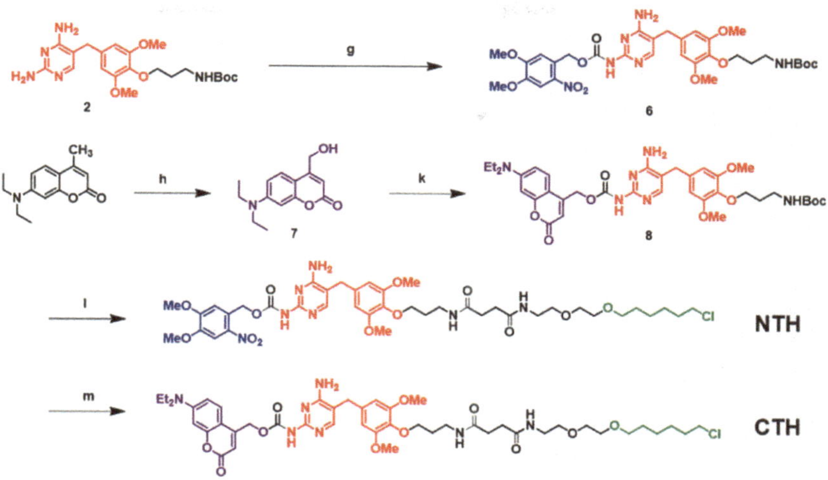

Scheme 2 *Chemical synthesis of CTH and NTH* (Coumarin and NVOC *Trimethoprim–Haloligand*) Reagents and conditions: (g) NVOC-Cl, DIEA, DCM, rt. (41% major isomer) (h) (i) DMF-DMA, DMF, reflux o.n, (ii) (4:1) THF:H_2O, 3 equiv. $NaIO_4$, 2h, (iii) $NaBH_4$, MeOH, 1h, 65% (3 steps) (k) **2**, CDI, DIEA, DMAP, DCM, reflux, 16h; (ii) **7**, DIEA, DMAP, DCM, rt., 16h, 12% (2 steps) (l, m) (i) TFA:DCM (1:3), rt., 2h; (ii) **6** or **8** (ii) **2,** HATU, DIEA, DMF, rt., 16h, 64% (2 steps).

(k) The trimethoprim was activated with CDI before addition of DEACM-OH.

(l,m) The caged trimethoprim was deprotected and dried in vacuo to remove TFA and solvent. The intermediate was then redissolved with DIEA and dry DMF. Compound **2** was activated with HATU, and then, the solution of caged TMP-NH_2 was added.

2.3.3 Synthesis of cleavable dimerizer TNH

2.3.3.1 Design strategy

The PEGylated Haloligand was chosen to condense synthesis. The removal of the amide did not exhibit significant changes in Halotag binding kinetics. TNH also utilizes Compound **2** as a building block (Scheme 3).

2.3.3.2 Synthetic notes

(n) A potassium permanganate stain was used to visualize the product on TLC. The starting material is water-soluble and removed thorough separation.

(p) The benzyl group is necessary to block formation of unwanted isomers.

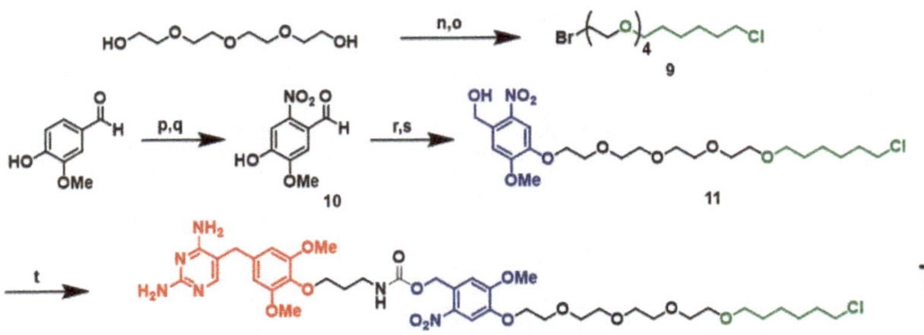

Scheme 3 *Chemical synthesis of TNH (Trimethoprim–NVOC-Haloligand)* Reagents and conditions: (n) NaH, 1-bromo-6-chlorohexane, THF:DMF (2:1), 0°C to 25°C, 16h, 40%; (o) PPh_3, CBr_4 THF, 25°C, 16h, 50%; (p) K_2CO_3, benzyl bromide, DMF, 80°C, 16h 99%; (q) (i) Fuming HNO_3, 1,2-dichloroethane −30°C to 25°C, 3h (ii) 48% HBr_{aq} 90°C, 3h, 23% (two steps); (r) **9**, K_2CO_3, DMF, 60°C, 16h, 30%; (s) $NaBH_4$, MeOH:Dioxane (1:1) 0°C to 25°C, 2h, quantitative yield; (t) (i) 4-Nitrophenyl chloroformate, DMAP, DCM, 25°C, 16h (ii) deprotected-**11** (TMP-NH_2TFA salt) DIEA, DCM, 25°C, 16h 16% (2 steps).

(q) Care should be taken when handling fuming nitric acid. Slow dropwise addition is necessary to avoid heating and over-nitration.

(t) The nitrophenyl carbonate of NVOC-OH is formed first, and then, deprotected TMP-NH_2 (**2**) is added.

2.3.3.3 Storage and handling of photocaged dimerizers

Care should be taken when working with photocaged compounds, strong ambient light (e.g., from brightfield microscopy) has been reported to remove coumarin photocages. Care was taken to minimize exposure of dimerizers to light and heat prior to experiments. In our hands, working quickly in low levels of white light used for differential interference contrast microscopy did not cause detectable premature cleavage of TNH or uncaging of NTH, CTH.

All compounds were dissolved in DMSO at 10mM and stored in amber plastic microcentrifuge tubes at −80°C for long-term storage. For experiments, a frozen aliquot was thawed and diluted into growth medium (DMEM, 10% FBS) to the final working concentration (μM) stored at −20°C and warmed to 37°C when ready for use. We recommend that once prepared, stock solutions are stored aliquoted in amber tubes at −20°C or below and used within 6months. In addition, freeze-thaw cycles should generally be avoided. All compounds are stable in growth and imaging media at 37°C for several hours. Cytotoxicity over 24h was only observed in TNH.

3. Methods

3.1 Generation and culture of transfected cell lines

3.1.1 Plasmids

All plasmids in this study are derived from pEM705, which contains a CAG promoter for constitutive expression, obtained from E. V. Makeyev (Khandelia et al., 2011). Additional information regarding the plasmid and usage can be found in the Supplementary section of Ballister et al. (2014, 2015).

- Halo-GFP-ActA (referred to as Halo-GFP-Mito) + mCherry-eDHFR includes the C-terminal 47 amino acids of the *Listeria monocytogenes* ActA gene, which consist of mitochondrial outer membrane targeting domain. mCherry-eDHFR is also expressed. (Addgene: Plasmid #61500)
- PEX3-GFP-Halo includes the N-terminal 42 amino acids of the human Pex3 gene, which target peroxisome. (Addgene: Plasmid #67764)
- CENPB-GFP-Halo + mCherry-eDHFR contains the full-length human kinetochore protein CENP-B and mCherry-eDHFR fusion (Addgene: Plasmid #61502)
- K560-mCherry-eDHFR includes residues 1–560 of human kinesin-1 heavy chain. (Addgene: Plasmid #67760)
- BICD-mCherry-eDHFR includes residues 1–572 of mouse BICD2. (Addgene: Plasmid #67763)

Equipment

1. Biosafety cabinet.
2. Incubator with CO_2 (for mammalian cell lines).
3. Pipettes (1000 μL, 100 μL, 10 μL).
4. Pipettors.

3.2 Materials

1. Six-well plates (CELLTREAT 229106).
2. Growth medium:
   Dulbecco's Modified Eagle's Medium (DMEM) (Invitrogen 11965084).
   1% Pen/Strep (Invitrogen 15140122).
   10% fetal bovine serum (FBS) (Life Technologies 16000-044).
3. Transfection medium: DMEM with 10% FBS.
4. Lipofectamine 2000 (Invitrogen).

5. Serum-free media such as Opti-MEM (Life Technologies 31985062).
6. DMSO (CAS: 67-68-5; Alfa Aesar 43998AE).

Procedure

1. Split the cell line and seed enough cells on poly-D-lysine coated coverslips placed in six-well plates (or an imaging dish) so that they reach 70–80% confluency after 24h of culture.
2. Lipofectamine 2000 is used as the transfection reagent to introduce the plasmid of choice into the cell. Wash each well with antibiotic-free media and replace with 2mL of transfection media. Antibiotics can impact cell viability during transfection.
3. Transfect each well with 1 μg of plasmid and 5 μL of Lipofectamine 2000 in OptiMEM 24–48h prior to imaging. The transfection complex can be left in the well overnight without noticeable impact on cell viability. However, the media should be exchanged with growth media 6–16h after transfection.
4. After 24–48h, set up microscope and begin preparation of cell chamber for imaging.

3.3 Preparation of cells and live-cell imaging

Equipment

1. Microscope suitable for live-cell fluorescent imaging and capable of photobleaching.

   Minimal required laser lines: 405nm with tunable output or 405nm and 444nm

   Imaging laser lines: 444nm may be used to image CFP, but will also initiate

   photoactivation within field of view. In this case, the pre-activation image will initiate dimerization. CTH is stable and compatible with standard laser lines that are red-shifted from 444nm (e.g., 488nm, 532nm, 561nm, etc.).
2. Environmental chamber (e.g., Tokai Hit Chamber).
3. Magnetic live-cell imaging chamber (Chamlide CM-S22-1, LCI).

   Alternative: 35mm Dish No. 1.5 Coverslip Poly-D-Lysine Coated (MatTek P35GC-1.5-14-C).
4. Incubator with CO_2 (for mammalian cell lines).
5. Biosafety cabinet.
6. Water bath (set to 37°C).
7. Pipettes.
8. Tweezers.

3.4 Materials

1. 22 × 22 mm glass Poly-D-lysine coated #1.5 coverslips (Millipore Sigma A-003-E).
2. Imaging medium:
   Leibovitz-15 (L15) CO2-independent media, no phenol red (Life Technologies 11415064).
   1% Pen/Strep (Invitrogen 15140122).
   10% FBS (Life Technologies 16000-044).
3. Growth medium:
   Dulbecco's Modified Eagle's Medium (DMEM) (Invitrogen 11965084).
   1% Pen/Strep (Invitrogen 15140122).
   10% Fetal bovine serum (FBS) (Life Technologies 16000-044).
4. 10 μM NTH in growth medium.
5. 10 μM CTH in growth medium.
6. 0.1 μM TNH in growth medium.
7. Trypsin 0.25% EDTA (Invitrogen 25200056).
8. Six-well plates (CELLTREAT 229106).

Procedure

1. After transfection, set up microscope. An environmental control apparatus is used for our experiments. L15 CO_2 independent media can be used in absence of a CO_2 line. However, the cells should be kept at 37 °C throughout the experiment.
2. Warm imaging medium to 37 °C.
3. Incubate cells with 10 μM dimerizer for 1 h.
4. Rinse each well three times with fresh growth medium to remove excess dimerizer. Incubate for 10 min.
5. Replace growth medium with prewarmed imaging medium, transfer coverslip to magnetic imaging chamber, and add 1 mL prewarmed imaging medium.
6. Mount chamber on microscope and locate cells with bright GFP signal (488/532 nm) on HaloTag fusion protein and mCherry-eDHFR fusion protein (593/625 nm) signal.
7. Obtain an initial image before uncaging. NVOC and DEACM can be orthogonally cleaved with different intensities of 405 nm light. Use the ROI tool to select the region of interest and activate with the 405 nm laser at 7% power for 40 repetitions at each of the imaging planes for NVOC or 6% for 15 repetitions for DEACM. Alternatively, DEACM can be removed with 444 nm light (used at maximum intensity with 2 s exposure). Take one image immediately, sequential GFP and mCherry

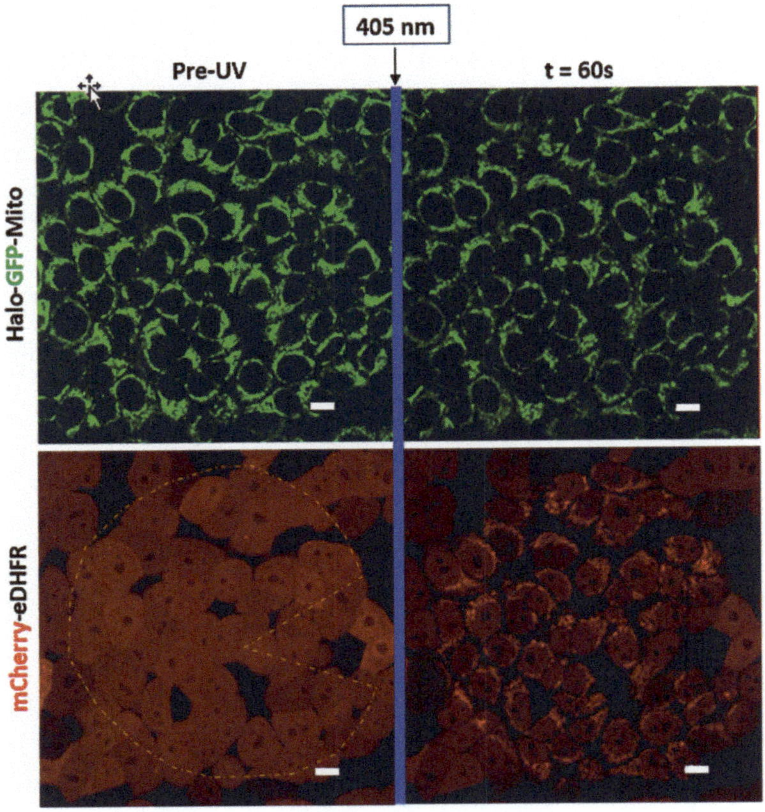

Fig. 2 *Spatiotemporal control of protein localization.* HeLa cells expressing Halo-GFP-Mito and mCherry-eDHFR are incubated with 10 μM of NTH followed by washout. Region of interest (yellow outline) is illuminated with 405 nm light. After 1 min, mCherry-eDHFR has localized to mitochondria within illuminated region of interest to reveal an iconic symbol of the 80s. Scale bar, 10 μm.

images should then be acquired at time intervals of your choice. We recommend taking images every minute for 30 min. Images were collected as maximum-intensity projections using 5–7 confocal imaging planes with 0.5 μm spacing.

8. Standard photobleaching effects should be considered (Figs. 2–5).

4. Analysis

Images were acquired using Nikon Elements—Advanced Research. Image analysis software can be accessed at no cost from the NIH at https://imagej.nih.gov/ij/. An open-source version of this software, called FIJI, includes several plugins that facilitate image analysis: https://fiji.sc/.

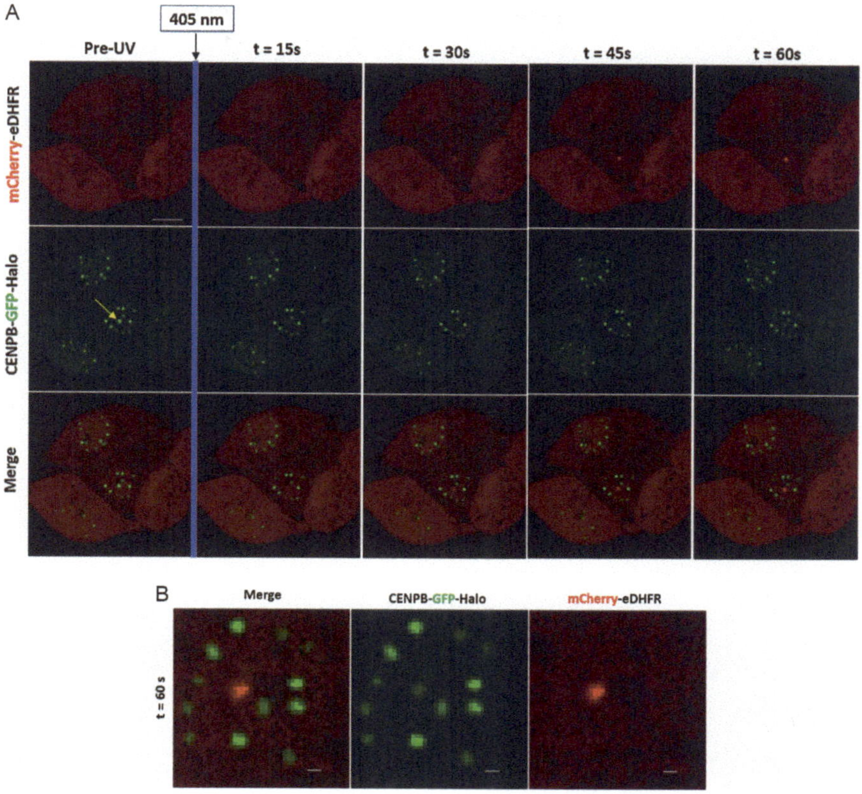

Fig. 3 *Protein recruitment at submicron resolution.* (A) (Top) HeLa cells expressing CENPB-GFP-Halo and mCherry-eDHFR incubated with 10 μM NTH followed by washout. A single centromere (yellow arrow) is pulsed with 405 nm light. After 1 min, mCherry-eDHFR has localized to target. Scale bar, 10 μm. (B) (Below) A magnified image at $t = 60$ s is shown below. Scale bar, 1 μm.

Images were analyzed using maximum intensity projection of the slices in a *Z*-stack. The fluorescence of mCherry-eDHFR localization was quantified by using the regions defined in the GFP channel with the threshold algorithm provided with ImageJ. All images shown are maximum intensity projections of all slices in a z-stack. For quantification of the mitochondrial mCherry-DHFR localization, the regions were defined in the GFP channel by thresholding using the ImageJ default algorithm.

After obtaining measurement values, the mCherry intensity at mitochondria was normalized such that the minimum value observed over the course of experiment for each field is 0 and the maximum value is 1 and then averaged at each timepoint over 10 fields of cells. For quantification of peroxisome density as a function of time, the average GFP intensity in

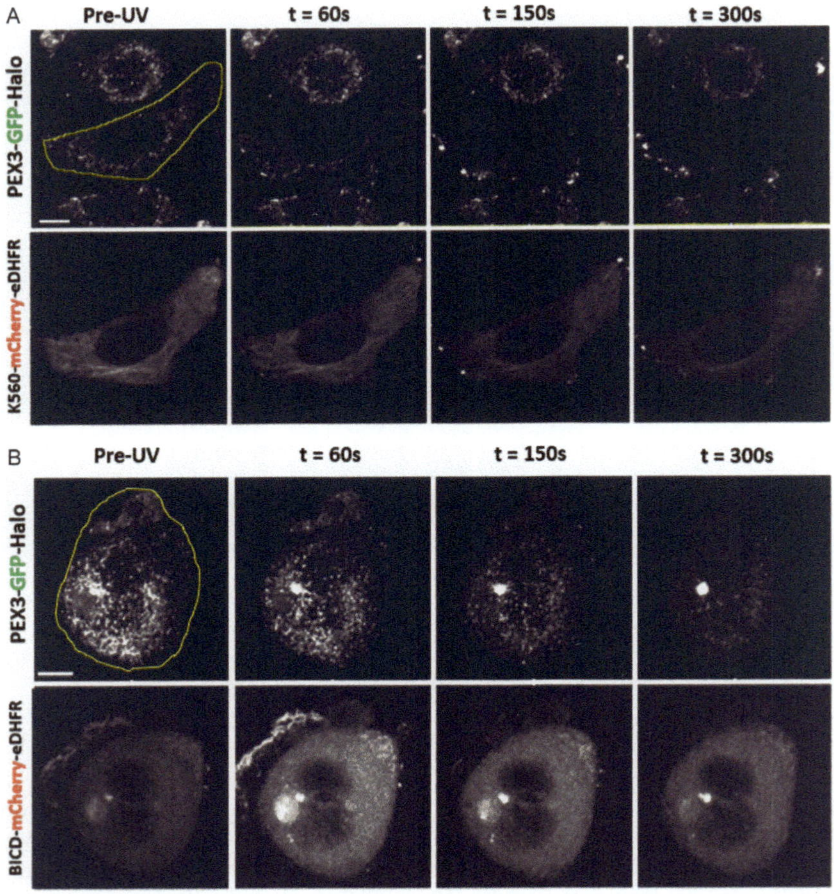

Fig. 4 *Optogenetic control of protein transport.* (A) Top panels show image of HeLa cells expressing PEX3-GFP-Halo and K560-mCherry-eDHFR incubated with 10 μM CTH followed by washout. Upon illumination, peroxisomes of target cell begin to travel anterograde. (B) Bottom panels show image of HeLa cells expressing PEX3-GFP-Halo and BICD-mCherry-eDHFR incubated with 10 μM CTH followed by washout. Upon illumination, peroxisomes of target cell begin to travel retrograde. Scale bar, 5 μm.

the regions of interest were measured at each time point. The background signal was estimated as the average intensity in large areas outside the cells and subtracted. To normalize between different regions, the background-subtracted measurements for each region at each time point were divided by the maximum intensity observed for each region over the course of the experiment.

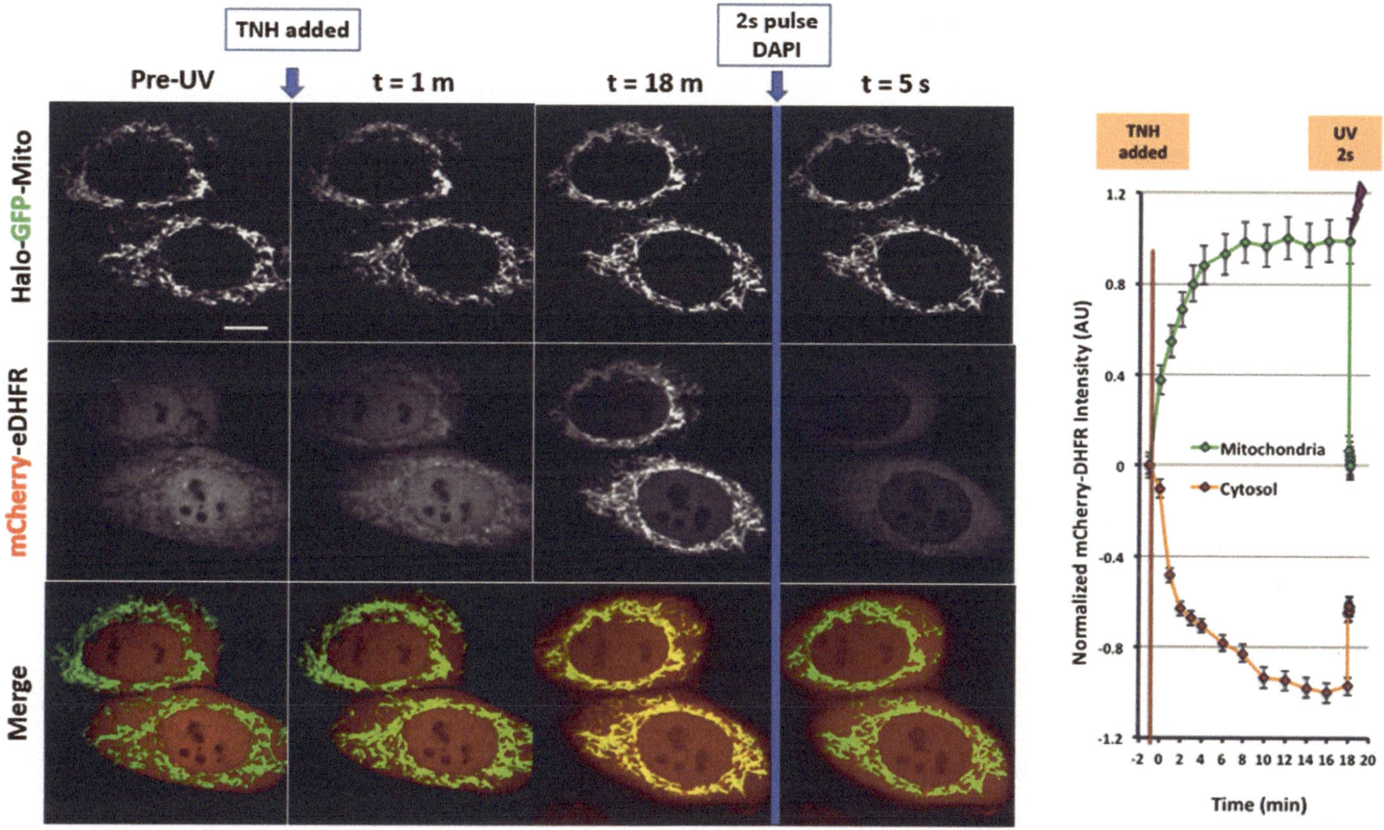

Fig. 5 *Spatiotemporal release of protein localization.* HeLa cells expressing Halo-GFP-Mito and mCherry-eDHFR are incubated with 0.1 μM of TNH and imaged for 18 m. Field of view is illuminated with 2 s DAPI light and imaged after 5 s. After DAPI illumination, mCherry-eDHFR is released back into the cytosol. Normalized mCherry intensity at mitochondria is graphed with cytosolic intensity throughout the experiment. Scale bar, 10 μm.

5. Notes

1. Proton nuclear magnetic resonance (^1H NMR) and carbon nuclear magnetic resonance (^{13}C NMR) spectra were recorded on a Bruker UNI 400 MHz, AVII 500 MHz and Biodrx 600 MHz NMR and processed by MestReNova or TopSpin software.
2. Low-resolution mass spectra were obtained using liquid chromatography-mass spectrometry (LCMS) on a Waters instrument with electrospray ionization in either positive or negative mode. High-resolution mass spectra (HRMS) were obtained at the University of Pennsylvania's Mass Spectrometry Service Center on Waters LC-TOF mass spectrometer (model LCT-XE Primer) using electrospray ionization in positive or negative mode depending on the analytes. HRMS data analysis was performed using the automated Waters software.
3. Spinning Disk Confocal Microscope Specifications: DM4000 Leica with a 100×1.4 NA objective, an XY Piezo-Z stage (Applied Scientific Instrumentation), a spinning disk (Yokogawa), an electron multiplier charge-coupled device camera (ImageEM; Hamamatsu Photonics), and a laser merge module equipped with 488 and 593 nm lasers (LMM5; Spectral Applied Research) controlled by MetaMorph software (Molecular Devices) equipped with an environmental chamber for temperature control (Incubator BL; PeCon GmbH). Targeted uncaging or cleavage was performed with an iLas2 illuminator system (Roper Scientific) equipped with a 405 nm laser (output of 27 mW after fiber coupling) controlled using the iLas2 software module within Metamorph.
4. We note that because the 405 nm laser power differs for each microscope setup, optimal laser conditions should be determined with a test cell or control experiment.

6. Summary and Conclusions

Our results demonstrate the utility of photocaged trimethoprim derivatives for manipulating biological processes with spatiotemporal control. Recruitment after uncaging is stable, leading to a sustained dimerization event after a single pulse of light. We demonstrate the modularity of the platform by synthesizing photoactivatable and photocleavable probes from the same building blocks. This chemi-optogenetic system enables prolonged activation or rapid reversal of protein interactions without the use of constant illumination. We envision these probes to be readily adaptable to dissect other processes where spatiotemporal activation and reversal are crucial, such as regulation of kinases, histone deacetylases, or genetic modifications (Chen et al., 2015; Duan et al., 2015; Funabiki & Wynne, 2013).

Acknowledgments

This work was supported by the National Institutes of Health (GM122475 to M.A.L. and GM118510 to D.M.C.) and the National Institutes of Health, National Cancer Institute (U54-CA193417). We thank Rachel Lackner, Chanat Aonbangkhen and Huaiying Zhang for helpful discussions, scientific contributions and image acquisition.

References

An, S., et al. (2008). Reversible compartmentalization of de novo purine biosynthetic complexes in living cells. *Science*, *320*(5872), 103.

Ballister, E. R., et al. (2014). Localized light-induced protein dimerization in living cells using a photocaged dimerizer. *Nature Communications*, *5*, 5475.

Ballister, E. R., et al. (2015). Optogenetic control of organelle transport using a photocaged chemical inducer of dimerization. *Current Biology*, *25*(10), R407–R408.

Benjamin, D., et al. (2011). Rapamycin passes the torch: A new generation of mTOR inhibitors. *Nature Reviews Drug Discovery*, *10*, 868.

Chen, H. P, et al. (2015). Histone deacetylases and mechanisms of regulation of gene expression. *Critical Reviews in Oncogenesis*, *20*(1–2), 35–47.

Coleman, K. G., & Crews, C. M. (2018). Proteolysis-targeting chimeras: Harnessing the ubiquitin-proteasome system to induce degradation of specific target proteins. *Annual Review of Cancer Biology*, *2*(1), 41–58.

Czlapinski, J. L., et al. (2008). Conditional glycosylation in eukaryotic cells using a biocompatible chemical inducer of dimerization. *Journal of the American Chemical Society*, *130*(40), 13186–13187.

Duan, L., et al. (2015). Optogenetic control of molecular motors and organelle distributions in cells. *Chemistry & Biology*, *22*(5), 671–682.

Encell, L. P., et al. (2012). Development of a dehalogenase-based protein fusion tag capable of rapid, selective and covalent attachment to customizable ligands. *Current Chemical Genomics*, *6*, 55–71.

Funabiki, H., & Wynne, D. J. (2013). Making an effective switch at the kinetochore by phosphorylation and dephosphorylation. *Chromosoma*, *122*(3), 135–158.

Geißler, D., et al. (2005). (Coumarin-4-yl)methyl esters as highly efficient, ultrafast phototriggers for protons and their application to acidifying membrane surfaces. *Angewandte Chemie International Edition*, *44*(8), 1195–1198.

Hüll, K., et al. (2018). In vivo photopharmacology. *Chemical Reviews*, *118*(21), 10710–10747.

Khandelia, P., et al. (2011). Streamlined platform for short hairpin RNA interference and transgenesis in cultured mammalian cells. *Proceedings of the National Academy of Sciences of the United States of America*, *108*(31), 12799.

Klán, P., et al. (2013). Photoremovable protecting groups in chemistry and biology: Reaction mechanisms and efficacy. *Chemical Reviews*, *113*(1), 119–191.

Li, J., et al. (2014). Rapamycin: One drug, many effects. *Cell Metabolism, 19*(3), 373–379.
Los, G. V., et al. (2008). HaloTag: A novel protein labeling technology for cell imaging and protein analysis. *ACS Chemical Biology, 3*(6), 373–382.
Mastro, A. M., et al. (1984). Diffusion of a small molecule in the cytoplasm of mammalian cells. *Proceedings of the National Academy of Sciences of the United States of America, 81*(11), 3414–3418.
Miyamoto, T., et al. (2012). Rapid and orthogonal logic gating with a gibberellin-induced dimerization system. *Nature Chemical Biology, 8*(5), 465–470.
Ohana, R. F., et al. (2015). Deciphering the cellular targets of bioactive compounds using a chloroalkane capture tag. *ACS Chemical Biology, 10*(10), 2316–2324.
Robinson, M. S., et al. (2010). Rapid inactivation of proteins by rapamycin-induced rerouting to mitochondria. *Developmental Cell, 18*(2), 324–331.
Spencer, D. M., et al. (1993). Controlling signal transduction with synthetic ligands. *Science, 262*(5136), 1019.
Stanton, B. Z., et al. (2018). Chemically induced proximity in biology and medicine. *Science, 359*(6380), eaao5902.
Weinrich, T., et al. (2017). Synthesis of a cytidine phosphoramidite with protected nitroxide spin label for EPR experiments with RNA. *European Journal of Organic Chemistry, 2017*(3), 491–496.
Zhang, H., et al. (2017). Optogenetic control of kinetochore function. *Nature Chemical Biology, 13*, 1096.

CHAPTER FIFTEEN

Site-specific antibody fragment conjugates for targeted imaging

Robert Maloney[†], **Zakey Yusuf Buuh**[†], **Yue Zhao**[†], **Rongsheng E. Wang***
Department of Chemistry, Temple University, Philadelphia, PA, United States
*Corresponding author: e-mail address: rosswang@temple.edu

Contents

Abstract

This work details the use of amber suppression-mediated genetic incorporation of unnatural amino acids (UAAs), specifically p-azido-L-phenylalanine (pAzF) and *p*-acetyl-L-phenylalanine (pAcF), to develop site-specifically labeled antibody Fab fragments. These antibody fragment conjugates represent a novel class of imaging agents with optimal stability, efficacy, and pharmacological properties, which have demonstrated

[†] These authors contributed equally.

Methods in Enzymology, Volume 638
ISSN 0076-6879
https://doi.org/10.1016/bs.mie.2020.02.023

promising potential for probing and understanding the *in vivo* bio-distributions of protein targets of interest. This chapter provides general guidelines for preparing these Fab conjugates, and details of follow-up bioassays such as single-agent based positron emission tomography (PET) imaging of immune-checkpoint protein PD-L1, and the use of GCN4-mediated switchable antibody conjugates for near-infrared fluorescent imaging of cancer-related biomarkers.

1. Introduction

Conjugation with small molecules (Hernandez et al., 2016; Lewis Phillips et al., 2008; Wang et al., 2015), oligomers (Sano, Smith, & Cantor, 1992), and proteins (Gill, Forouzandeh, Rahbarizadeh, Ramezani, & Rasaee, 2006), etc. has endowed versatile functions to antibodies, resulting in antibody conjugates capable of being used for broad applications such as immunoassay, imaging-based detection, diagnosis, and therapeutic development (Lyu et al., 2018). In addition to the development of antibody-drug conjugates (ADCs), which have emerged as promising immunotherapeutics (Lewis Phillips et al., 2008; Ricart, 2011; Wang et al., 2015), antibodies conjugated with radioisotopes or fluorophores are increasingly pursued for positron emission tomography (PET) or near-infrared fluorescence (NIRF) imaging, thereby becoming a critical component of modern precision medicine (Adumeau, Sharma, Brent, & Zeglis, 2016a, 2016b; Hernandez et al., 2016). While effective, most conjugates were based on full-length IgG antibodies which have limited tissue and tumor penetration (Chatterjee et al., 2016; England et al., 2018; Heskamp et al., 2015; Hettich, Braun, Bartholoma, Schirmbeck, & Niedermann, 2016). On top of this, common conjugates including FDA-approved ADCs have been prepared *via* random conjugation with naturally occurring lysine and cysteine residues, resulting in heterogeneous constructs with varying efficacies, stabilities, and suboptimal pharmacological properties (Adumeau et al., 2016a, 2016b; Schumacher, Hackenberger, Leonhardt, & Helma, 2016). Early studies, such as THIOMABs, utilized antibodies bearing engineered cysteine residues to prepare site-specific immunoconjugates (Adumeau et al., 2016a), and revealed higher signal-to-noise imaging intensity ratios *in vivo* in comparison to random conjugates, demonstrating the positive effects of the precise control of conjugation sites and stoichiometry. Nevertheless, most cysteine conjugation methods rely on maleimdyl thioester bonds that possess suboptimal biological stability, thereby limiting the broad applications of the THIOMAB approach (Adumeau et al., 2016a). Other site-specific strategies such as expressed protein ligation (EPL) could result in relatively stable constructs (Muir, 2003). Yet, the ligation site is usually limited to the C-terminal, which may not generate the most stable conjugates, and makes it challenging for multi-site conjugation (Muir, 2003; Wang, Xie, & Schultz, 2006).

Amber suppression-mediated genetic incorporation of unnatural amino acids (UAA), on the other hand, has recently emerged as a promising site-specific protein conjugation strategy, due to the stably bonded constructs and flexibility in site-selections it offers (Adumeau et al., 2016b). Through site-directed mutation, the genetic code of the target site can be changed to the amber stop codon (TAG) (Liu & Schultz, 2010; Wang et al., 2006). During the translation of recombinant proteins, a pair of orthogonal tRNA and aminoacyl tRNA synthetase were employed to charge the amber codon (TAG) with the desired UAA (Wals & Ovaa, 2014) (Fig. 1). This pair has been evolved to recognize UAA specifically, and function orthogonally to

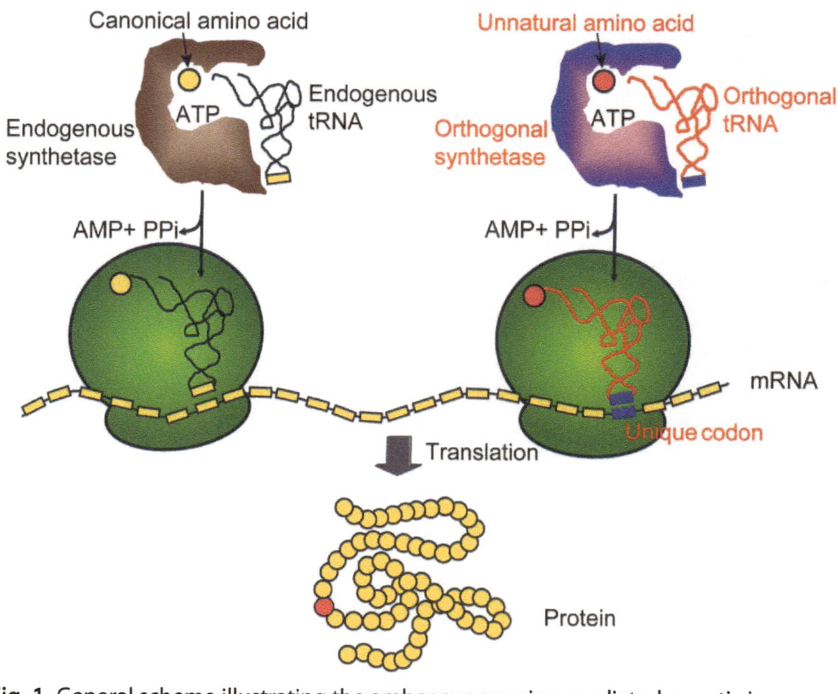

Fig. 1 General scheme illustrating the amber suppression-mediated genetic incorporation of unnatural amino acid. *Adapted with permission from Wang, Q., Parrish, A. R., & Wang, L. (2009). Expanding the genetic code for biological studies.* Chemical Biology, *16(3), 323–336. Copyright (2009) Elsevier Ltd.*

intrinsic tRNA/aminoacyl transferases (Liu & Schultz, 2010; Wang et al., 2006). This novel approach allows the generation of antibodies which site-specifically incorporated UAAs such as *p*-acetyl-L-phenylalanine (pAcF) and p-azido-L-phenylalanine (pAzF) that can be coupled to molecules of interest *via* a stable linkage (Adumeau et al., 2016b; Cao et al., 2015; Lyu et al., 2018; Wissler et al., 2019). Protein therapeutics based on the UAA conjugates have displayed better *in vivo* performance than the cysteine conjugates, in terms of efficacy, stability, and toxicology (Jackson et al., 2014; Liu & Schultz, 2010; Tian et al., 2014).

More recently, imaging studies using probes comprised of UAA-mediated conjugation have been increasingly reported (Lyu et al., 2018; Wissler et al., 2019; Wu, Zhu, et al., 2016), including our lab's development of UAA-based site-specific antibody Fab fragment conjugates. Along our efforts to develop a class of immunoconjugates based novel imaging agents, we were particularly interested in Fab fragments, which retain the binding of the antibody but are much smaller in size, with the Fc effector domain removed (Lyu et al., 2018; Wissler et al., 2019). The small size of Fab fragments led to shorter *in vivo* circulation half-lives, which incur less background signals for imaging (Badescu et al., 2014). Compared to other antibody fragments such as ScFv, Fab still possess constant regions that are ideal for modifications and conjugations without compromising Fab's binding with antigens (Lyu et al., 2018). We have since developed a site-specific αPD-L1 Fab conjugate with the isotope chelator 1,4,7-triazacyclononane-N,N′,N″-triacetic acid (NOTA), based on the incorporation of pAzF (Wissler et al., 2019) (Fig. 2). This immuno-PET tracer was used to capture the significant *in vivo* expression of PD-L1 in mouse tissues such as brown adipose and spleen (Wissler et al., 2019). Given the complications in antibody conjugation regarding yields, time, and labor (Young & Schultz, 2010), we have also developed a switchable antibody conjugate strategy (Lyu et al., 2018) (Fig. 2), which is focused on synthesizing and optimizing only one site-specific conjugate (αGCN4-Fab conjugate with Cy7 dye based on UAA pAcF) that is capable of recognizing GCN4 as a switch, thereby allowing for variability in the GCN4-fused IgG warhead. This GCN4-mediated switchable conjugate has been demonstrated to allow for NIRF imaging of different tumor types *in vivo* (Lyu et al., 2018). Here we describe the detailed protocols for the synthesis of UAA-based site-specific Fab conjugates with NOTA and Cy7, as well as a multitude of *in vitro* and *in vivo* assays to confirm the utility of this site-specific conjugation platform for molecular imaging.

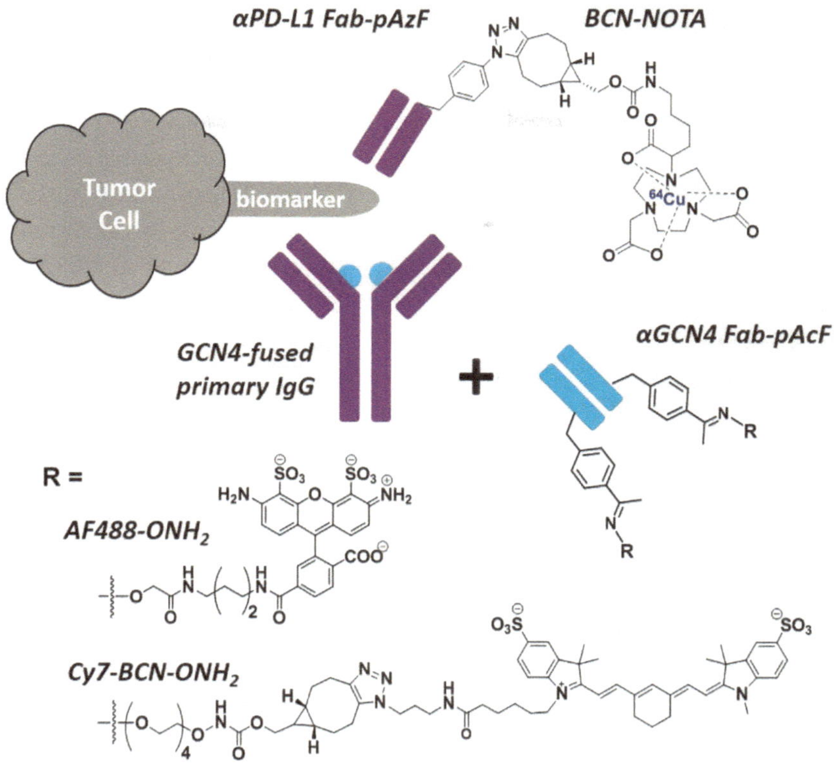

Fig. 2 Summary of developed site-specific Fab conjugates based on the incorporation of UAA (pAzF or pAcF). Blue dot: GCN4 peptide fused onto primary antibodies.

2. Genetic incorporation of UAA to fab fragments

2.1 Construction of plasmids for wild type and mutant Fab's

The sequence information for the variable and constant regions of αPD-L1 (avelumab) and αGCN4 antibodies can be obtained through literature (Zahnd et al., 2004) or patents (Nastri et al., 2014), which is the key for antibody plasmid construction. Plasmids for the expression of antibody Fab fragments can be feasibly synthesized *via* molecular cloning on the commercially available pBAD plasmid system that adopts a StII signaling peptide. The coding sequences can be directly synthesized from IDT, or quickly constructed by overlapping PCR of the fragments, followed by sub-cloning into the pBAD vector using a pair of restriction enzymes (*Nhe*I, *Apa*I) and T4-ligase-mediated ligation. Site-specific mutation to the amber codon can then be efficiently introduced *via* directed mutagenesis using the kit from New England Biolabs or Agilent Technologies. For UAA incorporation, the tyrosyl aminoacyl-tRNA synthetase (TyrRS)/ $tRNA^{Tyr}_{CUA}$ pair from the archaea *Methanococcus jannaschii* (Mj) was introduced to the *E. coli* expression machinery, which is orthogonal to other intrinsic tRNA synthetase/tRNA pairs (Wang, Brock, Herberich, & Schultz, 2001). The mutant tRNA synthetases have been evolved to selectively aminoacylate their cognate tRNAs with pAcF (Kim et al., 2013), or pAzF (Young et al., 2011). Thus, a separate pUltra plasmid that encodes the Mj-TyrRS/tRNA pair for pAcF (Kim et al., 2013) or pAzF (Young et al., 2011) was directly used, along with the pBAD plasmid that encodes the Fab fragment during protein expression.

2.1.1 Equipment and materials

- PCR Thermocycler (Bio-Rad)
- Sub-Cell GT Electrophoresis Cell (Bio-Rad)
- NanoDrop Lite Spectrometer (Thermo Scientific)
- Gene Pulser Xcell Electroporator (Bio-Rad)
- MaxQ 8000 Orbital Shaker (Thermo Scientific)
- Heratherm Microbiological Incubator (Thermo Scientific)
- Allegra Centrifuge (X-14R) (Beckman Coulter)
- pBAD plasmid, pULTRA plasmid (Thermo Scientific)
- Restriction enzymes, CutSmarter Buffer (New England BioLabs)
- T4 ligase, ligase buffer (New England BioLabs)
- DNA Clean & Concentrator (Zymo Research)
- Zymoclean Gel DNA Recovery Kit (Zymo Research)
- *Quick*-DNA Miniprep Kit (Zymo Research)
- Q5 Site-Directed Mutagenesis Kit (New England BioLabs)
- SOC Medium (Fisher Scientific)
- Luria-Bertani Broth and agar (Sigma Aldrich)
- Glycerol (Fisher Scientific)
- Antibiotics: ampicillin, spectinomycin (Sigma Aldrich)
- *Escherichia coli* strain TOP10 (Sigma Aldrich)

2.1.2 Protocol

1. Mix the pBAD template plasmid (~5 μg) with 2 μL *Apa*I in the 100 μL reaction solution containing the CutSmarter buffer. Incubate the mixture at room temperature for 2 h. Add approximately 2 μL of the other restriction enzyme, *Nhe*I, and incubate the mixture at 37 °C overnight.

2. Assemble the DNA strands corresponding to the coding sequences of the target proteins (αPD-L1 or αGCN4) in the order of "Light Chain Variable Domain (VL)—Light Chain Constant Domain (CL)—Heavy Chain Variable Domain (VH)—Heavy Chain Constant Domain 1(CH1)" by overlapping PCR, and enzyme digest the strands following the same procedure as step 1.
3. Purify products from steps 1–2 with DNA agarose gel electrophoresis, and extract the DNA *via* the Zymoclean Gel DNA recovery kit. Measure the concentrations accordingly on NanoDrop.
4. Mix approximately 100 ng each of the enzyme digested pBAD template and the antibody coding strand. To the final ligation solution (20 μL), add about 1 μL of T4 ligase and 2 μL of the 10 × ligation buffer. After ligation at room temperature for 1 h, extract the DNA mixture containing possible products by the DNA Clean & Concentrator kit.
5. Elute the ligated plasmid out from the DNA concentrator column with 6 μL water, and transform 1 μL of the elutes into competent cells made from TOP 10 *E. coli* cells (Warren, 2011) *via* gene pulser electroporator. Recover the cells in 500 μL SOC medium with continuous shaking at 250 rpm and 37 °C for 1 h, and plate 50 μL of the medium on an Ampicillin containing LB agar plate.
6. After overnight incubation at 37 °C, pick the colonies from the plate and amplify each of them in 10 mL of Ampicillin containing LB media. Mix 0.5 mL of each saturate culture with 0.5 mL of 50% glycerol, and store them at −80 °C as stock solutions. Extract the plasmids for the rest of the culture with the *Quick*-DNA Miniprep Kit and validate the sequences by Sanger sequencing.
7. For site-directed mutation to amber codon, design the primers containing the TAG codon *via* the online software (NEBaseChanger). Mix ~20 ng of the vector that has been correctly inserted the Fab protein sequence in the PCR tube with forward and reverse primers (final concentration 0.5 μM) in the Q5 hot start high-fidelity 2 × Master Mix. Perform 25 cycles of PCR amplification, followed by the KLD reaction.
8. Transform 5 μL of the KLD reaction mixture to competent cells, and perform the recovery, plating, and colony picking as described above. Validate the incorporation of TAG codon at the desired site by sanger sequencing.
9. If needed, steps 7–8 can be repeated to introduce more than one amber codon in the antibody sequence. With the incorporation of more than

one equivalent of UAA, a Fab fragment can be conjugated with multi-equivalents of payloads, which is the case for the αGCN4 switchable Fab conjugate (Lyu et al., 2018).

10. For future expression of mutant Fabs, transform both the correct pBAD plasmid (with amber codon mutation) and the pULTRA plasmid coding the Mj-TyrRS/tRNA pair for UAA incorporation into the *E. coli* competent cells (e.g., BL21). Plate the cells onto the agar plate that contains antibiotics for both plasmids. The next day, pick the colonies and amply them in the LB media with both antibiotics added. Extract the plasmids from part of the saturated culture by miniprep, and validate the two plasmids through DNA agarose gel electrophoresis. Mix the rest of the culture with 50% glycerol and store at −80 °C as the stock.

2.2 Fab expression and purification

All the engineered Fab fragments are expected to be readily expressible in *E. coli*. Compared to wild type, some modifications in the expression procedures are necessary in order to incorporate UAAs by amber suppression. Briefly, the nutritious cell culture media should be supplemented with UAAs. Antibiotic spectinomycin corresponding to the pULTRA plasmid should be added to the media, in addition to ampicillin that corresponds to the pBAD plasmid. Both antibiotics are required to selectively maintain the viability of cells that carry both pBAD and pUltra plasmids. During the induction of protein expression, IPTG is required to induce the T7 promoter in the pULTRA plasmid, while L-arabinose is needed to initiate the protein expression from the pBAD vector. The following protocol is focused on the expression of mutant Fab fragments. In practice, the UAA-based site-specific labeling also facilitates our exploration of various sites on the antibody that could lead to different stability, thereby resulting in varied expression yields (Lyu et al., 2018; Wissler et al., 2019). In that sense, experimental screening (Lyu et al., 2018; Wissler et al., 2019), along with *in silico* prediction by the RosettaBackrub algorithm (Wissler et al., 2019) should lead to the optimal mutation site for a given UAA.

2.2.1 Equipment and materials

- MaxQ 8000 Orbital Shaker (Thermo Scientific)
- Biowave CO 8000 Cell Density Meter (WPA)
- Allegra Centrifuge (X-14R) (Beckman Coulter)
- NanoDrop Lite Spectrometer (Thermo Scientific)

- Nalgene Rapid-Flow Sterile Filter Units with 0.2 μm PES Membrane (Fisher Scientific)
- CaptureSelect IgG-CH1 Affinity Matrix (Thermo Scientific)
- Amicon Ultra Centrifugal Filters, 10K MWCO (EMD Millipore)
- 2 × YT medium (Sigma Aldrich)
- Antibiotics: ampicillin, spectinomycin (Sigma Aldrich)
- L-(+)-Arabinose (Alfa Aesar)
- Isopropyl β-D-1-thiogalactopyranoside (IPTG) (VWR)
- Lysozyme (VWR)
- Dulbecco's phosphate-buffered saline (DPBS), without calcium and magnesium (Corning)

2.2.2 Protocol

To minimize the possibility of contamination, steps 1–4 should be performed in the area sterilized by 70% ethanol and close to the flame of natural gas. Protein extraction and purification (after step 5) should be performed in a cold room (4 °C) or on ice.

1. Inoculate the *E. coli* stock containing the desired plasmids in 8 mL 2 × YT media with 100 μg/mL ampicillin and 100 μg/mL spectinomycin. Incubate the mixture by shaking at 37 °C, 250 rpm for overnight.
2. Dilute the start culture in 800 mL 2 × YT media with 100 μg/mL ampicillin, 100 μg/mL spectinomycin, and 1 mM UAA (pAzF or pAcF). Continue to shake the cell culture at 37 °C, 250 rpm.
3. Monitor the cell growth phase by measuring OD_{600} using the Cell Density Meter, until the value reaches 0.8–1.0.
4. Induce the protein expression by adding 0.2% L-arabinose and 0.5 mM IPTG (all are final concentrations). Continue the expression at 30 °C, 250 rpm, for 24 h.
5. Harvest cells by centrifuging at 4 °C, 500 rpm, for 30 min. Discard the supernatant and freeze the cell pellets at −80 °C for at least 3 h.
6. Thaw the cell pellets and re-suspend them in 80 mL of the periplasmic lysis buffer (20% sucrose, 30 mM Tris (pH 8.0), 1 mM EDTA) that has been freshly dissolved with 0.2 mg/mL lysozyme. Mix thoroughly by pipetting up and down. Incubate the cell suspension at 37 °C, 250 rpm, for 20 min.
7. Extract the periplasmic lysates by centrifuging the suspension at 4 °C, 9000 rpm for 30 min.
8. Filter the supernatant through 0.2 μm membrane to remove debris.

9. Prepare the CH1 affinity column by equilibrating 1.5 mL IgG-CH1 Affinity Matrix beads slurry in PBS buffer. Allow the filtered periplasmic lysates to flow through the CH1 affinity column twice under gravity to capture Fab fragments.
10. Wash the CH1 affinity column twice with PBS buffer. Elute Fab fragments with 6 mL of 100 mM aqueous glycine solution (pH 2.8), and collect the eluent in a centrifuge tube with 600 μL 1 M Tris buffer (pH 8.0) pre-added into the bottom of the tube for neutralization.
11. Concentrate the protein containing eluent using an Amicon Ultra centrifugal filter and exchange the buffer into PBS. Determine the Fab concentration on NanoDrop.
12. Aliquot and snap-freeze the Fab proteins by liquid nitrogen. Store the aliquots at −80 °C for long-term storage.

3. Linker compound synthesis

A variety of payloads can be conjugated site-specifically to the mutant Fab fragments that possess the UAA as a handle. Depending on the chemical reactivity of the incorporated UAA, different bioorthogonal conjugation chemistries can be employed. For example, pAzF bears electron-deficient aryl azides, and is ideal for the strain-promoted azide-alkyne cycloaddition (SPAAC) conjugation with strained alkyne derivatized linker compound at neutral pH (Dommerholt et al., 2014; Wissler et al., 2019). In the process of developing a site-specific immuno-PET tracer, we have designed and synthesized a NOTA derivative that is capable of chelating ^{64}Cu, a metallic radionuclide for PET, and was tethered to the strained bicyclo[6.1.0]non-4-yne (BCN) linker for SPAAC conjugation with the pAzF containing αPD-L1 Fab mutant (Wissler et al., 2019). The NOTA derivative design was inspired by the enhanced chelation capabilities of hexadentate NOTA with branched substitution (3p-C-NOTA) (Wu, Kang, et al., 2016). On the other hand, pAcF is a typical UAA tag for coupling with molecules of interest *via* a stable oxime linkage (Axup et al., 2012; Hutchins et al., 2011; Kazane et al., 2012; Kularatne et al., 2014). To develop the switchable antibody conjugate, we have conjugated the pAcF containing αGCN4 Fab with alkoxy-amine derivatized Alexa Fluor 488 (AF488), or the alkoxy-amine modified PEGylated BCN linker in tandem with the sulfo-cyanine 7 (Cy7) azide (Lyu et al., 2018). Given that all alkoxy-amine related dyes and linkers are commercially available, this section will focus on the preparation of the NOTA-BCN payload (Fig. 3) for PET imaging.

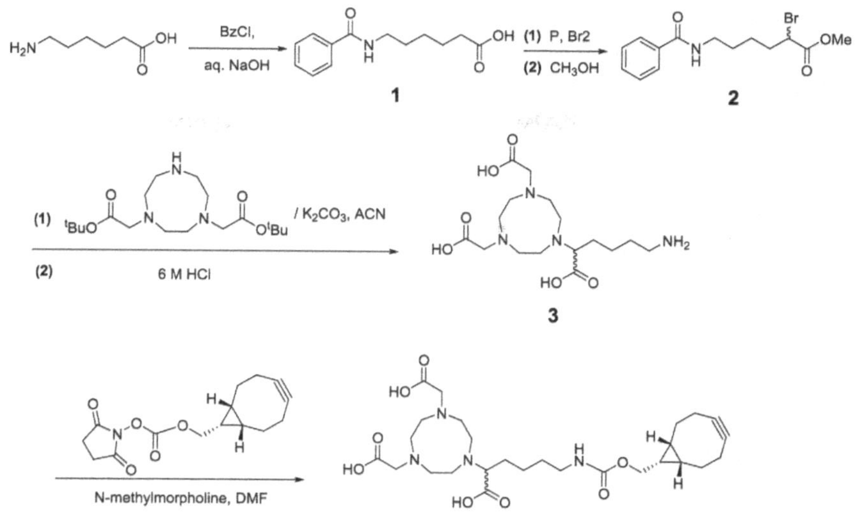

4

Fig. 3 Synthetic scheme of 1,4,7-triazacyclononane-N,N′,N″-triacetic acid (NOTA) derivative with a (1*R*, 8*S*, 9*S*)-bicyclo[6.1.0]non-4-yn-9-ylmethylcarbamate (BCN) linker. *Adapted with permission from Wissler, H. L., Ehlerding,E. B., Lyu, Z., Zhao, Y., Zhang, S., Eshraghi, A., et al. (2019). Site-specific immuno-PET tracer to image PD-L1,* Molecular Pharmaceutics, 16, *2028–2036. Copyright (2019) American Chemical Society.*

3.1 Equipment and materials

- Chemical Fume Hood (SafeAire)
- Stirring Hot Plate (Thermo Fisher)
- Handheld UV Lamp (Chemglass)
- Thin-layer chromatography plates (VMR International)
- Rotary Evaporator System (Buchi R-200)
- Agilent 1100 series, high-resolution LC–MS
- 400 and 500 MHz NMR (Bruker Advance)
- Waters 1525 series HPLC equipped with a XBridge Prep C18 column (5 μm, $19 \times 250\,mm^2$)
- LABCONCO FreeZone 4.5 Lyophilizer
- Silica Gel (Fisher Scientific 50–60 Å pore diameter)
- Round Bottom Flasks (ChemGlass)

3.2 Reagents

- Organic solvents (methanol, ethyl acetate, hexane, diethyl ether, acetonitrile, *N,N*-dimethylformamide) (Fisher Scientific/VWR International)
- HPLC-grade water (Fisher Scientific)

- Sodium thiosulfate (Fisher Scientific)
- Sodium bicarbonate (VWR International)
- Potassium carbonate (VWR International)
- N-methyl morpholine (VWR International)
- Bromine liquid (EMD Millipore)
- Phosphorus, red stabilized for synthesis (MilliporeSigma)
- 1,4,7-Triazacyclononane-1,4-bis(t-butyl acetate) (Macrocyclics, Inc)
- Hydrochloric acid, 36.5–38.0% (Fisher Scientific)
- BCN-succinimidyl ester (Sigma Aldrich)
- Benzamidohexanoic acid (Fisher Scientific)
- Benzoyl chloride (Fisher Scientific)
- Sodium hydroxide (VWR International)
- Hydrochloric acid, 36.5–38.0% (Fisher Scientific)

3.3 Protocol

The number of the following steps refers to the intermediates given in Fig. 3. Care should be taken when handling phosphorus (toxic, flammable) and liquid bromine (toxic, corrosive) in step 2. All steps should be performed in a chemical fume hood, with the wearing of appropriate personal protective equipment.

1. Add 6-aminohexanoic acid (4 g, 0.0341 mol) to a round-bottom flask and dissolve it with 1 M NaOH in water (41 mL). Place the reaction mixture into an ice bath. Add benzoyl chloride (4.76 mL, 0.0410 mol) dropwisely and stir the reaction mixture overnight. Once completed, wash the aqueous layer with diethyl ether (2 × 40 mL) to extract unreacted benzoyl chloride. Acidify the aqueous layer with 1 M HCl (~45 mL) to pH ~3. Extract the product with ethyl acetate (3 × 40 mL) and vacuum concentrate the organic layer to obtain compound **1** (70% yield, 5.29 g) as a white solid.
2. Mix compound **1** (3 g, 0.0128 mol) and dry red phosphorus (0.53 g, 0.0172 mol) in a flame-dried 100 mL round-bottom flask. Cool the mixture down to −15 °C in an ice-salt bath. Add liquid bromine (8.16 g, 0.102 mol) dropwisely, remove the ice bath and reflux the reaction mixture in a silicon oil bath overnight at 100 °C. Once completed, remove the flask from the oil bath and cool the mixture to room temperature and then 0 °C, followed by adding 20 mL of methanol dropwisely to quench the reaction. To form the desired methyl

ester (compound **2**), reflux the reaction mixture at 100 °C for 30 min. Upon reaction completion as monitored by TLC, vacuum concentrate, and re-suspend the mixture in 100 mL of ethyl acetate. Wash the organic phase with saturated sodium bicarbonate (1 × 30 mL), saturated sodium thiosulfate (2 × 30 mL), and brine (1 × 30 mL). After washing, concentrate the organic layer to a crude oil *via* rotary evaporation and purify the product *via* flash column chromatography (5:1, hexane:ethyl acetate) to obtain compound **2** at an 85% yield (3.57 g).

3. Add compound **2** (82.3 mg, 0.243 mmol) along with 1,4,7-triazacyclononane-1,4-bis(t-butyl acetate) (72.6 mg, 0.203 mmol), and K_2CO_3 (33.6 mg, 0.243 mmol) to 5 mL dry acetonitrile in a flame-dried round bottom flask under nitrogen atmosphere. Reflux the reaction for 48 h, and then concentrate the mixture under reduced pressure. Redissolve the crude mixture in 4 mL of 6 M aqueous HCl and reflux the solution for 24 h. Purify the hydrolyzed product *via* HPLC to yield compound **3** in ~50% yield (38 mg) over two steps.
4. To a flame-dried flask, dissolve compound **3** (6.9 mg, 0.0185 mmol) and BCN-Succinimidyl ester (5.4 mg, 0.0185 mmol) in anhydrous DMF. Add *n*-methylmorpholine (3.75 mg, 0.037 mmol) dropwisely and then stir the reaction mixture for 4 h at room temperature, at which point the succinimidyl ester should be fully converted. Concentrate the reaction mixture under vacuum, and purify the product *via* HPLC to afford 3.2 mg of final compound **4** (NOTA-BCN, 31% yield) as a white solid.

4. Site-specific fab conjugation and purification

Site-specific conjugation between the UAA on the Fab proteins and the linker compound is performed under mild reaction conditions and can be classified into two major categories: SPAAC for pAzF incorporated antibodies at neutral pH, or Oxime coupling for pAcF incorporated antibodies under mildly acidic pH. For both reactions, at most 10% of DMSO can be added into the final reaction buffer in order to promote the solubility of chemical linkers. Addition of any reducing agents may risk the stability of disulfide bonds within the protein structure. Lastly, the concentration of linker compound should be restrained (usually ~10 equivalents relative to the protein) to prevent non-specific side reactions, and to facilitate follow-up purification.

4.1 Equipment and materials

- Allegra Centrifuge (X-14R) (Beckman Coulter)
- Fisherbrand AccuSpin Micro 17R Microcentrifuges (Fisher Scientific)
- NanoDrop Lite Spectrometer (Thermo Scientific)
- Heratherm Microbiological Incubator (Thermo Scientific)
- Amicon Centrifugal Filter (EMD Millipore)
- AKTA fast protein liquid chromatography (FPLC) system (Amersham Biosciences)
- Superdex 200 Increase 10/300 GL (GE Healthcare)
- Zeba Spin Desalting Column (Thermo Scientific)
- Dulbecco's phosphate-buffered saline (DPBS), without calcium and magnesium (Corning)
- Sodium acetate (VWR International)

4.2 Protocol

1. Thaw the mutant Fab stock solution on ice. Measure the stock concentrations again by NanoDrop, as freezing and thawing sometimes may lead to protein loss. Concentrate and/or buffer exchange with the Fab fragments using an Amicon centrifugal filter. Since the storage buffer is PBS that is ideal for SPAAC conjugation, directly concentrate the Fab fragments to 2.5 mg/mL for coupling between pAzF and the NOTA-BCN compound. For conjugation of pAcF incorporated Fab with aminooxy-derivatized AF488 or BCN linker compound, exchange the PBS stock buffer into 50 mM NaOAc reaction buffer (pH 4.5).
2. Transfer the concentrated and/or buffer exchanged antibody into an Eppendorf tube. Add 10 equivalents of the desired linker compound. For oxime coupling, add to the mixture acetyl hydrazide (1000 equivalents) as a catalyst. Mix the reaction mixture gently by pipetting or flipping.
3. Incubate the reaction mixture at 37 °C for 6 h or 24 h, for pAzF or pAcF-mediated conjugation, respectively.
4. Before work-up, ESI-MS characterization of the crude mixture is necessary to ensure the complete reaction conversion, since it is challenging to separate conjugates with unconjugated antibodies. Purify the fully converted mixture by FPLC. Carefully collect the eluted fractions to avoid contamination by unreacted linker compounds. Alternatively, pass the mixture through at least two rounds of Zeba spin column with PBS as the equilibrium buffer. The purified conjugates will need to be revalidated by ESI-MS, followed with either subsequent *in vitro* and *in vivo* assays or snap-freezing in liquid nitrogen for long time storage.

5. *In vitro* characterization of the fab conjugates

With the site-specific Fab conjugates prepared in Section 4, a series of *in vitro* assays (Fig. 4) can be performed to evaluate their purity (by sodium dodecyl sulfate—polyacrylamide gel electrophoresis (SDS-PAGE)), identity (by electrospray ionization—mass spectrometry (ESI-MS)), and binding affinity (by enzyme-linked immunosorbent assay (ELISA) and/or flow cytometry). It is important to compare the antigen binding affinity between the wild type Fab and the mutants, as well as the Fab conjugates, in order to make sure that neither the incorporation of UAA nor the payload conjugation compromises the antibody's binding.

5.1 Equipment and materials

- Direct-Q 3 UV Water Purification System (Millipore Sigma)
- Mini Gel Tank (Thermo Fisher)
- KJ-201BD Orbital Shaker (KangJian)
- Q Exactive Plus Orbitrap Mass Spectrometer (Thermo Fisher)
- Synergy H1 Plate Reader (BioTek)
- Allegra Centrifuge (X-14R) (Beckman Coulter)
- LSR Fortessa Flow Cytometer (BD Biosciences)
- NuPAGE Precast Gels (Thermo Fisher)
- PageRuler Prestained Protein Ladder (Thermo Fisher)
- LDS Sample Buffer (Novex)
- MES SDS Running Buffer (Novex)
- Coomassie Brilliant Blue (CBB) R-250 (Thermo Scientific)
- Methanol (Fisher Scientific)
- Glacial acetic acid (Fisher Scientific)
- Flat-bottom 96-well plates (Corning)
- Phosphate-buffered saline (PBS) buffer (Corning)
- Bovine serum albumin (Thermo Fisher)
- QuantaBlu Fluorogenic Peroxidase Substrate Kit (Thermo Fisher)
- Anti-Human Lambda Light Chain Antibody HRP-Labeled (MilliporeSigma)
- Anti-Human Kappa Light Chain Antibody HRP-Labeled (MilliporeSigma)
- Recombinant Human ErbB2/Her2 Fc Antigen (R&D Systems)
- Recombinant Human CD19 Fc Antigen (R&D Systems)
- SK-BR-3 Breast Cancer Cell Line (ATCC)
- Ramos Burkitt's Lymphoma Cell Line (ATCC)
- DMEM Cell Culture Medium (Corning Cellgro)

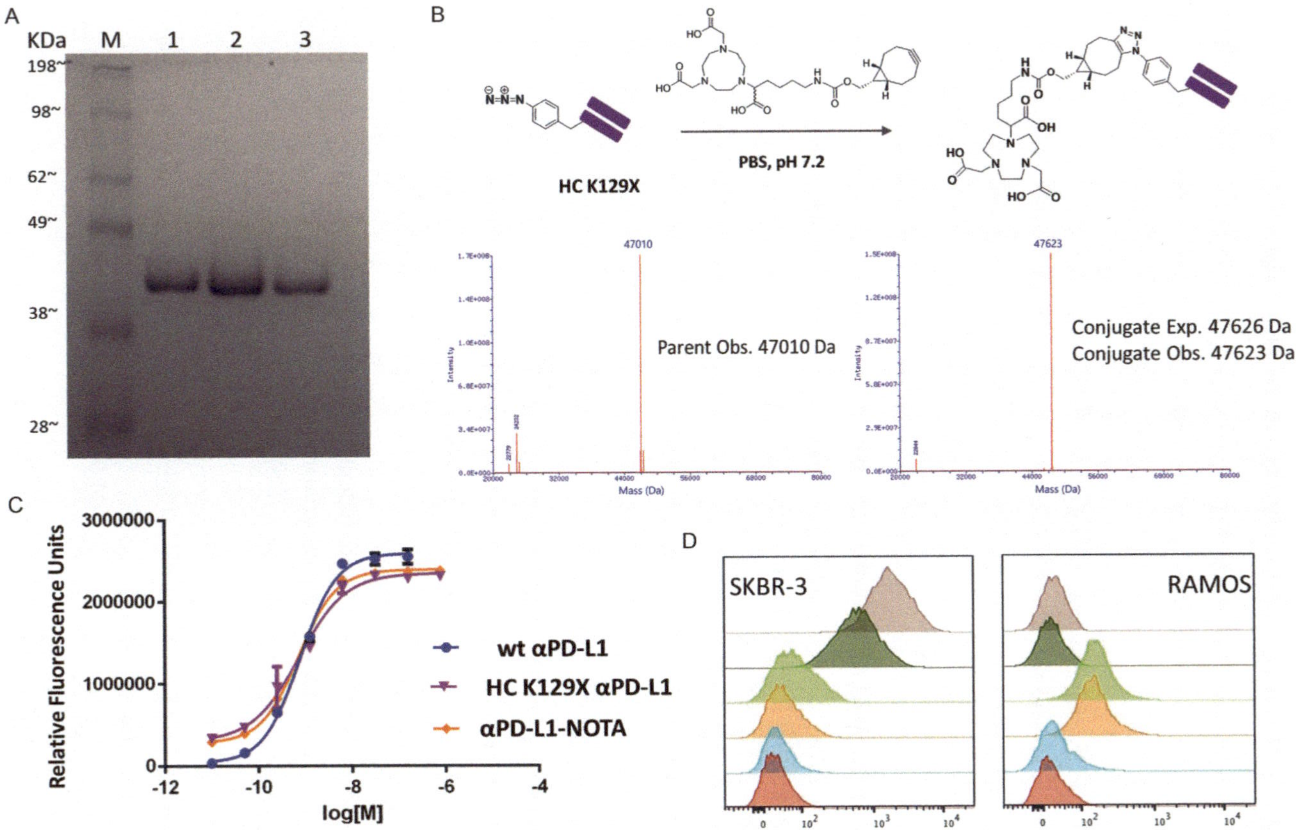

Fig. 4 See legend on opposite page.

- RPMI-1640 Cell Culture Medium (Corning Cellgro)
- Fetal bovine serum (Fisher Scientific)
- Antibiotic-Antimycotic 100 × Mix (Corning)
- 0.05% trypsin, 0.53 mM EDTA solution (Corning)
- 32% paraformaldehyde solution (Electron Microscopy Sciences)

5.2 Protocol

5.2.1 SDS-PAGE and ESI-MS

1. Prepare samples for SDS-PAGE in Eppendorf tubes. For samples without the use of Dithiothreitol (DTT) as a reductant, add 2 μg of the desired conjugate, 5 μL of LDS sample loading buffer (4 ×) and add 13 μL of Milli-Q water to dilute the sample to a total volume of 20 μL. For samples with DTT, add 3 μg of the desired Fab conjugate, 2 μL of 100 μM DTT, 5 μL of LDS sample loading buffer, and 10 μL of Milli-Q to dilute the sample to a total volume of 20 μL. All samples are mixed thoroughly *via* pipetting and heated at 90 °C for 2–5 min prior to loading.
2. Insert a piece of 4–12% Bis-Tris Plus, 10-well pre-cast gel into a Mini Gel Tank. Add 1 × MES SDS running buffer to fully submerge the gel.
3. Load the PageRuler™ Prestained protein ladder, followed by each antibody or antibody conjugate sample into individual wells.

Fig. 4 *In vitro* characterization of the site-specific Fab conjugates. (A) SDS-PAGE analysis of αPD-L1 Fab wild type (lane 1), αPD-L1 Fab mutant (HC K129X, X = pAzF) (lane 2), and αPD-L1 Fab-NOTA conjugate (HC K129X, X = pAzF-BCN-NOTA) (lane 3); (B) ESI-MS analysis of αPD-L1 Fab before (left) and after conjugation (right) with the BCN-NOTA linker; (C) ELISA analysis of the binding affinities of αPD-L1 Fab and Fab conjugates toward antigen PD-L1. EC_{50} was determined by GraphPad Prism to be around 7×10^{-10} M for Wt, the mutant (HC K129X), and the NOTA conjugate. (D) Flow cytometry analysis of cancer cell detection by GCN4-mediated switchable Fab conjugates. Peak traces from top to bottom represent cells treated with the mixture of the αGCN4 Fab AF488 conjugate and GCN4-fused αHER2 IgG at 25 nM (pink) and 5 nM (deep green); or the mixture of the same Fab conjugate but with GCN4-tagged αCD19 IgG at 25 nM (green) and 5 nM (orange). The cells without treatment were used as the negative control (red). *Adapted with permission from Wissler, H. L., Ehlerding,E. B., Lyu, Z., Zhao, Y., Zhang, S., Eshraghi, A., et al. (2019). Site-specific immuno-PET tracer to image PD-L1,* Molecular Pharmaceutics, *16, 2028–2036; Lyu, Z., Kang, L., Buuh, Z. Y., Jiang, D., McGuth, J. C., Du, J., et al. (2018). A switchable site-specific antibody conjugate,* ACS Chemical Biology, 13, *958–964. Copyright (2018, 2019) American Chemical Society.*

4. Close the lid to Mini Gel Tank and run gel electrophoresis at 150 V for approximately 50 min or until the target molecular weight on the ladder has moved halfway down the gel.
5. Remove the gel from the Mini Gel Tank and place it in a container for CBB gel staining.
6. Prepare the CBB staining solution that contains 1.0 g of CBB R-250 per 400 mL with 10% glacial acetic acid, 45% methanol, and 45% H_2O.
7. Fully submerge the gel in the CBB staining solution and incubate at room temperature for 20 min on the orbital shaker.
8. Decant the staining solution and rinse the gel with 100 mL of Milli-Q water. Incubate the gel in Milli-Q water with moderate shaking. Repeat the washing step until the gel has been fully destained. At this point, the CBB stained protein bands corresponding to Fab conjugates should be clearly observed. The non-reduced Fab or Fab conjugate should migrate as a single band, with >95% purity and a molecular mass of ~45–50 kDa (Fig. 4A). In the lane for samples treated by DTT, two distinct bands can be observed at ~20–30 kDa, corresponding to the reduced heavy chain and light chain, respectively.
9. After validation of purity, the protein conjugates are also analyzed by ESI-MS. This part is usually handled by the staff at the mass spectrometry core. Briefly, samples will be diluted with 0.1% formic acid. About 2 μL of the diluted Fab conjugates at a concentration of 1 pmol/μL will be injected onto a custom-made Poros 10R2 column, and reverse phased into a Q Exactive Plus Orbitrap Mass Spectrometer. The mass raw data will be extracted by Thermo Xtract program, and the protein deconvolution will be carried out using Thermo Promass for Xcalibur. A sample ESI-MS spectra is shown in Fig. 4B.

5.2.2 ELISA

ELISA was used to determine binding affinity, comparing the Fab conjugates and Fab mutations incorporating the UAA against the wild type. Briefly, the antigen is coated onto a well and the Fab or Fab conjugates specific to that antigen is added. A detection antibody conjugated with horse radish peroxidase (HRP) is then added, which will recognize the antigen captured Fab. The HRP on the detection antibody will then react with a highly-sensitive fluorogenic substrate to emit a stable and robust fluorescent signal (excitation: 325 nm/emission: 420 nm) that can be detected and analyzed. For measuring different classes of Fab antibodies (e.g., αGCN4 Fab has a kappa light chain (Lyu et al., 2018) while αPD-L1 Fab has a lambda light chain (Wissler et al., 2019)), different detection antibodies were used (e.g., anti-human kappa light chain antibody HRP-labeled for detecting αGCN4 Fab, anti-human lambda light chain antibody HRP-labeled for detecting αPD-L1 Fab).

1. Add 100 μL of antigen (1–10 μg/mL) in PBS solution to each well of a flat-bottom 96-well black plate and let them sit overnight for coating.
2. Wash with PBS buffer three times, and block each well with 300 μL of PBS buffer/1% BSA at room temperature for 1.5 h.
3. Serially dilute Fab/Fab mutants/Fab conjugates in the blocking buffer to desired concentrations. Then add these at 100 μL/well in triplicate and incubate the plate at room temperature for 2 h.
4. Wash all the wells with washing buffer (PBS/0.05% Tween) for three times.
5. Dilute HRP-labeled detection antibody at an appropriate concentration (usually 1000–10,000-fold) in blocking buffer, add them to wells and incubate for another 2 h.
6. Wash wells extensively with washing buffer a further five times.
1. Add 100 μL QuantaBlu fluorogenic ELISA substrate to each well. After 20 min of signal development, run fluorescent test on a plate reader (excitation: 325 nm/emission: 420 nm).
7. Process the data by Graphpad Prism software. Plot the results using the nonlinear regression equation "Log(agonist) vs. response—variable slope" to determine the EC_{50} values (half-maximal binding constant). A sample ELISA results was shown in Fig. 4C.

5.2.3 Flow cytometry

For certain applications (GCN4-mediated switchable antibody conjugate (Lyu et al., 2018)), we also gauged if the switch GCN4 can mediate Fab conjugates' detection of different cancer cell lines, which serves as a solid tool to predict the feasibility of follow-up *in vivo* experiments. Thus, flow cytometry experiments were carried out using GCN4 fused primary antibody (αHER2 or αCD19) and αGCN4-AF488 conjugate, which demonstrated the targeted detection of corresponding cancer cell lines (αHER2 combo mix bound HER2+ SK-BR-3 cell line, while αCD19 combo mix bound CD19+ Ramos cell line).

1. Harvest SK-BR-3 and Ramos cells. The adherent cell line SK-BR-3 was separated from the culture plate by 0.05% trypsin, 0.53 mM EDTA solution. Wash cells by repeated suspending in 4 °C PBS and spinning at 1000 rpm for 5 min.

2. Add 4% paraformaldehyde solution to the cell pellets. Pipette up and down to mix well. Incubate at room temperature for 15 min to allow for complete fixing of cells.
3. Wash cells with PBS and block with 2% BSA/PBS buffer for 1 h at 1×10^6 cells/mL.
4. Mix cells with a GCN4-tagged antibody (either αCD19 or αHER2) and the AF488 conjugated αGCN4 Fab at a molar ratio of 1:2 in the blocking buffer. Shake the tube or plate mildly at room temperature for 2 h.
5. Spin the cells down at 1000 rpm for 5 min and decant the solution.
6. Wash cells with cold PBS three times.
7. Analyze cell samples with a flow cytometer and detect the signals in the FITC channel. Process the results with FlowJo software (TreeStar) (Fig. 4D).

6. *In vivo* imaging using fab conjugates

Depending on the payloads, the site-specific Fab conjugates prepared in previous sections can be utilized for different *in vivo* imaging applications. Conjugation with the 1,4,7-triazacyclononane-N,N′,N″-triacetic acid (NOTA) chelator allows for the radiolabeling of the antibody using ^{64}Cu. Following procedures similar to literature reports (D'Souza et al., 2017; Hettich et al., 2016), the radiolabeled conjugates can be used for immunoconjugate-based positron emission tomography (immuno-PET) (Wissler et al., 2019) (Fig. 5A). Conjugation with Cy7 allows for a safe means for the optical imaging of organs and tumor tissues in the near-infrared region, without interference by background tissues' autofluorescence and light scattering (Yi, Wang, Qin, Yang, & Yuan, 2014; Zhang, Bloch, Akers, & Achilefu, 2012) (Fig. 5B). Note: All animal experiments were performed in accordance with the National Institute of Health (NIH) guidelines and conducted under the approval by University Institutional Animal Care and Use Committee.

6.1 Equipment and materials

- Inveon microPET/CT scanner (Siemens)
- Automated gamma counter (PerkinElmer)
- IVIS Spectrum for NIRF imaging (PerkinElmer)
- Orbital Shaker (Thermo Fisher)
- Athymic (nude) mice (Envigo)
- C57BL/6 mice (Envigo)

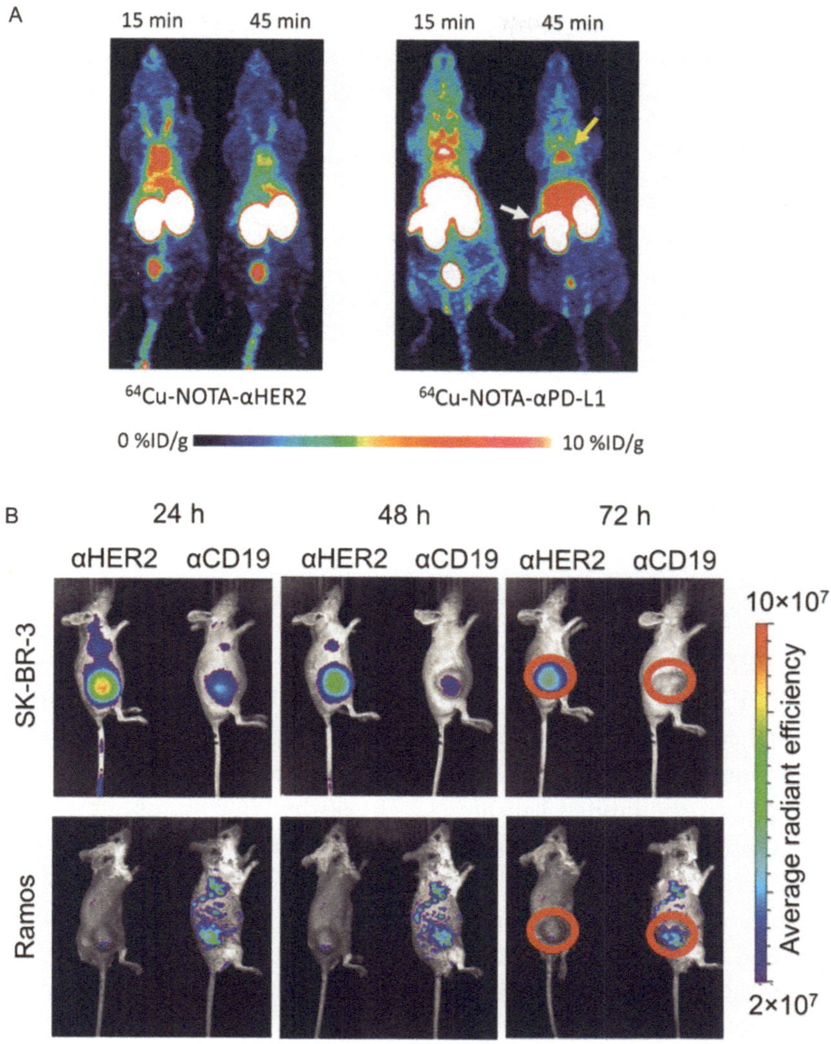

Fig. 5 Representative *in vivo* imaging using site-specific Fab conjugates. (A) PET imaging scans at 15 and 45 min post-injection of ^{64}Cu-αHER2-NOTA (left) or ^{64}Cu-α PD-L1-NOTA (right) in nude mice. The yellow arrow indicates adipose tissue while the white arrow points to the spleen; (B) NIRF imaging of SKBR3 and Ramos tumor bearing xenografts after the administration of premixed αHER2-GCN4 fusion: αGCN4-Cy7(LC-S155X, HC-K132X, X=pAcF) or αCD19-GCN4 fusion: αGCN4-Cy7(LC-S155X, HC-K132X, X=pAcF). *Adapted with permission from Wissler, H. L., Ehlerding,E. B., Lyu, Z., Zhao, Y., Zhang, S., Eshraghi, A., et al. (2019). Site-specific immuno-PET tracer to image PD-L1,* Molecular Pharmaceutics, 16, *2028–2036; Lyu, Z., Kang, L., Buuh, Z. Y., Jiang, D., McGuth, J. C., Du, J., et al. (2018). A switchable site-specific antibody conjugate,* ACS Chemical Biology, 13, *958–964. Copyright (2018, 2019) American Chemical Society.*

- CB17-SCID mice (Envigo)
- Matrigel matrix to re-suspend cancer cells (Invitrogen)
- Cell Biology Grade PBS buffer (Corning)
- 2% isoflurane (Millipore Sigma)
- $^{64}CuCl_2$ in 0.1 N HCl (University of Wisconsin-Madison)
- PD-10 column (GE Healthcare)
- DPBS Buffer, Cell Biology Grade (Corning)

6.2 Protocol

All mice were purchased between the ages of 4 and 6 weeks. For tumor implantation, harvest and re-suspend cancer cells in a matrigel matrix at 5×10^6 cells/well. Inject ~200 μL cancer cells into the lower right flank of each mouse. Monitor mice health (including body weight) and measure tumor volumes every other day. Once the tumor size grows to a diameter of ~1 cm, mice will be sorted, grouped ($n = 3$–5), and subjected to imaging experiments.

6.2.1 PET imaging

In vivo PET imaging has been performed with αPD-L1 Fab-NOTA conjugate and control conjugates such as αHER2 Fab-NOTA. Radiolabeling of the immunoconjugate is required before proceeding to the treatment of mice. Given the limited half-life (12.7 h) of the ^{64}Cu isotope, the radiolabeled protein conjugate is usually immediately administered to mice after purification. Caution should be taken when performing the series of experiments involved with radioisotope materials. The $^{64}CuCl_2$ stock should be stored behind 2 inch thick lead bricks. All the experiments should be performed in the lab area clearly outlined for radioactive material use. Disposable gloves, lab coats, and safety glasses as the minimum PPE are required. The persons working in the area should perform experiments behind L-block shields, and always wear radiation dosimetry monitoring badges (body and ring). All the experiment steps need to be planned ahead to minimize exposure to radiation. Lastly, potentially contaminated PPE should be removed and carefully disposed prior to leaving the area.

Radiolabeling of αPD-L1 Fab-NOTA was performed similar to previously reported methods (Ehlerding et al., 2017; Ferreira et al., 2018).

1. Dilute $^{64}CuCl_2$ in 150–300 μL of sodium acetate buffer (pH ~5).
2. Add 50 μg of αPD-L1 Fab-NOTA or αHER2 Fab-NOTA to 1 mCi (37 MBq) ^{64}Cu. Then incubate for 1 h at 37 °C with constant shaking.

3. Purify radiolabeled samples with a PD-10 column using PBS as the mobile phase. Note that all injectable solutions were prepared in sterile PBS.
4. Anesthetize mice with 3%/1% isoflurane *via* inhalation.
5. Intravenously inject the mice with 50–70 μCi of one of the radiolabeled ^{64}Cu-NOTA conjugates, either the αPD-L1 or αHer2.
6. Perform PET imaging with an Inveon microPET/CT scanner at 5, 15, 45, and 90 min post-injection. At each of the scanning time points, 20 million counts per mouse was obtained. Reconstruct the PET images using a 3D ordered subsets expectation maximization reconstructions algorithm (Fig. 5A).
7. To quantify the Fab conjugate accumulation in organs and tissues, use the Inveon Research Workspace software to analyze the mean percent injected dose per gram (%ID/g) at the regions-of-interest (ROI) following the equation: %ID/g = [activity in tissue (μCi/g)/administered activity (μCi)] × 100%

6.2.2 NIRF imaging

In vivo NIRF imaging has been performed with GCN4-mediated switchable antibody conjugates, during which either the GCN4-fused primary antibody and the αGCN4 Fab-Cy7 can be co-injected or the primary antibody can be injected hours ahead for pretargeted imaging.

1. Premix GCN4-tagged primary antibody (either αCD19 or αHER2) with αGCN4 Fab-Cy7 conjugate at a molar ratio of 1:2 for 2 h.
2. Intravenously inject the antibody mixture into three mice with a dosage of 60 μg of primary antibody per mouse.
3. Anesthetize mice using 2% isoflurane at set post-injection time intervals: 1, 4, 8, 16, 24, 48, and 72 h.
4. After each anesthetization, place the mouse on the IVIS Spectrum stage with the tumor side facing the CCD camera. Perform NIRF imaging using an excitation wavelength of 745 nm and an emission wavelength of 800 nm. To prevent image saturation, use automatic scanning time.
5. Place regions of interests (ROIs) on the 2D NIRF image to encompass the entire NIRF signal (Fig. 5B).

Pre-targeted NIRF Imaging

1. Inject a GCN4-tagged primary antibody (either αCD19 or αHER2) intravenously into the SK-BR-3 mouse model at a dosage of 60 μg of primary antibody per mouse for pre-targeting.
2. Intravenously inject the αGCN4 Fab-Cy7 conjugate at a dosage of 35 μg/mouse 4 h after the primary antibody injection.

3. The control group mice were injected with PBS buffer and then 4 h later with the same dosage of αGCN4 Fab-Cy7 conjugate.
4. Perform the NIRF imaging using the same procedures as above.

7. Summary and conclusions

The methods presented here provide an efficient and reliable approach to develop site-specific antibody Fab fragment conjugates. The UAA-based site-specific protein conjugation approach has been thereby illustrated as a general strategy to endow versatility to antibodies, making homogeneous conjugates with various types of small molecules of distinct functions. Further efforts may include site-by-side comparisons of the *in vivo* properties and activity of this class of conjugates with random conjugates and cysteine-based site-specific conjugates. Nevertheless, future applications of these imaging probes into clinical settings would be expected to facilitate image-guided therapy in multiple disease areas.

Acknowledgments

We thank Dr. Hsin-Yao Tang and Mr. Thomas Beer at the Wistar Institute, Proteomics and Metabolomics Facility for performing the ESI-MS analysis service. This work was supported by Temple University (start-up funds to R.E.W.), The American Cancer Society (#15-175-22 to R.E.W.), and the National Institutes of Health Grants (1R35GM133468-01 to R.E.W., P30CA014520 to collaborators at the University of Wisconsin-Madison). Support for the NMR facility at Temple University by a CURE grant from the Pennsylvania Department of Health is also gratefully acknowledged.

References

Adumeau, P., Sharma, S. K., Brent, C., & Zeglis, B. M. (2016a). Site-specifically labeled Immunoconjugates for molecular imaging—Part 1: Cysteine residues and Glycans. *Molecular Imaging and Biology*, *18*, 1–17.

Adumeau, P., Sharma, S. K., Brent, C., & Zeglis, B. M. (2016b). Site-specifically labeled Immunoconjugates for molecular imaging—Part 2: Peptide tags and unnatural amino acids. *Molecular Imaging and Biology*, *18*, 153–165.

Axup, J. Y., Bajjuri, K. M., Ritland, M., Hutchins, B. M., Kim, C. H., Kazane, S. A., et al. (2012). Synthesis of site-specific antibody-drug conjugates using unnatural amino acids. *Proceedings of the National Academy of Sciences of the United States of America*, *109*, 16101–16106.

Badescu, G., Bryant, P., Bird, M., Henseleit, K., Swierkosz, J., Parekh, V., et al. (2014). Bridging disulfides for stable and defined antibody drug conjugates. *Bioconjugate Chemistry*, *25*, 1124–1136.

Cao, Y., Axup, J. Y., Ma, J. S., Wang, R. E., Choi, S., Tardif, V., et al. (2015). Multiformat T-cell-engaging bispecific antibodies targeting human breast cancers. *Angewandte Chemie International Edition*, *54*, 7022–7027.

Chatterjee, S., Lesniak, W. G., Gabrielson, M., Lisok, A., Wharram, B., Sysa-Shah, P., et al. (2016). A humanized antibody for imaging immune checkpoint ligand PD-L1 expression in tumors. *Oncotarget*, *7*, 10215–10227.

Dommerholt, J., van Rooijen, O., Borrmann, A., Guerra, C. F., Bickelhaupt, F. M., & van Delft, F. L. (2014). Highly accelerated inverse electron-demand cycloaddition of electron-deficient azides with aliphatic cyclooctynes. *Nature Communications*, *5*, 5378.

D'Souza, J. W., Hensley, H., Doss, M., Beigarten, C., Torgov, M., Olafsen, T., et al. (2017). Cerenkov luminescence imaging as a modality to evaluate antibody-based PET radiotracers. *Journal of Nuclear Medicine*, *58*, 175–180.

Ehlerding, E. B., England, C. G., Majewski, R. L., Valdovinos, H. F., Jiang, D., Liu, G., et al. (2017). ImmunoPET imaging of CTLA-4 expression in mouse models of non-small cell lung cancer. *Molecular Pharmaceutics*, *14*, 1782–1789.

England, C. G., Jiang, D., Ehlerding, E. B., Rekoske, B. T., Ellison, P. A., Hernandez, R., et al. (2018). (89)Zr-labeled nivolumab for imaging of T-cell infiltration in a humanized murine model of lung cancer. *European Journal of Nuclear Medicine and Molecular Imaging*, *45*, 110–120.

Ferreira, C. A., Hernandez, R., Yang, Y., Valdovinos, H. F., Engle, J. W., & Cai, W. (2018). ImmunoPET of CD146 in a murine hindlimb ischemia model. *Molecular Pharmaceutics*, *15*, 3434–3441.

Gill, P., Forouzandeh, M., Rahbarizadeh, F., Ramezani, R., & Rasaee, M. J. (2006). Production of anti-digoxigenin antibody HRP conjugate for PCR-ELISA DIG detection system. *Journal of Immunoassay and Immunochemistry*, *27*, 303–318.

Hernandez, R., Sun, H., England, C. G., Valdovinos, H. F., Ehlerding, E. B., Barnhart, T. E., et al. (2016). CD146-targeted immunoPET and NIRF imaging of hepatocellular carcinoma with a dual-labeled monoclonal antibody. *Theranostics*, *6*, 1918–1933.

Heskamp, S., Hobo, W., Molkenboer-Kuenen, J. D., Olive, D., Oyen, W. J., Dolstra, H., et al. (2015). Noninvasive imaging of tumor PD-L1 expression using radiolabeled anti-PD-L1 antibodies. *Cancer Research*, *75*, 2928–2936.

Hettich, M., Braun, F., Bartholoma, M. D., Schirmbeck, R., & Niedermann, G. (2016). High-resolution PET imaging with therapeutic antibody-based PD-1/PD-L1 checkpoint tracers. *Theranostics*, *6*, 1629–1640.

Hutchins, B. M., Kazane, S. A., Staflin, K., Forsyth, J. S., Felding-Habermann, B., Smider, V. V., et al. (2011). Selective formation of covalent protein heterodimers with an unnatural amino acid. *Chemistry & Biology*, *18*, 299–303.

Jackson, D., Atkinson, J., Guevara, C. I., Zhang, C., Kery, V., Moon, S. J., et al. (2014). In vitro and in vivo evaluation of cysteine and site specific conjugated herceptin antibody-drug conjugates. *PLoS One*, *9*, e83865.

Kazane, S. A., Sok, D., Cho, E. H., Uson, M. L., Kuhn, P., Schultz, P. G., et al. (2012). Site-specific DNA-antibody conjugates for specific and sensitive immuno-PCR. *Proceedings of the National Academy of Sciences of the United States of America*, *109*, 3731–3736.

Kim, C. H., Axup, J. Y., Lawson, B. R., Yun, H., Tardif, V., Choi, S. H., et al. (2013). Bispecific small molecule-antibody conjugate targeting prostate cancer. *Proceedings of the National Academy of Sciences of the United States of America*, *110*, 17796–17801.

Kularatne, S. A., Deshmukh, V., Ma, J., Tardif, V., Lim, R. K., Pugh, H. M., et al. (2014). A CXCR4-targeted site-specific antibody-drug conjugate. *Angewandte Chemie International Edition*, *53*, 11863–11867.

Lewis Phillips, G. D., Li, G., Dugger, D. L., Crocker, L. M., Parsons, K. L., Mai, E., et al. (2008). Targeting HER2-positive breast cancer with trastuzumab-DM1, an antibody-cytotoxic drug conjugate. *Cancer Research*, *68*, 9280–9290.

Liu, C. C., & Schultz, P. G. (2010). Adding new chemistries to the genetic code. *Annual Review of Biochemistry*, *79*, 413–444.

Lyu, Z., Kang, L., Buuh, Z. Y., Jiang, D., McGuth, J. C., Du, J., et al. (2018). A switchable site-specific antibody conjugate. *ACS Chemical Biology*, *13*, 958–964.

Muir, T. W. (2003). Semisynthesis of proteins by expressed protein ligation. *Annual Review of Biochemistry*, *72*, 249–289.

Nastri, H. G., C. Iffland, O. Leger, Q. An, M. Cartwright, S. D. McKenna, et al. 2014. Anti-pd-l1 antibodies and uses thereof. United States, USPTO.

Ricart, A. D. (2011). Antibody-drug conjugates of calicheamicin derivative: Gemtuzumab ozogamicin and inotuzumab ozogamicin. *Clinical Cancer Research*, *17*, 6417–6427.

Sano, T., Smith, C. L., & Cantor, C. R. (1992). Immuno-PCR: Very sensitive antigen detection by means of specific antibody-DNA conjugates. *Science*, *258*, 120–122.

Schumacher, D., Hackenberger, C. P. R., Leonhardt, H., & Helma, J. (2016). Current status: Site-specific antibody drug conjugates. *Journal of Clinical Immunology*, *36*, S100–S107.

Tian, F., Lu, Y., Manibusan, A., Sellers, A., Tran, H., Sun, Y., et al. (2014). A general approach to site-specific antibody drug conjugates. *Proceedings of the National Academy of Sciences of the United States of America*, *111*, 1766–1771.

Wals, K., & Ovaa, H. (2014). Unnatural amino acid incorporation in E. coli: Current and future applications in the design of therapeutic proteins. *Frontiers in Chemistry*, *2*, 15.

Wang, L., Brock, A., Herberich, B., & Schultz, P. G. (2001). Expanding the genetic code of *Escherichia coli*. *Science*, *292*, 498–500.

Wang, R. E., Liu, T., Wang, Y., Cao, Y., Du, J., Luo, X., et al. (2015). An immunosuppressive antibody-drug conjugate. *Journal of the American Chemical Society*, *137*, 3229–3232.

Wang, L., Xie, J., & Schultz, P. G. (2006). Expanding the genetic code. *Annual Review of Biophysics and Biomolecular Structure*, *35*, 225–249.

Warren, D. J. (2011). Preparation of highly efficient electrocompetent *Escherichia coli* using glycerol/mannitol density step centrifugation. *Analytical Biochemistry*, *413*, 206–207.

Wissler, H. L., Ehlerding, E. B., Lyu, Z., Zhao, Y., Zhang, S., Eshraghi, A., et al. (2019). Site-specific immuno-PET tracer to image PD-L1. *Molecular Pharmaceutics*, *16*, 2028–2036.

Wu, N. J., Kang, C. S., Sin, I., Ren, S. Y., Liu, D. J., Ruthengael, V. C., et al. (2016). Promising bifunctional chelators for copper 64-PET imaging: Practical Cu-64 radiolabeling and high in vitro and in vivo complex stability. *Journal of Biological Inorganic Chemistry*, *21*, 177–184.

Wu, Y., Zhu, H., Zhang, B., Liu, F., Chen, J., Wang, Y., et al. (2016). Synthesis of site-specific radiolabeled antibodies for radioimmunotherapy via genetic code expansion. *Bioconjugate Chemistry*, *27*, 2460–2468.

Yi, X., Wang, F., Qin, W., Yang, X., & Yuan, J. (2014). Near-infrared fluorescent probes in cancer imaging and therapy: An emerging field. *International Journal of Nanomedicine*, *9*, 1347–1365.

Young, T. S., & Schultz, P. G. (2010). Beyond the canonical 20 amino acids: Expanding the genetic lexicon. *Journal of Biological Chemistry*, *285*, 11039–11044.

Young, D. D., Young, T. S., Jahnz, M., Ahmad, I., Spraggon, G., & Schultz, P. G. (2011). An evolved aminoacyl-tRNA synthetase with atypical polysubstrate specificity. *Biochemistry*, *50*, 1894–1900.

Zahnd, C., Spinelli, S., Luginbuhl, B., Amstutz, P., Cambillau, C., & Pluckthun, A. (2004). Directed in vitro evolution and crystallographic analysis of a peptide-binding single chain antibody fragment (scFv) with low picomolar affinity. *The Journal of Biological Chemistry*, *279*, 18870–18877.

Zhang, X., Bloch, S., Akers, W., & Achilefu, S. (2012). Near-infrared molecular probes for in vivo imaging. *Current Protocols in Cytometry*, *60*, 12.27.1–12.27.20, [Chapter 12: Unit12 27].

CHAPTER SIXTEEN

Quantifying protein-protein interactions of the acyl carrier protein with solvatochromic probes

Katherine Charov, Michael D. Burkart*
Department of Chemistry and Biochemistry, University of California, San Diego, La Jolla, CA, United States
*Corresponding author. e-mail address: mburkart@ucsd.edu

Contents

Abstract

Protein-protein interactions (PPIs) are universal to life and their study and understanding is critical to drug discovery and bioengineering efforts. Historically, X-ray crystallography, isothermal titration calorimetry and other biophysical methods have been used to study PPIs, but can be costly and are low throughput, hindering progress towards rapid evaluation of these interactions. Recent interest in targeting PPIs and in engineering biosynthetic pathways in which PPIs play a critical role has driven innovation in their evaluation but a universal screen is still needed. One of the best characterized systems relying upon PPIs is *Escherichia coli* type II fatty acid biosynthesis in which the central acyl carrier protein (EcACP) shuttles substrates to a series of partner enzymes. Here we present a method by which EcACP is labeled with a solvatochromic dye, 4-DMN, and then allowed to interact with its various partner enzymes. Upon interaction, there is a large

Methods in Enzymology, Volume 638
ISSN 0076-6879
https://doi.org/10.1016/bs.mie.2020.03.016

increase in fluorescence intensity which is easily monitored via fluorometer or plate reader. This method is useful in the study of known PPI, hypothetical PPI and in evaluation of inhibitors of both partner enzyme active site and of the PPI itself.

1. Introduction

Protein-protein interactions (PPIs) are ubiquitous to all domains of life, playing roles in biosynthetic pathways, signaling and regulation. While the surfaces of each protein involved in interaction are diffuse and large, typically PPIs are mediated by a small sub-set of residues on the surface of each protein involved, termed "hot-spot" residues. Each interaction is transient and of low affinity, making PPIs a challenging phenomenon to study. In the past, crosslinking strategies have been employed to trap the interaction between interacting proteins for X-ray crystallography and solution NMR experiments (Nguyen et al., 2014). Isothermal titration calorimetry and surface plasmon resonance have also helped to characterize the thermodynamics behind binding, but these methods are limited by cost and are low-throughput.

One of the best characterized proteins engaging in numerous PPI is the *Escherichia coli* acyl carrier protein (EcACP), which was first identified for its centrality to fatty acid biosynthesis (FAS). In FAS, EcACP carries the growing acyl chain as it is elongated by a series of nine different partner enzymes (Fig. 1). EcACP sequesters intermediates within its alpha helical bundle to prevent cross-reactivity and degradation. To present the correct fatty acyl intermediate to each different FAS partner enzyme, EcACP must engage in correct PPI with high specificity. Beyond FAS, EcACP is known to interact with at least 30 different enzymes from biotin and lipid A biosynthesis, to proteins involved in chromosome segregation and iron cluster formation, among others (Brozek & Raetz, 1990; Issartel, Koronakis, & Hughes, 1991; Jordan & Cronan, 1997; Lin, Hanson, & Cronan, 2010). Each interaction is critical to cell survival and have gained recent interest in antibiotic drug discovery. Already, many well-known antibiotics such as cerulenin and thiolactomycin target FAS and others have been developed to inhibit lipid A biosynthesis (Price et al., 2001). Targeting the PPI between EcACP and one or more of its many partner enzymes presents a new mode of inhibition.

EcACP is a model protein to study other carrier protein mediated biosynthetic pathways including polyketide (PKS) and nonribosomal peptide

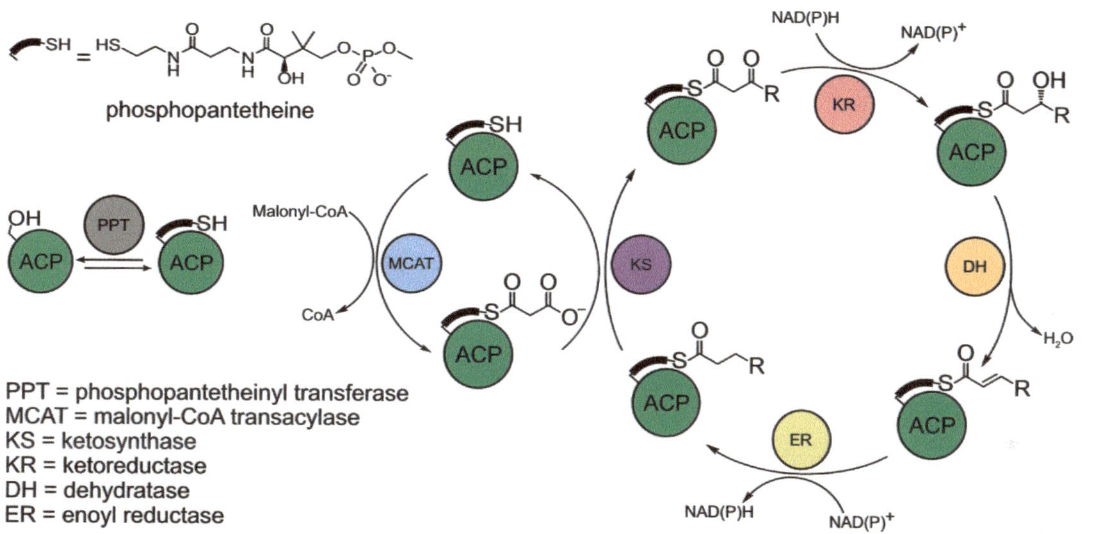

Fig. 1 (A) Structure of phosphopantetheine arm responsible for tethering cargo to the acyl carrier protein, (B) cartoon representation of the type II fatty acid biosynthesis cycle: *apo*-ACP is first post-translationally modified by a phosphopantetheine arm to form *holo*-ACP. To begin the cycle, *holo*-ACP is loaded with a malonyl starter unit by a malonyl-CoA acyl transferase (*E. coli:* FabD) to form malonyl-ACP. A ketosynthase (FabH, FabF or FabB in *E. coli)* condenses malonyl-ACP with another unit of acyl-ACP, which is then reduced by a ketoreductase (FabG, *E. coli*). A dehydratase (FabA/FabZ) forms an α/β-unsaturated ketone which undergoes reduction by an enoyl reductase (FabI) to form the fully saturated acyl chain which can re-enter the cycle by condensation with another unit of malonyl-ACP. Each of these reactions requires ACP to engage in protein-protein interaction with the various enzymes in the pathway.

(NRPS) synthases. In both PKS and NRPS, a central carrier protein shuttles growing intermediates to partner enzymes during extension of the growing polyketide or peptide chains. NRPS and PKS pathways are the source of many highly prized natural products such as vancomycin and erythromycin. The study of the PPI behind FAS, PKS and NRPS is of high interest to research into recombinant biosynthetic production of these compounds.

By labeling a carrier protein with a synthetically prepared coenzyme A analogue bearing a terminal solvatochromic dye, we have developed a method by which multiple protein-protein interactions between carrier proteins and partner enzymes can be rapidly visualized (Charov & Burkart, 2019). We have demonstrated the utility of this method to study previously characterized PPI between EcACP and partner enzymes from a variety of pathways and have also validated proposed partner enzymes with EcACP. This method can also be used to study PPI and active site inhibitors.

In this chapter, we describe procedures for the preparation of EcACP labeled with the solvatochromic dye, 4-dimethylnaphthalene (4-DMN) and the procedure for assaying PPI and inhibitors.

2. Preparation of 4-DMN-EcACP

2.1 Purification of recombinant EcACP from *Escherichia coli*

EcACP is obtained by gene expression and protein purification from *E. coli* BL21 (DE3). EcACP can be purified without the use of a His-tag by exploiting the rare solubility and stability of the protein in isopropanol.

2.1.1 Cloning of the acpP gene into an expression vector

acpP was cloned using the following primers into a pET11a-pLysS vector, as previously described (Hill, MacKenzie, Flanagan, Cronan, & Prestegard, 1995).

5'-GTCGGTATCATATGTCTACTATCGAAGAACG.

NdeI site underlined.

5'-TCATGCGGATCCTTAAGCCTGGTGACCGTTG.

BamHI site underlined.

2.1.2 Expression of acpP in E. coli BL21 (DE3) and purification of recombinant EcACP protein

2.1.2.1 Materials

- Competent *E. coli* BL21 (DE3) cells (New England Biolabs)
- *acpP* expression vector (pET11a-pLysS)

- Luria-Bertani (LB) agar (Sigma Aldrich)
- LB Broth (Sigma Aldrich)
- Terrific broth (TB) medium (Sigma Aldrich)
- Isopropyl-β-D-thiogalactoside (IPTG) (Sigma Aldrich)
- Isopropanol
- Lysis buffer: 50 mM Tris, 150 mM NaCl, 1 mM TCEP, 5% glycerol (pH 7.4), and 1 mg/mL lysozyme
- HiTrap Q HP anion exchange chromatography column 5 mL (GE Healthcare Life Sciences)
- Low salt buffer: 50 mM Tris, 1 mM TCEP, 5% glycerol (pH 7.4)
- High salt buffer: 50 mM Tris, 1 M NaCl, 1 mM TCEP, 5% glycerol (pH 7.4)
- Clear polystyrene 96-well microplate (Fisher Scientific)
- Bradford reagent (Bio-Rad)
- Disposable polypropylene column (Qiagen)
- SDS-PAGE (12% Tris-Glycine Novex gel with Novex Tris-Glycine SDS running buffer, ThermoFisher)

2.1.2.2 Procedure

1. Transform competent BL21 (DE3) cells with *acpP* expression vector according to manufacturer's instructions. Plate transformed cells on LB agar containing 50 mg/L kanamycin and incubate overnight at 37 °C.
2. Inoculate 2 × 5 mL of sterile LB broth containing 50 mg/L kanamycin with a single colony for each culture from the transformed cells. Incubate overnight at 37 °C on rotisserie or shaker.
3. Prepare long-term storage of transformed cells from one overnight culture by mixing in a 1:1 ratio the *E. coli* BL21(DE3) culture containing the *acpP* plasmid with sterile 40% glycerol in water. Store these stocks at −80 °C.
4. Use the other overnight culture to inoculate 1 L of TB media containing 50 mg/L kanamycin. Incubate the flask at 125 rpm at 37 °C, monitoring the A_{600} of the growth. Once an A_{600} of 0.6–0.8 is reached, chill the cells on ice for 20 min to arrest growth. After chilling, induce cells with 1 mM of sterile IPTG (final concentration) and incubate overnight at 16 °C at 125 rpm.
5. Harvest cells by centrifugation at 2000 rpm for 30 min. Transfer pellet to 50 mL Falcon tube. This pellet can be stored at −20 °C for up to 6 months or used immediately for protein purification.
6. Suspend pellet in ~25 mL of lysis buffer and chill on ice for 20 min.

7. Lyse cells by sonication by 1 s on, 3 s off for 10 min total. Alternatively, a French press can be used to lyse cells. Keep cell samples on ice during lysis.
8. Transfer lysed cell solution to beaker and slowly add equal volume of isopropanol to the lysed cells with constant stirring.
9. Clarify isopropanol-lysis solution by high-speed centrifugation (e.g. 12,000 rpm) for 45 min at 4 °C. Save supernatant and discard pellet.
10. Purify EcACP from supernatant by anion exchange. Follow manufacturer's protocol. A brief description of purification is described below:
11. Equilibrate HiTrap Q HP 5 mL (GE Healthcare Life Sciences) with low salt buffer.
12. Load full volume of supernatant from step 9 onto HiTrap Q HP column.
13. Elute using the following gradient: 100.0 mL total length, 0% low salt buffer to 70% high salt buffer. Collect 3.0 mL fractions. EcACP appears at 300 mM NaCl concentrations, between 40.0 mL and 50.0 mL of gradient.
14. Use Bradford reagent to access in which fractions protein is present. To analyze via Bradford: dispense 90 μL of Bradford reagent into a well of a clear-bottomed 96-well plate and mix with 10 μL of protein solution. A deep blue color indicates high protein concentration. A NanoDrop instrument can also be used to locate fractions with high protein concentration.
15. Analyze fractions for EcACP content and purity using 12% SDS-PAGE.
16. Pool fractions containing EcACP and freeze at −78 °C and store at −20 °C. If necessary, EcACP can be purified further by re-subjecting impure fractions to anion exchange chromatography or size exclusion chromatography.

2.2 Purification of recombinant acpH from *Escherichia coli*

AcpH from *Pseudomonas aeruginosa* is obtained by gene expression and protein purification from *E. coli* BL21 (DE3).

2.2.1 Cloning of the acpH gene into an expression vector

acpH is amplified from genomic DNA of *P. aeruginosa* using the following primers:

5'-CGAGGATCCATGAACTACCTCGC-3′.
5'-GTACTCGAGTCAGCGCTGGCTCAG-3′.

and ligated into pET-28b(+) (Novagen) between BamHI and *Xho*I sites, as previously described (Murugan, Kong, Sun, Rao, & Liang, 2010).

2.2.2 Expression of *acpH* in *E. coli* BL21 (DE3) and purification of recombinant acpH protein

2.2.2.1 Materials

- Competent *E. coli* BL21 (DE3) cells (New England Biolabs)
- *acpH* expression vector (pET-28b(+))
- Luria-Bertani (LB) agar (Sigma Aldrich)
- LB Broth (Sigma Aldrich)
- Terrific broth (TB) medium (Sigma Aldrich)
- Isopropyl-β-D-thiogalactoside (IPTG) (Sigma Aldrich)
- Ni^{+2}-NTA resin (Qiagen)
- Lysis buffer: 50mM Tris, 100mM NaCl, 0.5mM TCEP, 15mM $MgCl_2$, 1mM $MnCl_2$, 10% glycerol (pH 8.0), and 1mg/mL lysozyme
- Wash buffer: 50mM Tris, 100mM NaCl, 0.5mM TCEP, 15mM $MgCl_2$, 1mM $MnCl_2$, 10% glycerol, 25mM imidazole (pH 8.0)
- Elution buffer: 50mM Tris, 100mM NaCl, 0.5mM TCEP, 15mM $MgCl_2$, 1mM $MnCl_2$, 10% glycerol, 250mM imidazole (pH 8.0)
- Storage buffer: 50mM Tris, 100mM NaCl, 0.5mM TCEP, 15mM $MgCl_2$, 1mM $MnCl_2$, 10% glycerol (pH 8.0)
- Clear polystyrene 96-well microplate (Fisher Scientific)
- Bradford reagent (Bio-Rad)
- Disposable polypropylene column (Qiagen)
- SDS-PAGE (12% Tris-Glycine Novex gel with Novex Tris-Glycine SDS running buffer, ThermoFisher)
- Spin concentrator 10kDa molecular weight cut-off (Millipore Sigma)
- PD-10 column (Sephadex G-25)

2.2.2.2 Procedure

Note: AcpH is an unstable protein and it is imperative that all reagents and protein purification be kept and performed at 4°C. If possible, it is recommended that the purification be performed in a cold room.

1. Transform competent BL21 (DE3) cells with *acpH* expression vector according to manufacturer's instructions. Plate transformed cells on LB agar containing 50mg/L kanamycin and incubate overnight at 37°C.

2. Inoculate 2 × 5 mL of sterile LB broth containing 50 mg/L kanamycin with a single colony for each culture from the transformed cells. Incubate overnight at 37 °C on rotisserie or shaker.
3. Prepare long-term storage of transformed cells from one overnight culture by mixing in a 1:1 ratio the *E. coli* BL21(DE3) culture containing the *acpH* plasmid with sterile 40% glycerol in water. Store these stocks at −80 °C.
4. Use the other overnight culture to inoculate 1 L of TB media containing 50 mg/L kanamycin. Incubate the flask at 125 rpm at 37 °C, monitoring the A_{600} of the growth. Once an A_{600} of 0.6–0.8 is reached, chill the cells on ice for 20 min to arrest growth. After chilling, induce cells with 1 mM of sterile IPTG (final concentration) and incubate overnight at 16 °C at 125 rpm.
5. Harvest cells by centrifugation at 2000 rpm for 30 min. Transfer pellet to 50 mL Falcon tube. This pellet can be stored at −20 °C for up to 6 months or used immediately for protein purification.
6. Suspend pellet in ~25 mL of lysis buffer and chill on ice for 20 min.
7. Lyse cells by sonication by 1 s on, 3 s off for 2 min total. Rest for 1 min. Repeat sonication two more times, resting on ice for 1 min in between each sonication. Alternatively, a French press can be used to lyse cells. Take care to keep all cell samples on ice.
8. Clarify lysate by high-speed centrifugation, recommended 12,000 rpm for 45 min at 4 °C.
9. Purify cleared lysate by Ni^{2+}-NTA resin (Qiagen) according to manufacturer's instructions. A brief protocol is outlined in the following steps.
10. First, equilibrate 1 mL of Ni^{2+}-NTA resin with lysis buffer. If purifying a volume of cells from greater than 1 L growth, use 1 mL extra Ni^{2+}-NTA resin for each additional liter of cell culture.
11. Transfer equilibrated Ni^{2+}-NTA resin and cleared lysate to fresh 50 mL Falcon tube or autoclaved beaker and incubate at 4 °C with gentle shaking for 30 min.
12. Transfer the Ni^{2+}-NTA resin—lysate solution to a clean polypropylene with frit column and collect flow through liquid. Store the flow through on ice.
13. Wash the remaining Ni^{2+}-NTA resin with ~25 mL of lysis buffer and collect fraction. Store this fraction on ice.
14. Wash the remaining Ni^{2+}-NTA resin with ~25 mL of wash buffer and collect fraction. Store this fraction on ice.

15. Note: it is recommended to monitor nonspecific protein elution in lysis and wash buffer fractions by Bradford assay. In a clear bottomed 96-well plate, dispense 90 μL of Bradford reagent and mix with 10 μL of protein solution to be tested. A deep blue color indicates high concentration of protein. If at the end of 25 mL of wash buffer the Bradford test indicates a high concentration of protein wash the Ni^{2+}-NTA resin with ~25 mL of optional wash buffer 2 until Bradford test no longer indicates protein.
16. Elute AcpH using 10 mL of elution buffer, collecting in 1 mL fractions.
17. Analyze elution fractions by 12% SDS-PAGE.
18. Pool and concentrate fractions containing AcpH using a 10 kDa spin concentrator to a total volume of 2.5 mL. Immediately desalt concentrated AcpH using a PD-10 column (Sephadex G-25): remove the bottom tip of the PD-10 column and discard storage liquid. Equilibrate the column with 25 mL of storage buffer and load concentrated AcpH (maximum volume of 2.5 mL per PD-10 column). Discard the flow through. Elute with 3.5 mL storage buffer and immediately flash freeze at −78 °C. Store at −20 °C.

2.3 Purification of recombinant CoaA, CoaD and CoaE from *Escherichia coli*

CoaA, CoaD and CoaE are obtained by gene expression and protein purification from *E. coli* BL21 (DE3). Coa enzymes are cloned and expressed as fusion constructs with maltose binding protein (MBP) to enhance stability and solubility (Haushalter, Worthington, Hur, & Burkart, 2008). These enzymes are used in catalytic amounts and typically a single 1 L growth is sufficient for many large scale preparations of 4-DMN-EcACP.

2.3.1 Cloning of the coaA, coaD and coaE genes into expression vectors

The following primers are used to PCR amplify the open reading frame from a pCA24N construct for *E. coli* coaA:

CoaA_F: 5′-ggggacaagtttgtacaaaaaagcaggcttcatgagtataaaagagcaaacg-3’.

CoaA_R: 5′-ggggaccactttgtacaagaaagctgggtcttatttgcgtagtctgacctcttctac-3’.

The following primers are used to PCR amplify the open reading frame from a pET24b construct for *E. coli* coaD:

CoaD_F: 5′-ggggacaagtttgtacaaaaaagcaggcttcatgcaaaaacgggcgatttat-3’.

CoaD_R: 5′-ggggaccactttgtacaagaaagctgggtcctacgctaacttcgccatcagcgc-3’.

The following primers are used to PCR amplify the open reading frame from a pCA24N construct for *E. coli* coaE:

CoaE_F: 5′-ggggacaagtttgtacaaaaaagcaggcttcatgaggtatatagttgccttaacg-3’.

CoaE_R: 5′-ggggaccactttgtacaagaaagctgggtcttacggtttttcctgtgagacaaactg-3’.

Following PCR amplification, a vector coding for an N-terminal maltose binding fusion is combined with the vectors obtained above using LR Clonase (Invitrogen).

2.3.2 *Expression of coaA, coaD and coaE in* E. coli *BL21 (DE3) and purification of recombinant CoaA, CoaD and CoaE proteins*

2.3.2.1 Materials

- Competent *E. coli* BL21 (DE3) cells (New England Biolabs)
- *coaA, coaD* and *coaE* expression vector
- Luria-Bertani (LB) agar (Sigma Aldrich)
- LB Broth (Sigma Aldrich)
- Terrific broth (TB) medium (Sigma Aldrich)
- Isopropyl-β-D-thiogalactoside (IPTG) (Sigma Aldrich)
- Maltose
- Amylose resin (New England Biolabs)
- Lysis buffer: 20 mM Tris, 200 mM NaCl, 1 mM EDTA (pH 7.4) and 1 mg/mL lysozyme
- Wash buffer: 20 mM Tris, 200 mM NaCl, 1 mM EDTA (pH 7.4)
- Elution buffer: 20 mM Tris, 200 mM NaCl, 1 mM EDTA (pH 7.4) and 10 mM maltose
- Storage buffer: 20 mM Tris, 200 mM NaCl, 1 mM EDTA (pH 7.4)
- Clear polystyrene 96-well microplate (Fisher Scientific)
- Bradford reagent (Bio-Rad)
- Disposable polypropylene column (Qiagen)
- SDS-PAGE (12% Tris-Glycine Novex gel with Novex Tris-Glycine SDS running buffer, ThermoFisher)
- Spin concentrator 10 kDa molecular weight cut-off (Millipore Sigma)
- PD-10 column (Sephadex G-25)

2.3.2.2 Procedure

1. Transform 3 stocks of competent BL21 (DE3) cells with *coaA, coaD* and *coaE* expression vectors according to manufacturer’s instructions. Plate transformed cells on LB agar containing ampicillin and incubate overnight at 37 °C.

2. Inoculate 2 × 5 mL of sterile LB broth containing ampicillin with a single colony for each culture from the transformed cells. Incubate overnight at 37 °C on rotisserie or shaker.
3. Prepare long-term storage of transformed cells from one overnight culture by mixing in a 1:1 ratio the *E. coli* BL21(DE3) culture containing the *coaA, coaD* or *coaE* plasmid with sterile 40% glycerol in water. Store these stocks at −80 °C.
4. Use the other overnight culture to inoculate 1 L of TB media containing 100 mg/L ampicillin. Incubate the flask at 125 rpm at 37 °C, monitoring the A_{600} of the growth. Once an A_{600} of 0.6–0.8 is reached, chill the cells on ice for 20 min to arrest growth. After chilling, induce cells with IPTG (1 mM final concentration) and incubate overnight at 16 °C at 125 rpm.
5. Harvest cells by centrifugation at 2000 rpm for 30 min. Transfer pellet to 50 mL Falcon tube. This pellet can be stored at −20 °C for up to 6 months or used immediately for protein purification.
6. Suspend pellet in ~25 mL of lysis buffer and chill on ice for 20 min.
7. Lyse cells by sonication by 1 s on, 3 s off for 2 min total. Rest for 1 min. Repeat sonication two more times, resting on ice for 1 min in between each sonication. Alternatively, a French press can be used to lyse cells. Take care to keep all cell samples on ice.
8. Clarify lysate by high-speed centrifugation, recommended 12,000 rpm for 45 min at 4 °C.
9. Purify cleared lysate by amylose resin (NEB) according to manufacturer's instructions. A brief protocol for purification follows.
10. Mix cleared lysate and 1 mL of amylose resin (NEB) for 1 h at 4 °C with light shaking.
11. Load resin onto a column and wash with 20 mL of wash buffer.
12. Elute Coa-MBP fusion enzyme with 10 mL of elution buffer collecting in 1 mL fractions.
13. Analyze elution fractions by 12% SDS-PAGE.
14. Pool fractions containing Coa-MBP fusion enzyme and concentrate using a 10 kDa spin concentrator to a final concentration of 10 mg/mL. Freeze at −78 °C and store at −20 °C.

2.4 Conversion of *holo*-EcACP to *apo*-EcACP

EcACP is expressed and purified as a roughly 1:1 mix of *holo*-EcACP and *apo*-EcACP and it is necessary to remove the phosphopantetheine arm to generate 100% *apo*-EcACP prior to attachment of the 4-DMN-pantetheine. Removal of phosphopantetheine from *holo*-EcACP is accomplished using purified diesterase, AcpH.

2.4.1 *Removal of phosphopantetheine to convert* holo-*EcACP to* apo-*EcACP*

2.4.1.1 Materials

Proteins and enzymes

- Purified EcACP (see Section 2.1.)
- Purified His_6-AcpH (see Section 2.2)

Stock solutions and buffers

- Magnesium chloride dissolved in water, 600 mM
- Manganese chloride dissolve in water, 40 mM
- Tris(2-carboxyethyl)phosphine hydrochloride dissolved in water, 250 mM (TCEP)
- Apofication buffer: 50 mM Tris, 150 mM NaCl, 10% glycerol (pH 7.4)
- 40% acrylamide (Millipore Sigma)
- Urea 10 M
- Tris base 3 M (pH 8.8) (Millipore Sigma)
- Tris-HCl 1 M (pH 6.8)
- Ammonium persulfate 10 wt% in water (10% APS)
- Tetramethylethylenediamine (TEMED) (Thermo Scientific)
- 10 × concentrated native gel running buffer (144 g glycine/L, 30 g Tris base/L, pH 8.3)

Equipment

- 1.7 mL Eppendorf tubes (Eppendorf)
- 37 °C rotating incubator
- Power source and gel box for PAGE

2.4.1.2 Procedure

1. In a 1.7 mL Eppendorf tube, combine all reagents in the following concentrations: EcACP 1 mg/mL, His_6-AcpH 0.1 mg/mL, $MgCl_2$ 12.5 mM, $MnCl_2$ 5 mM, TCEP 5 mM. Fill remaining volume with apofication buffer.
2. Seal Eppendorf tube with parafilm and place in rotating incubator at 37 °C for 16 h.
3. Analyze conversion from *holo*-EcACP to *apo*-EcACP by 20% urea-PAGE. A recipe for a resolving urea gel is as follows:

Water	2 mL
40% acrylamide	6.25 mL
10 M urea	2.5 mL
Tris (3 M pH 8.8)	1.6 mL
10% APS	0.125 mL
TEMED	0.010 mL

A recipe for the companion stacking gel is as follows:

Water	2.95 mL
40% acrylamide	0.5 mL
Tris (1.0 M pH 6.8)	0.5 mL
10% APS	0.040 mL
TEMED	0.004 mL

4. Place gel in electrophoresis chamber and fill chamber with 1 × diluted running buffer.
5. Run gel for 2 h and 15 m at 150 V (Fig. 2).

2.5 Chemoenzymatic labeling of *apo*-EcACP with 4-DMN-pantetheine

2.5.1 One-pot labeling

2.5.1.1 Materials

Enzymes

- CoaA, 10 mg/mL (Section 2.3)
- CoaD, 10 mg/mL (Section 2.3)
- CoaE, 10 mg/mL (Section 2.3)
- Sfp, 10 mg/mL (expression and purification as previously described (Lambalot et al., 1996))
- *apo*-EcACP, 10 mg/mL

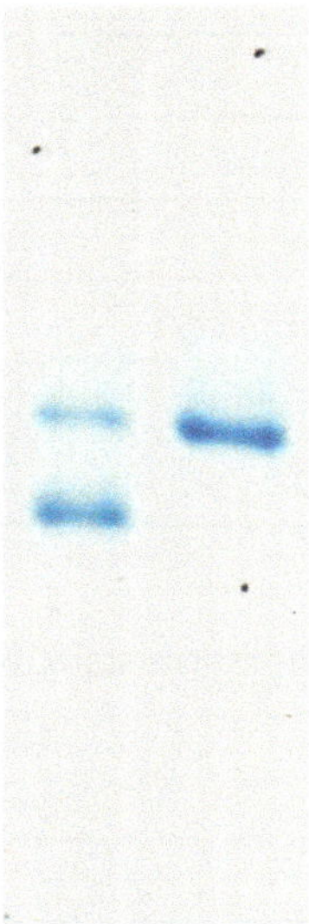

Fig. 2 20% urea-PAGE analysis of apofication reaction. Left band: *apo/holo*-EcACP mixture, right band: *apo*-EcACP. A change in conformation due to the presence of the phosphopantetheine arm of *holo*-EcACP is responsible for shrinking the apparent radius of EcACP causing it to move further in the gel in the *holo* state. In the *apo* state, EcACP does not travel as far in the gel.

Buffers and stock solutions

- Adenosine triphosphate (ATP), 130 mM, pH 7.5 (Millipore Sigma)
- Magnesium chloride, 600 mM
- Dimethyl sulfoxide
- 4-DMN-pantetheine, stored neat (synthesized as previously described (Beld, Cang, & Burkart, 2014))
- Triton-X, 1 wt% in water (Millipore Sigma)
- One-pot buffer: 50 mM phosphate, 5% glycerol (pH 8.0)
- Low salt buffer: 50 mM phosphate, 5% glycerol (pH 8.0)
- High salt buffer: 50 mM phosphate, 1 M NaCl, 5% glycerol (pH 8.0)
- Lyophilization buffer: 20 mM NH_4HCO_3

Equipment

- HiTrap Q HP anion exchange column (GE Healthcare Life Sciences)
- 1.7 mL Eppendorf tubes
- 37 °C rotating incubator
- Power source and gel box for PAGE
- SnakeSkin dialysis tubing 3 kDa molecular weight cut-off (ThermoFisher Scientific)
- Lyophilizer

2.5.1.2 Procedure

1. Make 100 mM stock solution of 4-DMN-pantetheine dissolved in dimethyl sulfoxide.
2. In a 1.7 mL Eppendorftube, mix all enzymes and reagents in the following final concentrations: CoaA (0.1 mg/mL), CoaD (0.1 mg/mL), CoaE (0.1 mg/mL), Sfp (0.1 mg/mL), ATP (8 mM), $MgCl_2$ (12.5 mM), Triton-X (0.01 wt%), 4-DMN-pantetheine (1 mM), *apo*-EcACP (1 mg/mL). Fill remaining volume with one-pot buffer.
3. Seal Eppendorf tube with parafilm and place in rotator incubator at 37 °C overnight.
4. Analyze conversion of *apo*-EcACP to 4-DMN-EcACP using 20% urea-PAGE (see Section 2.4.1.2. for urea-PAGE protocol) (Fig. 3).

2.5.2 Purification

1. Equilibrate 5 mL HiTrap Q HP anion exchange column with low salt buffer.
2. Load crude one-pot reaction from Section 2.5.1.2 onto column.
3. Elute using the following gradient: 100.0 mL total length, 0–70% high salt buffer, collecting 3.0 mL fractions. 4-DMN-EcACP will appear at an NaCl concentration of 300 mM, after 40.0–45.0 mL.
4. Analyze purity of 4-DMN-EcACP containing fractions by 12% SDS-PAGE.

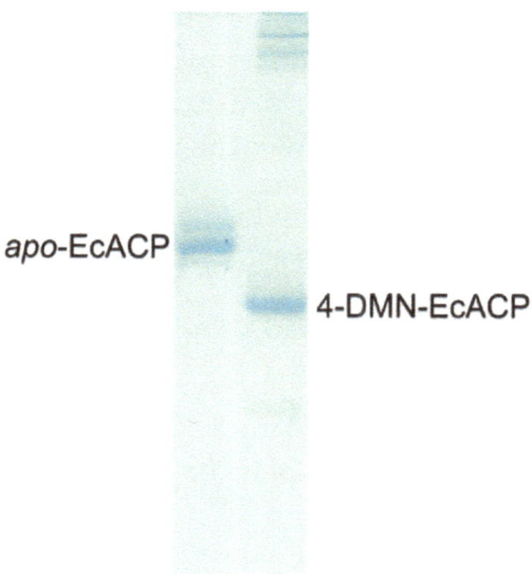

Fig. 3 20% urea-PAGE analysis of one-pot conversion of *apo*-ECACP (right band) to 4-DMN-ECACP (left band). A conformational change due to labeling causes 4-DMN-EcACP to travel further in the gel. Coa enzymes (CoaA, CoaD, CoaE, Section 2.3) and phosphopantetheinyl transferase (Sfp) appear at the top of the gel.

2.5.3 *Taking concentration of 4-DMN-EcACP*

1. Make 2 L of 50 mM NH_4HCO_3 buffer.
2. Place fractions containing pure 4-DMN-EcACP in appropriate length of SnakeSkin dialysis tubing.
3. Dialyze 4-DMN-EcACP into 50 mM NH_4HCO_3 buffer overnight at 4 °C.
4. Transfer 4-DMN-EcACP into pre-weighed 1.7 mL Eppendorf tubes.
5. Freeze at −78 °C and lyophilize to remove NH_4HCO_3 buffer. Store 4-DMN-EcACP powder at −20 °C.

3. Protein-protein interaction studies using 4-DMN-EcACP

The role of the acyl carrier protein (ECACP) in the cell is to shuttle substrates to different partner enzymes. To deliver cargo to a partner enzyme, ACP chain flips the cargo into the active site of the partner enzyme, a mechanism which is believed to be universal in acyl carrier protein interactions (Cronan, 2014). To monitor protein-protein interactions using 4-DMN labeled ACP (4-DMN-EcACP), this universal chain flipping mechanism is exploited. When labeled with the 4-DMN dye, EcACP sequesters the dye within its alpha helical core. Upon interaction with a partner enzyme, 4-DMN-EcACP flips the 4-DMN dye into a pocket on the partner enzyme, which is usually more hydrophobic than the alpha helical core of EcACP, leading to a large increase in fluorescence intensity. By monitoring fluorescence intensity, it is easy to monitor protein-protein interactions between 4-DMN-EcACP and its many partner enzymes.

3.1 Partner enzyme titrations

3.1.1 *Materials*

Enzymes

- 4-DMN-EcACP
- EcACP partner enzymes (40 mM stock concentration, in 50 mM sodium phosphate, 0.01 wt% Triton-X pH 8.0)

Equipment

- Plate reader
- Corning 96-well solid black polystyrene microplate (Fisher Scientific)

Buffer

- 50 mM sodium phosphate, 0.01 wt% Triton-X (pH 8.0)

3.1.2 Procedure

1. Dissolve lyophilized 4-DMN-EcACP to 1 mg/mL concentration in 50 mM sodium phosphate, 0.01 wt% Triton-X (pH 8.0).
2. Transfer 50 μL of 4-DMN-EcACP solution to wells of a black 96-well plate.
3. Read fluorescence using the following parameters: $\lambda_{ex} = 408$ nm, $\lambda_{em} = 512$ nm for each well of 4-DMN-EcACP over the course of 1 min with light shaking (60 rpm) in between each fluorescence measurement.
4. Take the average fluorescence value over 1 min of 4-DMN-EcACP. This becomes the baseline fluorescence to which future readings are normalized.
5. Add 5 μL of desired partner enzyme to each well of 4-DMN-EcACP.
6. Read fluorescence using same parameters over the course of 3 min with light shaking (60 rpm) in between each fluorescence reading. An initial jump in fluorescence is expected; the fluorescence value will plateau after a minute.
7. Take the average fluorescence value over the final 1 min of reading, after value has plateaued.
8. Divide this average fluorescence value of 4-DMN-EcACP plus partner enzyme by the average fluorescence value of 4-DMN-EcACP alone (step 4) to normalize fluorescence value.
9. Repeat steps 5–8 until 50 μL of partner enzyme has been added, representing an equimolar amount of 4-DMN-EcACP and partner enzyme in each well.
10. Graph fluorescence vs concentration of partner enzyme to generate a titration curve.

3.2 Inhibitor studies using 4-DMN-EcACP

3.2.1 Materials

Enzymes

- 4-DMN-EcACP
- EcACP partner enzymes (40 mM stock concentration, in 50 mM sodium phosphate, 0.01 wt% Triton-X pH 8.0)

Equipment

- Plate reader
- Corning 96-well solid black polystyrene microplate (Fisher Scientific)
- 1.7 mL Eppendorf tubes (Eppendorf)

Buffers and stock solutions

- 50 mM sodium phosphate, 0.01 wt% Triton-X (pH 8.0)
- Inhibitor(s) dissolved in DMSO at 100 mM concentration

3.2.2 Procedure

1. Dissolve lyophilized 4-DMN-EcACP to 1 mg/mL concentration in 50 mM sodium phosphate, 0.01 wt% Triton-X (pH 8.0).
2. Make dilutions of stock solution of inhibitor according to following chart:

Final concentration of inhibitor (mM)	Volume of 100 mM inhibitor stock solution (μL)	Volume DMSO (μL)
50	25	25
25	25	75
5	25	100

3. Mix 4-DMN-EcACP, partner enzyme of interest and inhibitor according to the following chart

Tube number	Volume 4-DMN-EcACP (μL)	Volume partner protein (μL)	Stock [inhibitor] (μM)	Final [inhibitor] (μM)	Volume of inhibitor stock solution (μL)	Volume buffer (μL)
1	50	50	0	0	0	20
2	50	50	5	0.05	1.2	18.2
3	50	50	5	0.1	2.4	17.6
4	50	50	25	0.25	1.2	18.2
5	50	50	25	0.5	2.4	17.6
6	50	50	50	0.75	1.8	18.2
7	50	50	50	1.25	3	17
8	50	50	100	2.5	3	17
9	50	50	100	5	6	14

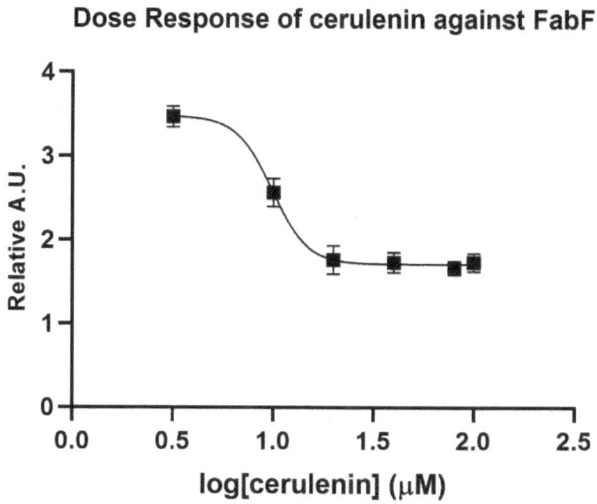

Fig. 4 Dose response curve generated using solvatochromic method described in this section. Cerulenin forms a covalent bond within the active site of ketosynthase II (FabF) and blocks EcACP from chain flipping its 4-DMN cargo into the active site. This is easily read as a decrease in fluorescence intensity relative to uninhibited FabF. The calculated IC_{50} value using this method is 9.87 (8.52, 10.82) μM for cerulenin (literature value of $IC_{50} = 0.4$–9 μM) (Kodali et al., 2005). The IC_{50} value was calculated using GraphPad Prism 8 and data was reported with 95% confidence intervals.

Make three samples for each inhibitor concentration.

4. Dispense 100 μL of each 4-DMN-EcACP, partner enzyme, inhibitor sample into wells of a 96-well black bottomed plate.
5. Read fluorescence using same parameters over the course of 10 min with light shaking (60 rpm) in between each fluorescence reading.
6. To process data: take the average fluorescent value over time for each well after fluorescence value reaches a plateau. Normalize the fluorescence response in the presence of inhibitor by dividing by the fluorescence response in the absence of inhibitor. Find the average and standard deviation of the fluorescence response for each triplicate data point. Graph average fluorescence response vs log[inhibitor] to obtain an IC_{50} curve (Fig. 4).

4. Summary and conclusions

We have developed a rapid and highly sensitive assay to monitor protein-protein interactions between *E. coli* acyl carrier protein and its many partner enzymes. This assay is useful in its ability to identify new partner enzymes and towards drug discovery applications wherein new inhibitors of the partner enzyme active site, or the protein-protein interactions themselves, can be detected.

Acknowledgments

Support was provided by NIH RO1 GM095970. K.C. was supported by the National Science Foundation Graduate Assistance in Areas of National Need (NSF GAANN) Fellowship.

References

Beld, J., Cang, H., & Burkart, M. D. (2014). Visualizing the chain-flipping mechanism in fatty-acid biosynthesis. *Angewandte Chemie (International Ed in English)*, *53*, 14456–14461.

Brozek, K. A., & Raetz, C. R. (1990). Biosynthesis of lipid A in Escherichia coli. Acyl carrier protein-dependent incorporation of laurate and myristate. *The Journal of Biological Chemistry*, *265*, 15410–15417.

Charov, K., & Burkart, M. D. (2019). A single tool to monitor multiple protein-protein interactions of the Escherichia coli acyl carrier protein. *ACS Infectious Diseases*, *5*, 1518–1523.

Cronan, J. E. (2014). The chain-flipping mechanism of ACP (acyl carrier protein)-dependent enzymes appears universal. *The Biochemical Journal*, *460*, 157–163.

Haushalter, R. W., Worthington, A. S., Hur, G. H., & Burkart, M. D. (2008). An orthogonal purification strategy for isolating crosslinked domains of modular synthases. *Bioorganic & Medicinal Chemistry Letters*, *18*, 3039–3042.

Hill, R. B., MacKenzie, K. R., Flanagan, J. M., Cronan, J. E., & Prestegard, J. H. (1995). Overexpression, purification, and characterization of Escherichia coli acyl carrier protein and two mutant proteins. *Protein Expression and Purification*, *6*, 394–400.

Issartel, J. P., Koronakis, V., & Hughes, C. (1991). Activation of Escherichia coli prohaemolysin to the mature toxin by acyl carrier protein-dependent fatty acylation. *Nature*, *351*, 759–761.

Jordan, S. W., & Cronan, J. E. (1997). A new metabolic link. The acyl carrier protein of lipid synthesis donates lipoic acid to the pyruvate dehydrogenase complex in Escherichia coli and mitochondria. *The Journal of Biological Chemistry*, *272*, 17903–17906.

Kodali, S., et al. (2005). Determination of selectivity and efficacy of fatty acid synthesis inhibitors. *The Journal of Biological Chemistry*, *280*, 1669–1677.

Lambalot, R. H., et al. (1996). A new enzyme superfamily - the phosphopantetheinyl transferases. *Chemistry & Biology*, *3*, 923–936.

Lin, S., Hanson, R. E., & Cronan, J. E. (2010). Biotin synthesis begins by hijacking the fatty acid synthetic pathway. *Nature Chemical Biology*, *6*, 682–688.

Murugan, E., Kong, R., Sun, H., Rao, F., & Liang, Z.-X. (2010). Expression, purification and characterization of the acyl carrier protein phosphodiesterase from Pseudomonas aeruginosa. *Protein Expression and Purification*, *71*, 132–138.

Nguyen, C., et al. (2014). Trapping the dynamic acyl carrier protein in fatty acid biosynthesis. *Nature*, *505*, 427–431.

Price, A. C., et al. (2001). Inhibition of beta-ketoacyl-acyl carrier protein synthases by thiolactomycin and cerulenin. Structure and mechanism. *The Journal of Biological Chemistry*, *276*, 6551–6559.

Kolibri 360 GmbH
Ferdinand-Jühlke-Straße 7
99095 Erfurt, Deutschland
produktsicherheit@kolibri360.de